Y0-BYB-331

The Adrenal in Toxicology

Target organ and modulator of toxicity

The Adrenal in Toxicology

Target organ and modulator of toxicity

Edited by **PHILIP W. HARVEY**

AgrEvo UK Limited, Chesterford Park, Saffron Walden, UK

UK Taylor & Francis Ltd, 1 Gunpowder Square, London EC4A 3DE
USA Taylor & Francis Inc., 1900 Frost Road, Suite 101, Bristol, PA 19007

British Library Cataloguing in Publication Data

A catalogue record for this book is available from the British Library.

ISBN 0 7484 0330 2

Library of Congress Cataloguing Publication Data are available

Cover design by Youngs Design in Production, London

Typeset in Times 10/12pt by Santype International Ltd, Salisbury, Wilts

Printed in Great Britain by T. J. Press (Padstow) Ltd, Cornwall

In memory of my wife Denise who died from breast cancer on 7 October 1994, and for our daughters Jessica and Rebecca

Contents

Preface

This is the first book to consider the role of the adrenal gland in toxicology and, although it is a contribution to the endocrine toxicology literature, it differs from the usual target organ approach to consider whole body responses to adrenal pharmacology and toxicity, and instead offers a novel discussion of the modulation of toxicological responses by altered adrenal hormones. This approach significantly broadens the relevance of the volume; to this end the generic involvement of the adrenal gland in a wide variety of toxicological responses is considered in detail. Indeed, the book has been designed for the specialist endocrine toxicologist in addition to the general toxicologist who may be unfamiliar with the significant involvement of the adrenal gland in toxicological responses.

The volume is structured to provide an introduction to the types of endocrine toxicity, the role of stress in toxic insult, how the stress response modulates toxicity, and why the adrenal gland is the most common toxicological target organ in the endocrine system. The second section deals with the endocrinology and pharmacology of the adrenal gland of most relevance to toxicology and this is related to species, age and sex. This section also introduces the anatomical, cellular and molecular features of the adrenal that predispose the gland to direct toxic insult and indirect secondary change. The third section considers the adrenal cortex and adrenal medulla as toxicological target organs and the fourth section details adrenocortical and glucocorticosteroid modulation of toxicity. This latter section covers the role of adrenal glucocorticosteroids in neurotoxicity, developmental toxicity and immunotoxicity and also how glucocorticosteroids can both enhance toxicity or ameliorate toxicity in a variety of other organs such as brain, liver, kidney or gastrointestinal tract. The final section of the book covers clinical interfaces such as adverse drug reactions and the adrenal gland, and corticosteroid–drug interactions. A chapter is specifically devoted to the significance to man of adrenal changes in pre-clinical or experimental toxicity studies. Most of the available information on the adrenal gland and toxicity centres on the cortex and this is reflected in this volume.

The book is designed to be accessible to both specialist and generalist, and to develop the concept that the adrenal gland is vital to tolerance to, and modulation of, toxicity at the whole-body level. The molecular and mechanistic basis of this is explored and extrapolation is made from natural hormone to synthetic analogues. Finally, the book is designed to be a contribution to a variety of fields including endocrinology, pharmacology, toxicology, pathology, pharmacy and clinical science, and this is apparent in the scope of its contents.

P. W. Harvey

Contributors

J. A. Baldwin
Formerly of SmithKline Beecham Pharmaceuticals, The Frythe, Welwyn, Hertfordshire, UK

H. D. Colby
Albany College of Pharmacy, Union University, 106 New Scotland Avenue, Albany, NY12208, USA

P. F. D'Arcy
School of Pharmacy, Medical Biology Centre, The Queen's University of Belfast, 97 Lisburn Road, Belfast BT9 7BL, Northern Ireland

A. D. Dayan
D H Department of Toxicology, St Bartholomew's Hospital Medical College, 59 Bartholomew Close, London EC1 7ED, UK

M. Gumbleton
Welsh School of Pharmacy, University of Wales, Cardiff CF1 3XF, UK

P. W. Harvey
AgrEvo UK Limited, Toxicology, Chesterford Park, Saffron Walden, Essex CB10 1XL, UK

J. P. Hinson
Department of Biochemistry, Faculty of Basic Medical Sciences, Queen Mary and Westfield College, Mile End Road, London E1 4NS, UK

I. Kimber
Zeneca Central Toxicology Laboratory, Alderley Park, Macclesfield, Cheshire SK10 4TJ, UK

R. D. Mann
Drug Safety Research Unit, Bursledon Hall, Southampton SO31 1AA, UK

L. J. McIntosh
Department of Biological Sciences, Stanford University, Stanford, CA 94305, USA

Contributors

P. J. Nicholls
Welsh School of Pharmacy, University of Wales, Cardiff CF1 3XF, UK

P. W. Raven
Metabolic Studies Section, Institute of Psychiatry, De Crespigny Park, London SE5 8AF, UK

R. M. Sapolsky
Department of Biological Sciences, Stanford University, Stanford, CA 94305, USA

M. J. Tucker
Zeneca Pharmaceuticals, Alderley Park, Macclesfield, Cheshire SK10 4TG, UK

SECTION ONE

Introduction to Adrenal Gland Involvement in Toxicology

1

An Overview of Adrenal Gland Involvement in Toxicology

From target organ to stress and glucocorticosteroid modulation of toxicity

PHILIP W. HARVEY

AgrEvo UK Limited, Chesterford Park, Saffron Walden

1.1 An Introduction to Endocrine Toxicology

Toxicology usually focuses on the adverse effects of drugs and chemicals at the level of an organ or tissue and it is a research goal to identify 'the target organ'. Techniques are available at the cellular and molecular levels to identify target organs and to assist in elucidating mechanisms of action. Toxicology, with foundations in many disparate disciplines in the chemical and biomedical sciences, has developed specialities which are so focused in approach that there is a danger that too great a divergence of knowledge may occur, at the expense of considering whole body interactions and consequences.

Endocrine toxicology is one such speciality and the limited literature that has been devoted to it has predominantly adopted the target organ approach, with considerations of the actions of chemicals on the structure and function of a particular gland. However, the endocrine system, more than any other, is homeostatically balanced and the endocrine system as a whole is sensitive to changes in the function of its constituent glands and to non-endocrine organs such as the liver. The actions of the various hormonal products of endocrine glands provide the route by which toxic actions on an endocrine gland can fundamentally affect whole body physiology and biochemistry and, as is often neglected, the function of distal non-endocrine target tissues. For example, toxic action on the adrenal cortex could disturb the synthesis of steroids important for glucose regulation or water and sodium balance, thereby extending the impact of the toxic insult to general metabolism or effects on the kidney. Conversely, chemically induced changes in non-endocrine organs can affect the endocrine system (eg the primary effects of phenobarbital on the liver and secondary effects on thyroid pathology reviewed by Atterwill *et al.* 1992). Furthermore, toxicology often focuses on the damage induced by a chemical to an organ or tissue, leading to total or subtotal failure in function. In endocrine toxicology, as with other comparatively rare special cases (eg immunotoxicology), chemically induced increases in function are just as harmful as is loss of function. Indeed, many examples of non-genotoxic endocrine carcinogenesis result from pituitary over-stimulation of peripheral endocrine targets. This chapter aims to provide a brief overview of adrenal gland involvement in endocrine and general toxicology and explore why the adrenal gland, and in particular the cortex, is such a commonly affected organ (Ribelin 1984). To this end, basic types of endocrine toxicology are reviewed, followed by specific examples of direct, secondary and indirect toxicity involving the adrenal gland. The concept that the adrenal gland is a generic endogenous modulator of toxicological responses is reviewed in relation to the literature on stress and toxicity and to the literature on glucocorticosteroid interactions with drugs and chemicals. Where possible, stress interactions with toxicological responses are ascribed to adrenomedullary or adrenocortical mechanisms. Finally, the importance of glucocorticosteroids as endogenous modulators of toxicity is introduced.

1.2 Types of Endocrine Toxicity

There are various classifications of the different types of endocrine toxicity which are derived largely from clinical endocrine function and pathology (eg Capen & Martin 1989, with regard to the thyroid) or pharmacological properties of a compound (eg Atterwill & Flack 1992). The most recent analysis of the types of endocrine toxicology is by Atterwill & Flack (1992), and in their analysis endocrine toxicological effects are grouped into 'classes' based on the pharmacological activity of the compound. Thus, 'class 1' effects are those which can be readily predicted from the known pharmacology of the compound at the *pharmacological dose range* (eg sex steroids). A 'class 2' agent is one which has predicted endocrine pharmacology but which requires high dose levels 'well in excess of the therapeutic dose range' to induce toxicity (eg glucocorticoids and adrenal suppression and excessive catabolism). 'Class 3' compounds are those producing effects which could not be predicted from the known pharmacology of the compound. These can be subdivided into agents producing direct effects on a gland (eg ketoconazoles, on adrenal and testicular function) or secondary effects on a gland (eg phenobarbital effects on the thyroid mediated by liver enzyme induction). Atterwill & Flack's (1992) final class of agents ('class 4') are those which cause entirely unpredicted effects (from a pharmacological viewpoint) on the endocrine system and are described as 'idiosyncratic'. This scheme has the primary pharmacological action of a compound as its basis and is an extremely useful model. However, current models of endocrine toxicology fail to recognize the modulation of toxicity by hormones which is an important but neglected area in endocrine toxicology. Thus, an additional model of classifying endocrine toxicological effects other than those reported to be based on clinical functional and pathological parameters (Capen & Martin 1989) or predicted pharmacological actions of compounds (Atterwill & Fleck 1992) needs to be developed to cover all aspects of endocrine toxicology. The model developed here and in subsequent chapters has fundamental endocrinology as its basis and, whilst there is overlap in some examples, differs from previous models in allowing the recognition of 'indirect' toxic actions to include the *modulation of toxicity in non-endocrine* target organs. The latter is a particular feature of pituitary–adrenal involvement in toxicology and recognizes the fundamental importance of stress–toxicity interactions and chemical–hormone interactions which are not predictable from the 'pharmacology' of the toxic agent alone.

1.3 A Model of Endocrine Toxicity Incorporating Primary, Secondary and Indirect Endocrine Toxicity and Hormonal Modulation of Toxicity

This model of endocrine toxicology encompasses three major mechanistic principles by which endocrine toxicity can be manifested (see Table 1.1). Firstly, *primary endocrine toxicity* (type 1) where a compound is directly damaging to a gland [eg [*o,p*-DDD(1,1-dichloro-2,2-bis(4-chlorophenyl)-ethane; mitotane] which causes lesions in the adrenal cortex (see Colby & Longhurst 1992 and later chapters)]. Secondly, there is commonly evidence of *secondary endocrine toxicity* (type 2) where effects can be detected in an endocrine gland as a result of toxicity elsewhere in the endocrine axis (eg pituitary 'castration cells' that develop as a consequence of exposure to compounds causing testicular atrophy and impaired testosterone secretion (see Walker & Cooper 1992) or indeed changes in the adrenal cortex due to chemically-induced stress). The third aspect of endocrine toxicity in this model concerns *indirect endocrine toxicity* of which there are two types. The first type (type 3a) concerns the *initial site of toxic action in a non-endocrine organ*, the disturbance in function of which then has an impact on the endocrine system. Examples of this are compounds which induce hepatic cytochrome P450s and uridine diphosphate glucuronyltransferase, such as phenobarbital, which can indirectly cause thyroid pathology in the rat as a result of increased liver enzyme activity and thyroid hormone clearance and excretion, ultimately resulting in excessive thyroid stimulating hormone drive of the thyroid (reviewed in Atterwill *et al.* 1992). The second type (type 3b) of *indirect endocrine toxicity* concerns the *modulation of toxic responses in non-endocrine organs* (eg liver, kidney, brain) by hormones interacting with another co-administered chemical. An example of this is the modulation of kainic acid toxicity to the hippocampus as a function of adrenal corticosterone status in the rat, where high, physiologically relevant systemic concentrations of corticosterone exacerbate neurotoxicity, and corticosterone depletion is protective (Sapolsky 1986a).

It is with this latter type of mechanism in endocrine toxicology (ie the hormonal modulation of toxicological responses) that the adrenal gland more than any other endocrine gland is particularly predisposed to involvement. One explanation for this is that toxic insult can be stressful, particularly at the maximum tolerated dose (eg Miller 1992), and thus compounds inherently toxic to liver, kidney or brain may also cause stress–related secretion of adrenal hormones, particularly corticosterone, and coincidence of toxic agent with changed adrenal function allows common opportunity for interactions and the modulation of toxic response. Some compounds, such as the methylxanthines, may also pharmacologically stimulate corticosterone secretion (eg Spindel *et al.* 1983) in addition to any other target organ effects these compounds have, and this illustrates the importance of knowledge of the pharmacology of a compound, as in Atterwill & Flack's (1992) Schema, in explaining mechanisms behind some types of endocrine toxicity. The endocrinology of the adrenal gland according to species, age and sex and the importance of the adrenal in the adaptation and response to stressful conditions (environment and chemical insult) is fully reviewed by Hinson & Raven in the chapters that follow.

1.4 The Adrenal Gland in Toxicology

In examining the scope of the adrenal gland in toxicology it is worthwhile considering Ribelin's (1984) report that the most frequently reported endocrine lesions occur in the adrenals, and employing a structure of approach as outlined in the model of types of general endocrine toxicology developed in the preceding section of this chapter. The discussion will therefore cover the scope of the adrenal gland in toxicology in terms of primary toxicity (type 1 – the adrenal as a target organ), secondary toxicity (type 2 – effects on a gland due to pharmacotoxicological insult elsewhere in the endocrine axis or in certain cases changes in adrenal structure and function due to chemically-induced stress) and indirect toxicity (type 3) focusing on how stress (ie initial toxic insult outside the endocrine system which then activates the pituitary–adrenal axis) can alter the general toxicological response to chemical insult in non-endocrine target organs. The progress that has been made in identifying which adrenal products (both adrenocortical and adrenomedullary) actually modulate specific types of toxicity is discussed separately within the framework of indirect toxicity. For the purposes of illustration only, therefore, the type 3a response is discussed in terms of agents causing generalized toxicity which then invokes the adrenal stress response and how a primary action on the liver can alter adrenal steroid metabolism which may then affect the adrenal. It should be noted that in specifically discussing the adrenal gland in toxicology, type 2 and type 3a examples can be indistinguishable because of lack of mechanistic information. The type 3b response is discussed in terms of how the subsequent changes in adrenal activity then modulate target organ and system response to toxic insult (ie hormonal modulation of general toxicity). In discussing the type 3b response, the literature relating to the co-administration of natural and synthetic corticosteroids and, to a more limited extent, adrenaline in combination with a second compound used as a toxic challenge is introduced. These studies of adrenal hormone interactions in toxic response provide some evidence of the role of the adrenal *in vivo* in the modulation to toxicity.

1.5 Examples of Primary Endocrine Toxicity (Type 1) Involving the Adrenal Gland

There are several reviews of the range of compounds that directly affect the adrenal gland, the most recent and briefest focusing on the adrenal cortex is by Thomas (1993), with a more thorough analysis by Colby & Longhurst (1992). Compounds affecting the adrenal gland directly can be grouped into those with clearly expected pharmacological activity (class 1 compounds in Atterwill & Flack's 1992 Schema) and those that unpredictably affect the adrenal (class 3 and 4 compounds in Atterwill & Flack's 1992 Schema). Of those with clearly predicted or understood pharmacological activity, further subgrouping can be made according to the mechanism of pharmacological action (eg steroidogenesis–inhibiting agents such as aminoglutethimide, metyrapone, etomidate; or adrenal–stimulating agents such as the methylxanthines, caffeine, theophylline and denbufylline producing effects via phosphodiesterase inhibition and cAMP accumulation provided that the effect is direct; or even hormones and analogues, eg spironolactone) and the site in the adrenal gland at which the compound's activity is targeted (eg zona fasiculata and zona reticularis which produce steroids from the precursal compound pregnenolone through progesterone, glucocorticoids, androgens and oestrogens; zona glomerulosa which produces the mineralocorticoid, aldosterone, and finally the adrenal medulla producing catecholamines, such as adrenaline, from the precursor tyrosine).

Although covered in detail in the next chapter by Hinson & Raven, it is worthwhile showing the range of hormones produced in the adrenal gland and this is illustrated in Figure 1.1. Also, as the compounds causing effects on the adrenal are given detailed consideration in later chapters (from a pharmacological viewpoint by Hinson & Raven and from a toxicological viewpoint by Colby & Tucker), there is presented in Table 1.2 a selected list of drugs and chemicals producing direct effects on the adrenal with examples specifically selected to illustrate effects based on pharmacological activity, in contrast to unpredicted effects, and effects in different regions of the adrenal gland (cortical zones and medulla separately).

The endocrine system in experimental species, especially the rat, is one of the most common targets in toxicity studies (Heywood 1984). Further analysis has revealed that within the endocrine system, the adrenal gland is by far the most commonly affected target organ (Ribelin 1984), and within the adrenal gland the majority of effects have been noted to be in the cortical regions (eg Ribelin 1984, Thomas 1993). This raises the question of why the adrenal gland, and particularly the cortex, is so commonly implicated in endocrine toxicity. In terms of primary toxicity (type 1) there are various factors that predispose the adrenal gland to toxic insult, and these have been discussed by Colby & Longhurst (1992). The adrenal gland receives a disproportionately large fraction of cardiac output in relation to its tissue mass. Such a rich blood supply is not only important for the rapid delivery of adrenal hormones into the blood, but also for the delivery of potentially toxic drugs and chemicals to the gland. Due to the chemical composition of the adrenal cortex (ie high lipid content partly derived from high concentrations of cholesterol and steroids), the adrenal cortex is consequently hydrophobic and is particularly vulnerable to lipophilic drugs and chemicals that readily accumulate in the cortical tissue.

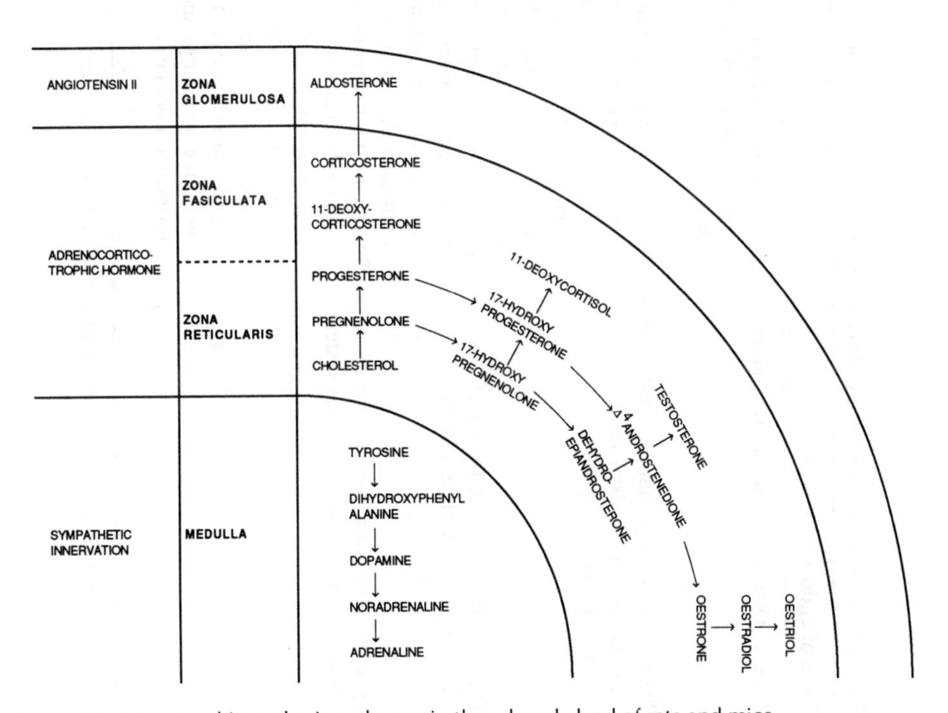

Figure 1.1 Hormone biosynthesis pathways in the adrenal gland of rats and mice

Table 1.1 A model of endocrine toxicity incorporating primary, secondary and indirect endocrine toxicity and hormonal modulation of toxicity

Type of endocrine toxicity	Example	Comment
Type 1: Direct toxic action to a gland	Mitotane (*o,p*-DDD) and adrenocortical destruction. Propylthiouracil inhibition of thyroid function	This is primary or direct toxicity of an agent to an endocrine tissue. Pharmacology may predict toxicity (see Colby & Longhurst 1992 and Atterwill *et al.* 1992, respectively)
Type 2: Secondary changes induced in a gland: toxicity is elsewhere in the endocrine axis or system	Compounds causing testicular atrophy (eg phthalates, cadmium) may cause secondary changes in pituitary gonadotrophs. Methylxanthines (eg denbufylline) or chemical 'stressors' stimulating ACTH secretion which then causes adrenocortical hypertrophy	This is secondary endocrine toxicity where an effect within an axis influences a second component tissue (see Walker & Cooper 1992, Hadley *et al.* 1990 and Spindel 1984, respectively)
Type 3a: Initial site of toxic insult is outside the endocrine system	Specific effect is generated in a non-endocrine tissue (eg phenobarbital induction of corticosterone metabolism in liver) which then influences the pituitary adrenocortical system	This is indirect toxicity where the principal site of toxicity lies outside the endocrine system (eg Rehulka & Kraus 1987). This may also appropriately describe compounds that induce toxicity and then the stress response
Type 3b: Hormonal modulation of toxicity outside the endocrine system	Specific hormonal modulation of toxicity (eg kainate or aflatoxin toxicity by corticosterone). Following on from the type 3a response, compounds that indirectly alter hormonal function may also interact with hormones in the expression of toxicity. This classification of endocrine toxicology specifically covers hormonal modulation of toxicity in the context of the whole body	The hormonal modulation of toxicity has been reviewed in general terms of stress by Vogel (1993) and specifically concerning glucocorticosteroid modulation of toxicity of natural toxins by Harvey *et al.* (1994). It is also discussed in detail in Chapter 7.

Note: Compounds can fall into more than one category dependent on multiple pharmacological and toxicological effects.

Table 1.2 Selected examples of compounds producing primary toxicity to the adrenal gland organized by target site and mechanism of action

Anatomical and functional region of adrenal gland		Compound	Reference	Comment
Cortex	Zona glomerulosa	Spironolactone	Sherry *et al.* (1986)	Aldosterone antagonist
	Zona fasiculata and zona reticularis	Aminoglutethimide	Dexter *et al.* (1967)	Enzyme inhibitors with known pharmacological effect on steroidogenesis
		Etomidate	Preziosi & Vacca (1988)	
		Spironolactone	Sherry *et al.* (1986)	
		Ketoconazole[a]	Feldman (1986)	Unpredicted steroidogenesis enzyme inhibitor
		Carbon tetrachloride[a]	Brogan *et al.* (1984)	Produces necrosis specifically to zona reticularis (not predicted from 'pharmacology')
		7-12-Diemethylbenz(a)-anthracene (DMBA)[a]	Hallberg (1990)	Necrosis in inner two zones possibly related to prostaglandin metabolism (not predicted from 'pharmacology')
Medulla		Growth hormone[a]	Moon *et al.* (1950)	Produces hypertrophy and hyperplasia (not predicted from 'pharmacology')
		Nicotine	Boelsterli *et al.* (1984)	Hypertrophy and hyperplasia, increased catechloamine content. Neuropharmacologically mediated effect
		Xylitol[a]	Boelsterli & Zbinden (1985)	Biochemical and morphological changes in medulla not predicted from 'pharmacology'

Note: [a] Examples of compounds that have effects which are not readily predicted by known pharmacological activity of the compound [ie class 3 or 4 compounds in Atterwill & Flack's (1992) classification].

This may be one reason why the majority of compounds that are known to affect the adrenal appear to induce toxicity in the cortex as noted by Thomas (1993). Once compounds are delivered and accumulate in the adrenal cortex, the rich concentration of enzymes present, normally catalysing steroidogenesis, may utilize foreign drugs and chemicals as substrate thereby increasing local exposure to reactive metabolic products.

1.6 Examples of Secondary Endocrine Toxicity (Type 2) Involving the Adrenal Gland

A second reason why the adrenal is so commonly implicated in endocrine toxicological responses (Ribelin 1984) concerns its potential for manifesting secondary toxicity, which in turn revolves around the gland's unique position in the endocrinology of the stress response. The endocrine and pathological phenomena comprising the stress syndrome and the role of the hypothalamo-pituitary-adrenocortical (HPA) axis in the adaptive response to adverse conditions was first described by Selye (1936, 1950). In toxicology, secondary changes to adrenal gland morphology often comprise adrenocortical hypertrophy and hyperplasia which is characteristic of prolonged stimulation by pituitary adrenocorticotrophic hormone (ACTH), the secretion of which is stimulated by noxious chemical insult.

The role of pituitary–adrenal activation in response to adverse or noxious stimuli has been well researched and more recent endocrinological reviews are available elsewhere (eg Smelik & Vermes 1980 and Hinson & Raven in the following chapters). The stress response, however, comprises the activation of the adrenal to produce a rapid phase increased secretion of adrenomedullary catecholamines, and a secondary more prolonged output of adrenocortical steroids. Although the adrenal cortex has the capability to synthesize and secrete a wide range of steroids (eg glucocorticosteroids, androgens, oestrogens, progestogens) in response to pituitary ACTH, it is the glucocorticosteroids that are important to the adaptive stress response, through a variety of pharmacological, physiological, metabolic and endocrinological actions. The predominant glucocorticosteroid in rats and mice is corticosterone but in other mammals such as the guinea pig, dog and man it is cortisol. This raises the first point concerning the endocrinology of the pituitary–adrenocortical system and the stress response, namely that there are significant species differences not only affecting which glucocorticosteroids are produced but also the sensitivity of different species to stress. Species differences and the detailed endocrinology of the adrenal gland, and the pharmacological and metabolic actions of the glucocorticosteroids of relevance to toxicology, are covered in the following chapters. Species differences can manifest in toxicology in terms of the relative frequency of adrenal changes occurring via 'secondary toxicity', with rats and mice showing the greater degree of sensitivity. In the context of this classification scheme secondary toxicity occurs where there are non-specific changes in adrenal morphology or where the original insult lies within an endocrine axis, that is the hypothalamo-pituitary component of the adrenal control axis. Concerning the latter there is often little information available to distinguish a true type 2 response from an indirect type 3a adrenal effect (see Table 1.1). This is relevant for compounds that induce the stress response and whilst the type 2 and type 3 effects may be readily distinguishable in general endocrine toxicology, there is overlap when considering the adrenal in the absence of additional information.

Some examples of compounds which induce toxicity and cause a generalized stress response, as evidenced by increased corticosterone secretion in rodents, are: natural toxins such as *Escherichia coli* endotoxin (Hawes *et al.* 1992) and *Rhodobacter sphaeroides* toxin (Zuckerman & Qureshi 1992); chemicals, eg alcohol (Moerland *et al.* 1985), chlordecone (Rosencrans *et al.* 1985) and carbon disulfide (Benes *et al.* 1985); neuropharmacologically active agents such as 9-tetrahydrocannabinol (Kumar & Chen 1983, Patel *et al.* 1985), cocaine (Levy *et al.* 1992) and lithium, amitryptylline and mianserine (Storlien *et al.* 1985); and, of course, cytotoxic drugs such as hydroxyurea (Preziosi *et al.* 1985), 5-fluorouracil (Preziosi & Vacca 1983) and vinblastine sulfate (Chung & Gabourel 1971). Additionally, some drugs such as methylxanthines (eg caffeine, theophylline and denbufylline) exert powerful direct pharmacological effects on the endocrine system and are especially known for their potent stimulatory effects on the adrenal cortex as evidenced by markedly increased secretion of corticosterone in rodents (eg Spindel *et al.* 1983, see also Hadley *et al.* 1990). The methylxanthines cause true secondary type 2 endocrine changes in the adrenal since the proximal pharmacological target is the hypothalamo-pituitary axis (Hadley *et al.* 1990). Other compounds affecting the adrenal may operate through a more indirect mechanism involving non-endocrine organs where toxic insult to these organs may be stressful, and may therefore be classified as inducing a type 3 effect. It should be noted that whilst caffeine potently stimulates corticosterone secretion in the rat (Spindel *et al.* 1983) it does not stimulate cortisol secretion in man (Spindel 1984), at least not at dose levels of 8 mg/kg body weight. This may be an example of a genuine species difference however other factors, including dose considerations and prior exposure and tolerance, may be involved. These examples are not intended to be exhaustive, but rather to illustrate that a wide variety of drugs, chemicals and natural toxins with diverse structures and actions can induce stress-related secretion of corticosterone in laboratory rodents. This response may well occur with the majority of chemicals administered to rodents at the maximum tolerated dose (eg Miller 1992) and this generic physiological response can provide the mechanism for hormone interactions and the modulation of toxicity.

1.7 Examples of Indirect Toxicity (Type 3a and 3b) Involving the Adrenal Gland

Two types of indirect toxicity mechanisms involving the adrenal gland can be identified. The first involves a general toxic action on a non-endocrine target organ which then specifically affects the adrenal (type 3a). An example of this is the induction of hepatic microsomal enzymes by phenobarbital in the rat, which subsequently affects corticosterone metabolism (Rehulka & Kraus 1987). A similar finding has been reported in patients where treatment with diphenylhydantoin can affect cortisol metabolism (Werk *et al.* 1971). The significance of this observation is that alteration of the rate of hepatic corticosteroid metabolism can cause imbalance in the pituitary-adrenocortical axis which, if sustained, can produce changes in adrenal morphology quite indirectly.

Thus the type 3a response involves the initial *specific* action to a non-endocrine target organ which then affects the adrenal. Other examples, mediated through stress, for example the increased corticosterone secretion occurring as a result of the general toxicity of compounds such as carbon disulfide (Benes *et al.* 1985) or hydroxyurea (Preziosi *et al.* 1985), or indeed the administration of compounds at the maximum tolerated dose (Miller 1992), may bear some similarity to a type 2 secondary response. Table 1.1 summarizes the criteria for classification.

The second type of indirect toxicity involving the adrenal gland concerns the modulation of toxicity by adrenal hormones (type 3b) and, in the context of this discussion, specifically by glucocorticosteroids. However, the modulation of toxicity by hormones of other endocrine axes is known (eg the protection of cyclophosphamide-induced testicular damage in mice by administration of a gonadotrophin-releasing hormone analogue; Glode *et al.* 1981) which makes this classification scheme relevant to endocrine toxicology generally, although most examples of hormonal modulation of toxicity involve the adrenal. Examples of hormonal modulation of toxicity (type 3b) are derived from studies on induced stress interactions in toxicity where animals are subjected to stressful conditions and their response to toxic insult is studied (note this is quite different from the secondary type 2 changes in adrenal function and morphology brought about by noxious chemical insult). In the type 3b response, the HPA axis is deliberately activated as an experimental variable by exposure to stressful conditions (eg footshock, restraint) with the intent of altering the hormonal milieu to examine how this then alters or modulates toxicological response to a chemical insult. A complimentary literature has developed where hormones have been administered in combination with a toxic agent to explore hormonal modulation of toxicity in more carefully controlled situations. Both the literature on stress interactions in toxicity and adrenal hormone interactions in toxicity (comprising the type 3b classification in endocrine toxicology) is reviewed in the following sections and related to the concept that the adrenal gland is a modulator of toxicity.

1.8 The Adrenal Gland as a Modulator of Toxicity

In developing the argument that the adrenal gland is an important modulator of toxicity, information illustrating this point has been gathered from a number of diverse sources. First, in experimental toxicology the information on stress interactions and stress modulation of toxicity is reviewed. Additionally, the more valuable studies of hormone administration (primarily natural glucocorticosteroids and synthetic analogues but also adrenomedullary hormones) and subsequent interactions with a second toxic agent is reviewed. Second, in clinical pharmacology, the literature on adverse drug interactions in patients prescribed glucocorticosteroids with other medicines is considered, as is the response of patients with adrenal dysfunction (eg Cushing's or Addision's patients) to prescribed drugs. Although these concepts are introduced in this chapter, the reviews on these topics by Harvey, D'Arcy and Mann in Sections 4 and 5 of this volume cover these subjects in detail, and this review will only cover the experimental toxicology literature, ie stress interactions in toxicity and the effects of adrenal hormone administration (particularly natural and synthetic glucocorticosteroids) on the modulation of toxic responses.

One of the main criticisms of the literature on stress and toxicity is that it is often unclear which of the many physiological elements of the stress response is actually responsible for modulating subsequent tolerance or sensitivity to a toxic chemical insult. For example, in many studies of stress interactions with toxicity it is difficult to establish whether the observed changes in drug or chemical response are due to adrenocortical or adrenomedullary hormone mechanisms or even more general effects on physiology. Nevertheless, it is worthwhile reviewing briefly the literature on stress and toxicity interactions in order to develop the argument that the adrenal gland is a modulator of toxic insult.

1.9 The Adrenal Gland as a Modulator of Toxicity: Evidence from Studies on Stress and Toxicity

A wide variety of drugs, chemicals and natural toxins (eg *E. coli* endotoxin, Hawes *et al.* 1992; alcohol, Moerland *et al.* 1985; chlordecone, Rosencrans *et al.* 1985; carbon disulfide, Benes *et al.* 1985; cocaine, Levy *et al.* 1992; hydroxyurea, Preziosi *et al.* 1985; caffeine, Spindel *et al.* 1983) have been shown to induce the HPA stress response in rodents as evidenced by increased corticosterone secretion. This response to toxic insult takes on greater significance in the light of suggestions that administration of compounds at the maximum tolerated dose level is stressful (Miller 1992), and at the very least implicates the adrenocortical stress response as a potentially integral part of many compound toxicity profiles.

Stress is known to alter xenobiotic oxidative metabolism in rats (Pollack *et al.* 1991). Additionally, glucocorticosteroids are known to induce various cytochrome P450 isoenzymes (Ortiz de Montellano 1986), inhibit aryl hydrocarbon hydroxylase (Bogdanffy *et al.* 1986), are involved in detoxification pathways (Bogdanffy *et al.* 1984), and alter the activity of specific liver enzymes such as tryptophan oxygenase (Moerland *et al.* 1985) and γ-glutamyltransferase (Barouki *et al.* 1982). Stress-induced alteration of metabolic capacity is an obvious prerequisite for the expression of toxicity, particularly in the liver. An example of this is that footshock stress accelerates carbon tetrachloride-induced liver injury in rats (Iwai *et al.* 1986). A more general example is of social stress (group housing) markedly increasing the acute toxicity of metamphetamine in mice (Greenblatt & Osterberg 1961).

The glucocorticosteroids have a range of pharmacological actions including effects on DNA, RNA and protein synthesis, carbohydrate metabolism and protein catabolism, and are immunosuppressive and possess anti-inflammatory actions (Haynes & Murad 1991). It is not surprising, therefore, that stress can alter the expression of toxicological endpoints through a variety of mechanisms. For example, restraint stress is reported to facilitate dimethylbenz-α-anthracene–induced mammary turmorigenesis in the rat (Tejwani *et al.* 1991). In this case, the augmentation of a carcinogenic response may be due to molecular mechanisms, metabolic mechanisms, hormonal mechanisms or immunosuppression and lack of immunosurveillance. Furthermore, although the discussion so far has centred on the potential involvement of adrenocortical steroids in the modulation of toxic response by stress, stress also alters the secretion of a wide variety of other hormones (eg adrenomedullary and indeed extra-adrenal hormones) such that it is not possible with any certainty to identify which factor, altered by the stressful procedure, is actually responsible for altering toxic response and tolerance to the test chemical. For example, stress has been reported to markedly increase cardiotoxicity resulting from isoproterenol and this has been attributed to markedly elevated blood levels of adrenaline (eg Vogel 1993). However, it is also known that corticoids can markedly increase isoproterenol cardiotoxicity through a mineralcorticoid action causing blood electrolyte imbalance (eg Guideri *et al.* 1974). Thus in this particular example of stress–isoproterenol interaction, it is not possible to distinguish between a mechanism of stress-induced adrenomedullary catecholamine secretion or stress-induced adrenocortical gluocorticosteroid secretion. The difficulties of attribution of mechanisms behind stress–toxicity interactions has been discussed by Harvey (1994). Vogel (1987, 1993) has reviewed in detail the general literature concerning stress and toxicity.

1.10 The Adrenal as a Modulator of Toxicity: Evidence from Glucocorticosteroid Interaction Studies with Drugs and Chemicals

The literature related to glucocorticosteroid toxicity interactions is broad and falls into two main categories. The first concerns experimental toxicology where natural and synthetic glucocorticosteroids have been administered before, during or shortly after a toxic insult with a drug or chemical. The literature relating to this type of research, but confined to glucocorticosteroid interactions with natural toxins (kainate, aflatoxins) rather than drugs or chemicals, has been reviewed recently by Harvey *et al.* (1994). The second body of literature relates to clinical pharmacology and covers adverse drug interactions in patients being prescribed glucocorticosteroids in addition to other medicines, and altered sensitivity to prescribed drugs in patients with concurrent adrenal dysfunction conditions such as Cushing's syndrome or Addison's disease.

In introducing the concept of the adrenal gland as a modulator of toxicity (citing experimental toxicology literature of glucocorticosteroid interactions with drugs and chemicals) it is worthwhile focusing here only on the reported studies that have employed natural adrenocortical steroids (eg corticosterone, cortisol/hydrocortisone). This allows closer comparison of what may be happening in the intact animal *in vivo*. The wider literature concerning synthetic drugs frequently used in medicine (eg prednisolone, dexamethasone) is reviewed in a later chapter.

Studies with corticosterone (the primary natural glucocorticosteroid in rats and mice) have shown that administration of this steroid, at dose levels designed to mimic the endogenous stress response in rats, can modulate neurotoxicity and hepatotoxicity. Sapolsky (1986a) studied the interaction of systemic corticosterone with central kainic acid-induced neurotoxicity: rats were either adrenalectomized or treated with 10 mg/day corticosterone, and then both groups received a microinfusion of kainic acid into the hippocampus. Corticosterone pre-treatment resulted in a marked increase in the volume of hippocampal damage following kainic acid microinfusion, whereas adrenalectomy reduced the degree of damage compared with controls treated with kainic acid only. In an elegant programme of research, Sapolsky and co-workers have shown that corticosterone also exacerbates 3-acetylpyridine hippocampal damage (Sapolsky 1985), that corticosterone exacerbation of kainic acid neurotoxicity can be demonstrated *in vitro* using cultured hippocampal neurons (Sapolsky *et al.* 1988), and that the mechanism behind corticosterone modulation of neurotoxicity involves both glucose and metabolic fuel utilization (Sapolsky 1986b) and neuronal calcium regulation (Elliott & Sapolsky 1992). As the central role of corticosterone and other glucocorticosteroids involves glucose metabolism, it is not surprising that the mechanism of modulation of neurotoxicity involves glucose utilization. Such a fundamental mechanism allows extrapolation to other organs where manipulation of natural glucocorticosteroids may modulate cellular toxicity, however glucocorticosteroids have a variety of other physiological and pharmacological effects which may be variably operative in other organs, and this is reviewed in detail in later chapters.

Corticosterone, administered at high but physiologically relevant dose levels, has been reported to exacerbate carbon tetrachloride hepatoxicity in rats (Lloyd & Franklin 1991). Corticosterone has also been reported to enhance the subacute general toxicity and lethality of Vitamin D_2 in rats (Kunitomo *et al.* 1989) but ameliorate dextran-induced shock in rats (Kogure *et al.* 1986). Additionally, English *et al.* (1987) have reported that the general toxicity of methotrexate varies in the rat according to time of day and endogenous corticosterone rhythm. Similar modulation of toxicity has been reported to occur with cortisol/hydrocortisone (the natural glucocorticosteroid in man): whilst this compound exacerbated aflatoxin B1 hepatoxicity in rats (Chentanez *et al.* 1988), it ameliorated hepatoxicity resulting from *Naja nigricollis* snake venom (Mohamed *et al.* 1974) thus providing evidence that the modulation of toxicity of different toxicants in the same target organ may involve different mechanisms.

1.11 General Conclusions

The involvement of the adrenal gland in the expression of toxicological responses is wide-ranging. The adrenal gland is a target for direct toxic insult, and more than any other endocrine gland has a propensity for developing secondary changes during toxic insult largely due to its focal involvement in the stress response. Indeed, the most frequent endocrine lesions in toxicology are reportedly in the adrenals (Ribelin 1984) and particularly the cortex, which although not precluding direct toxicity, is suggestive of non-specific, secondary stress response to generalized chemical insult.

The scope of the involvement of the adrenal gland in toxicology is further broadened when the modulation of toxic response by adrenal hormones and analogues is considered. The generic concept that hormones can modulate toxicity is an important, but often neglected, aspect of endocrine toxicology and the adrenal gland features prominently in examples of this phenomenon. The general literature on how experimentally-induced stress can alter tolerance to toxic insult (eg Vogel 1993) illustrates how the endocrine system as a whole can modulate the response to toxicity, as it is often unclear which adrenal products, cortical or medullary, are directly responsible for the observed effects on toxicity (eg Harvey 1994). The literature on stress and toxicity is particularly prone to be confounded because stress is associated with multiple extra-adrenal endocrine changes and can also produce a variety of non-endocrine physiological changes which may influence the response to a toxic insult. Studies on specific hormone–chemical interactions provide a more reliable insight into the role the adrenal gland plays in modulating toxic responses. The effects of co-administration of adrenaline, or natural glucocorticosteroids with drugs or chemicals, implicates the adrenal in the intact animal as a modulator of toxicity and identifies the pharmacological basis for observed interactions. The use of synthetic corticosteroids, with differing potencies of glucocorticoid or mineralcorticoid actions, further defines pharmacological mechanisms operative in adrenal gland modulation of toxicity in a variety of target organs.

The main conclusion from this review is that the adrenal gland is a modulator of toxicity. Endocrinologists have long been aware of the adrenal stress response and the functions that this serves in conditions of adversity (Selye 1936, 1950). Whilst chemical insult *per se* may not be overtly stressful in every case, there is clear evidence that administration of a wide range of compounds at dose levels that could well be considered a toxic challenge, ie the maximum tolerated dose, may also induce the adrenal stress response in addition to any other direct target organ effects the chemical may possess. The adrenal gland and the hormones secreted from it forms part of a fundamental 'whole-body'–based mechanism of adaptation to, and modulation of, chemical insult, the competency of which may be as important as that of the liver to metabolize a chemical or the kidney to excrete it. Initial adrenal secretions may form part of the first–line response to chemical insult, for example, corticosteroids have anti-inflammatory actions and alter liver metabolic enzyme systems which may have a beneficial 'adaptive' effect. Finally, it is clear that adrenal modulation of toxicity can also take the form of exacerbation of toxic responses; this is not surprising given the range of specific pharmacological and toxicological properties of drugs and chemicals which have been employed to generate these findings. The pharmacological and biochemical bases for this is explored in later sections, together with evidence of adrenal gland modulation of toxicity observed clinically in humans.

References

ATTERWILL, C. K. & FLACK, J. D. (1992) Introduction to endocrine toxicology. In ATTERWILL, C. K. & FLACK, J. D. (Eds), *Endocrine Toxicology*, pp. 3–11, Cambridge: Cambridge University Press.

ATTERWILL, C. K., JONES, C. & BROWN, C. G. (1992) Thyroid gland II – mechanisms of species-dependent thyroid toxicity, hyperplasia and neoplasia induced by xenobiotics. In ATTERWILL, C. K. & FLACK, J. D. (Eds), *Endocrine Toxicology*, pp. 137–182, Cambridge: Cambridge University Press.

BAROUKI, R., CHOBERT, M. N., BILLON, M. C., FINIDORI, J., TSAPIN, R. & HANOUN, J. (1982) Glucocorticoid hormones increase the activity of gamma-glutamyltransferase in a highly differentiated hepatoma cell line. *Biochimica et Biophysica Acta*, **721**, 11–21.

BENES, V., FRANTIK, E. & HORVATH, M. (1985) Interaction of chemical and psychogenic stress: biochemical response to repeated conditioned fear and chronic carbon disulfide poisoning. *Activitas Nervosa Superior*, **27**, 23–25.

BOELSTERLI, V. A. & ZBINDEN, G. (1985) Early biochemical and morphological changes of the rat adrenal medulla induced by xylitol. *Archives of Toxicology*, **57**, 25–30.

BOELSTERLI, U. A., CRUZ-ORIVE, L.-M. & ZBINDEN, G. (1984) Morphometric and biochemical analysis of adrenal medullary hyperplasia induced by nicotine in rats. *Archives of Toxicology*, **56**, 113–116.

BOGDANFFY, M. S., SCHATZ, R. A. & BROWN, D. R. (1984) Adrenal mediation of ethanol's inhibition of benzo(a)pyrene metabolism. *Journal of Toxicology and Environmental Health*, **13**, 799–810.

BOGDANFFY, M. S., ROBERTS, A. E., SCHATZ, R. A. & BROWN, D. R. (1986) Regioselective inhibition of benzo(a)pyrene metabolism by corticosterone in comparison with metyrapone and alpha-naphthoflavone. *Toxicology Letters*, **31**, 57–64.

BROGAN, W. C., HINTON, D. E. & COLBY, H. D. (1984) Effects of carbon tetrachloride on adrenocortical structure and function in guinea pigs. *Toxicology and Applied Pharmacology*, **75**, 188–227.

CAPEN, C. C. & MARTIN, S. L. (1989) The effects of xenobiotics on the structure and function of thyroid follicular and c-cells. *Toxicologic Pathology*, **17**, 266–293.

CHENTANEZ, T., PATRADILOK, P., GLINSUKON, T. & PIYACHATURAWAT, P. (1988) Effects of cortisol pretreatment on the acute hepatoxicity of aflatoxin B1. *Toxicology Letters*, **42**, 237–248.

CHUNG, L. W. & GABOUREL, J. D. (1971) Adrenal steroid release by vinblastine sulfate and its contribution to vinblastine sulfate effects on rat thymus. *Biochemical Pharmacology*, **20**, 1749–1756.

COLBY, H. D. & LONGHURST, P. A. (1992) Toxicology of the adrenal gland. In ATTERWILL, C. K. & FLACK, J. D. (Eds), *Endocrine Toxicology*, pp. 243–281. Cambridge: Cambridge University Press.

DEXTER, R. N., FISHMAN, L. M., NEY, R. L. & LIDDLE, G. W. (1967) Inhibition of corticosteroid synthesis by aminoglutethimide; studies on the mechanism of action. *Journal of Clinical Endocrinology and Metabolism*, **27**, 473–480.

ELLIOTT, E. M. & SAPOLSKY, R. M. (1992) Corticosterone enhances kainic acid-induced calcium elevation in cultured hippocampal neurons. *Journal of Neurochemistry*, **59**, 1033–1040.

ENGLISH, J., AHERNE, G. W., ARENDT, J. & MARKS, V. (1987) The effects of abolition of the endogenous corticosteroid rhythm on the circadian variation in methotrexate toxicity in the rat. *Cancer Chemotherapy and Pharmacology*, **19**, 287–290.

FELDMAN, D. (1986) Ketoconazole and other imidazole derivatives as inhibitors of steroidogenesis. *Endocrine Reviews*, **7**, 409–420.

GLODE, L. M., ROBINSON, J. & GOULD, S. F. (1981) Protection from cyclophosphamide-induced testicular damage with an analogue of gonadotrophin releasing hormone. *Lancet*, **1**, 1132–1134.

GREENBLATT, E. N. & OSTERBERG, A. C. (1961) Correlations of activating and lethal effects of excitatory drugs in grouped and isolated mice. *Journal of Pharmacology and Experimental Therapeutics*, **131**, 115–119.

GUIDERI, G., BARLETTA, M. A. & LEHR, D. (1974) Extraordinary potentiation of isoproterenol cardiotoxicity by corticoid pretreatment. *Cardiovascular Research*, **8**, 775–786.

HADLEY, A. J., FLACK, J. D. & BUCKINGHAM, J. C. (1990) Modulation of corticotrophin release *in vitro* by methylxanthines and adenosine anologues. *British Journal of Pharmacology*, **100**(Suppl.), 337.

HALLBERG, E. (1990) Metabolism and toxicity of xenobiotics in the adrenal cortex, with particular reference to 7,12-dimethylbenz(a)anthracene. *Journal of Biochemical Toxicology*, **5**, 71–90.

HARVEY, P. W. (1994) Stress and toxicity. *Human & Experimental Toxicology*, **13**, 275–276.

HARVEY, P. W., HEALING, G., REES, S. J., EVERETT, D. J. & COCKBURN, A. (1994) Glucocorticosteroid interactions with natural toxins: a mini review. *Natural Toxins*, **2**, 341–346.

HAWES, A. S., ROCK, C. S., KEOGH, C. V., LOWRY, S. F. & CALVANO, S. E. (1992) *In vivo* effects of the antiglucocorticoid RU486 on glucocorticoid and cytokine responses to *Escherichia coli* endotoxin. *Infection and Immunity*, **60**, 2641–2647.

HAYNES, R. C. & MURAD, E. (1991) Adrenocorticotrophic hormone; adrenocortical steroids and their synthetic analogues; inhibitors of adrenocortical steroid biosynthesis. In GOODMAN GILMAN, A., RALL, T. W., NIES, A. S. & TAYLOR, P. (Eds), *Goodman and Gilman's – The Pharmacological Basis of Therapeutics*, 8th Edn, pp. 1431–1462, New York: McGraw-Hill.

HEYWOOD, R. (1984) Prediction of adverse drug reactions from animal safety studies. In BOSTRUN, H. & LJUNGSTEDT, N. (Eds), *Detection and Prevention of Adverse Drug*

Reactions, pp. 173–189, Stockholm: Almquist and Wiksell Int.

IWAI, M., SAHEKI, S., OHTA, Y. & SHIMAZU, T. (1986) Footshock stress accelerates carbon tetrachloride-induced liver injury in rats. *Biomedical Research*, **7**, 145–154.

KOGURE, K., ISHIZAKI, M. & NEMOTO, M. (1986) Antishock effects of corticosterone on dextran-induced shock in rats. *American Journal of Physiology: Endocrinology and Metabolism*, **251**, 569–575.

KUMAR, M. S. & CHEN, C. L. (1983) Effects of an acute dose of delta 9-THC on hypothalmic luteinizing hormone releasing hormone and met-enkephalin content and serum levels of testosterone and corticosterone in rats. *Substance and Alcohol Actions and Misuse*, **4**, 37–43.

KUNITOMO, M., FUTAGAWA, Y., TANAKA, Y., YAMAGUCHI, Y. & BANDO, Y. (1989) Cholesterol reduces and corticosteroids enhance the toxicity of vitamin D in rats. *Japanese Journal of Pharmacology*, **49**, 381–388.

LEVY, A. D., LI, Q., ALVAREZ-SANZ, M. C., RITTENHOUSE, P. A., BROWNFIELD, M. S. & VAN DE KAR, L. D. (1992) Repeated cocaine modifies the neuroendocrine responses to the 5-HT (IC)/5-HT sub (2) receptor agonist DOI. *European Journal of Pharmacology*, **221**, 121–127.

LLOYD, S. A. & FRANKLIN, M. R. (1991) Modulation of carbon tetrachloride hepatotoxicity and xenobiotic-metabolizing enzymes by corticosterone pre-treatment, adrenalectomy and sham surgery. *Toxicology Letters*, **55**, 65–75.

MILLER, D. B. (1992) Caveats in hazard assessment – stress and neurotoxicity. In ISAACSON, R. L. & JENSEN, K. F. (Eds), *The Vulnerable Brain and Environmental Risks*, Vol. 1, *Malnutrition and Hazard Assessment*, pp. 239–266, New York: Plenum.

MOERLAND, J., STOWELL, L. & GJERDE, H. (1985) Ethanol increases rat liver tryptophan oxygenase: evidence for corticosterone mediation. *Alcohol*, **2**, 255–259.

MOHAMED, A. H., NAWAR, N. N. & MOHAMED, F. A. (1974) Influence of hydrocortisone on the microscopic changes produced by *Naja nigricollis* venom in kidney, liver and spleen. *Toxicon*, **12**, 45–48.

MOON, H. D., SIMPSON, M. E., LI, C. H. & EVANS, H. M. (1950) Neoplasms in rats treated with pituitary growth hormone II. Adrenal glands *Cancer Research*, **10**, 364–370.

ORTIZ DE MONTELLANO, P. R. (1986) *Cytochrome P-450 – Structure, Mechanism and Biochemistry*, New York: Plenum.

PATEL, V., BORYSENKO, M., KUMAR, M. S. S. & MILLARD, W. J. (1985) Effects of acute and subchronic delta 9-tetrahydrocannabinol on the plasma catecholamine, beta-endorphin and corticosterone levels and splenic natural killer cell activity in rats. *Proceedings of the Society for Experimental Biology and Medicine*, **180**, 400–404.

POLLACK, G. M., BROWNE, J. L., MARTON, J. & HABERER, L. J. (1991) Chronic stress impairs oxidative metabolism and hepatic excretion of model xenobiotic substrates in the rat. *Journal of Pharmaceutical Sciences*, **84**, 130–134.

PREZIOSI, P. & VACCA, M. (1983) Adrenocortical activation by 5-fluorouracil and its possible reversal by thymidine. *Archives of Toxicology*, **6**(Suppl.), 374–376.

PREZIOSI, P. & VACCA, M. (1988) Adrenocortical suppression and other endocrine effects of etomidate. *Life Sciences*, **42**, 477–489.

PREZIOSI, P., NUZIATA, A., MACRAE, S., RAGAZZONI, E. & VACCA, M. (1985) Possible adrenal involvement in hydroxyurea toxicity defence mechanisms. *Archives of Toxicology*, **8**(Suppl.), 380–384.

REHULKA, J. & KRAUS, M. (1987) Regulation of corticosterone metabolism in liver cell fractions in young and adult rats: cofactor requirements, effects of stress and phenobarbital treatment. *Physiologia Bohemoslovenica*, **36**, 21–32.

RIBELIN, W. E. (1984) The effects of drugs and chemicals upon the structure of the adrenal gland. *Fundamental and Applied Toxicology*, **4**, 105–119.

ROSENCRANS, J. A., SQUIBB, R. E., JOHNSON, J. H., TILSON, H. A. & HONG, J. S. (1985) Effects of neonatal chlordecone exposure on pituitary–adrenal function in adult

Fischer-344 rats. *Neurobehavioural Toxicology and Teratology*, **7**, 33–37.

SAPOLSKY, R. M. (1985) Glucocorticoid toxicity in the hippocampus: temporal aspects of neuronal vulnerability. *Brain Research*, **359**, 300–305.

SAPOLSKY, R. M. (1986a) Glucocorticoid toxicity in the hippocampus: temporal aspects of synergy with kainic acid. *Neuroendocrinology*, **43**, 440–444.

SAPOLSKY, R.M., (1986b) Glucocorticoid toxicity in the hippocampus: reversal by supplementation with brain fuels. *Journal of Neuroscience*, **6**, 2240–2244.

SAPOLSKY, R. M., PACKAN, D. R. & VALE, W. W. (1988) Glucocorticoid toxicity in the hippocampus: *in vitro* demonstration. *Brain Research*, **453**, 367–371.

SELYE, H. (1936) A syndrome produced by diverse nocuous agents. *Nature*, **138**, 32.

SELYE, H. (1950) *The Physiology and Pathology of Exposure to Stress*, Montreal: Acta Inc.

SHERRY, J. H., FLOWERS, L., O'DONNELL, J. P., LA CAGNIN, L. B. & COLBY, H. D. (1986) Metabolism of spironolactone by adrenocortical and hepatic microsomes: relationship to cytochrome P-450 destruction. *Journal of Pharmacology and Experimental Therapeutics*, **236**, 675–683.

SMELIK, P. G. & VERMES, I. (1980) The regulation of the pituitary–adrenal system in mammals. In CHESTER-JONES, I. & HENDERSON, I. W. (Eds), *General Comparative and Clinical Endocrinology of the Adrenal Cortex*, Vol 3, pp. 1–55, London: Academic Press.

SPINDEL, E. (1984) Action of methylxanthines on the pituitary and pituitary-dependent hormones. *Progress in Clinical and Biological Research*, **158**, 355–363.

SPINDEL, E., GRIFFITHS, L. & WURTMAN, R. J. (1983) Neuroendocrine effects of Caffeine. II. Effects on thyrotropin and corticosterone secretion. *Journal of Pharmacology and Experimental Therapeutics*, **225**, 346–350.

STORLIEN, L. H., HIGSON, F. M. & GLEESON, R. M. (1985) Effects of chronic lithium, amitriptylline and mianserin on glucoregulation, corticosterone and energy balance in the rat. *Pharmacology, Biochemistry and Behavior*, **22**, 119–125.

TEJWANI, C. A., GUDCHITHLU, K. P., HANISSIAN, S. H., GIENAPP, L. E., WHITEACRE, C. C. & MALARKEY, W. B. (1991) Facilitation of dimethylbenz-alpha-anthracene-induced rat mammary tumorigenesis by restraint: role of endorphins, prolactin and naltraxone. *Carcinogenesis*, **12**, 637–643.

THOMAS, J. A. (1993) Toxicology of the adrenal, thyroid and endocrine pancreas. In BALLANTYNE, B., MARRS, T. & TURNER, P. (Eds), *General & Applied Toxicology*, pp. 807–820, Basingstoke: Macmillan.

VOGEL, W. H. (1987) Stress – the neglected variable in experimental pharmacology and toxicology. *Trends in Pharmacological Sciences*, **8**, 35–38.

VOGEL, W. H. (1993) The effect of stress on toxicological investigations. *Human & Experimental Toxicology*, **12**, 265–271.

WALKER, R. F. & COOPER, R. L. (1992) Toxic effects of xenobiotics on the pituitary gland. In ATTERWILL, C. K. & FLACK, J. D. (Eds), *Endocrine Toxicology*, pp. 51–82, Cambridge: Cambridge University Press.

WERK, E. E., THRASHER, K., SHOLITON, L. J., OLINGER, C. & CHOI, Y. (1971) Cortisol production in epileptic patients treated with diphenylhydantoin. *Clinical Pharmacology and Therapeutics*, **12**, 698–703.

ZUCKERMAN, S. H. & QURESHI, N. (1992) *In vivo* inhibition of lipopolysaccharide-induced lethality and tumour necrosis factor synthesis by *Rhodobacter sphaeroides* diphosphoryl lipid A is dependent on corticosterone induction. *Infection and Immunity*, **60**, 2581–2587.

SECTION TWO

General Endocrinology, Pharmacology and Pathology of the Adrenal Gland Related to Species, Sex and Age and Glucocorticosteroid Actions

2

Adrenal Morphology and Hormone Synthesis and Regulation

J. P. HINSON

Queen Mary and Westfield College, London

P. W. RAVEN

Institute of Psychiatry, London

2.1 Introduction

There are many aspects of adrenal function which render this gland particularly susceptible to damage by toxins. In addition, the hormones of the adrenal gland, especially the steroids produced by the adrenal cortex, are an essential component of the physiological response to stress, and if adrenal function is impaired the consequences of toxic or other insult to the body are likely to be severe. This and the following chapter provide a summary of normal adrenal function, while giving an indication of why this tissue is vulnerable to toxicological assault, showing the importance of the corticosteroids in maintenance of homeostasis and describing the consequences of impaired adrenal function.

2.2 Functional Anatomy of the Mammalian Adrenal

In most mammalian species, including the rat, dog and human, the adrenal glands are paired organs, located close to the cephalic pole of the kidneys. The degree of proximity to the kidneys varies between species: in humans, for example, the adrenals are in direct contact with the kidneys, while in the rat they are removed from the kidneys, embedded in the perirenal fat. The adrenals comprise two endocrine tissues which are embryologically and functionally distinct: the adrenal cortex which is derived from the mesodermal lining of the coelom, and the adrenal medullary chromaffin tissue which originates in the neural crest and migrates into the centre of the cortical tissue during foetal development. The adrenal cortex secretes steroid hormones derived from cholesterol, while the principal secretion of the adrenal medulla is catecholamines, produced by metabolism of the amino acid, tyrosine. The unusual anatomical relationship between these very different tissue types

Figure 2.1 Cross-section through the normal rat adrenal, perfusion fixed under conditions of operative stress. G, Zona glomerulosa; F, zona fasciculata; R, zona reticularis; M, medulla. Figure reproduced with permission from Pudney *et al.* (1984).

strongly suggests some form of linked function, and indeed there is evidence both that steroids secreted by the adrenal cortex affect the biosynthesis of catecholamines (Ungar & Phillips 1983), and also that the adrenal medulla may influence cortical function (Hinson 1990). The interaction between these tissues is more of the nature of fine tuning of their respective functions however, rather than a major physiological regulatory mechanism.

2.2.1 Anatomy of the Adrenal Cortex

The division of the mammalian adrenal cortex into three distinct concentric zones was first described by Harley in 1858, and the terms zona glomerulosa, zona fasciculata and zona reticularis were introduced by Arnold in 1866 (see Neville & O'Hare 1982). Since that time, further species-specific and age-related divisions of the adrenal cortex have been described, including the foetal zone of the embryonic human adrenal cortex, and the zona intermedia in the rat.

The outermost layer of the adrenal cortex, immediately below the connective tissue capsule, is the zona glomerulosa. In the rat, this comprises four to six layers of cells, arranged in clusters forming a continuous band beneath the capsule (Figure 2.1), and occupies 10–15% of the gland volume (Nussdorfer 1980). In the human, the cells of the zona glomerulosa form discontinuous islets beneath the capsule (Figure 2.2), and comprise no more than 5% of the total cortical volume (Neville & O'Hare 1982). The zona glomerulosa in the dog has a very different appearance, consisting of large, flattened cells which stain palely and are stacked in large loops (Figure 2.3; Vinson *et al.* 1992).

The zona fasciculata of the rat adrenal consists of 30–40 layers of cells arranged in radial cords with parallel blood capillaries, and comprises the greater part of the adrenal cortex. A similar arrangement is seen in the adrenals of other mammals, including humans. The cells of this zone are larger than those of the zona glomer-

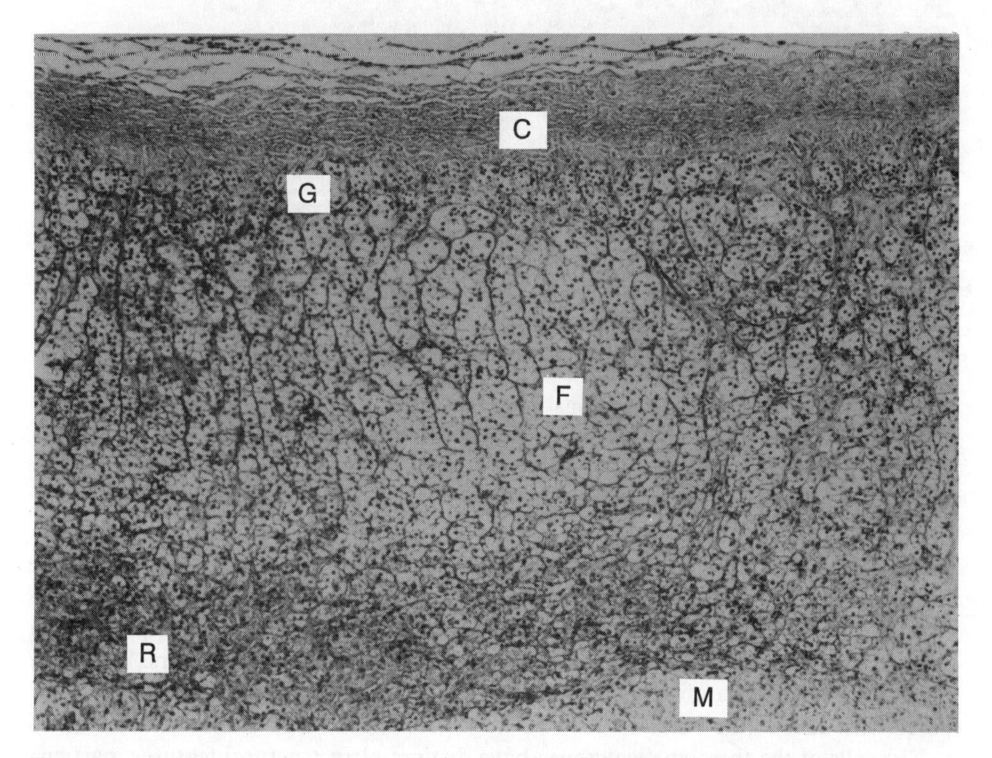

Figure 2.2 Cross-section through a normal human adrenal gland. G, Zona glomerulosa; F, zona fasciculata; R, zona reticularis; M, medulla; C, capsule. Figure reproduced with permission from Neville & O'Hare (1982).

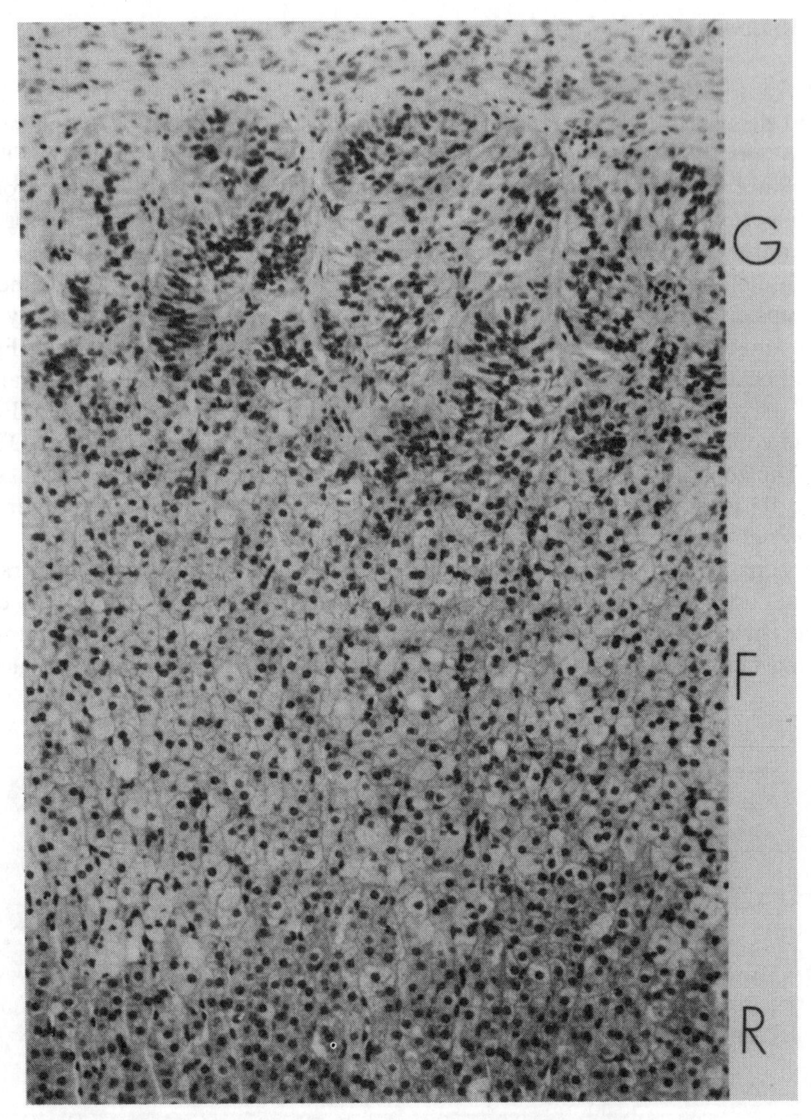

Figure 2.3 Cross-section through a normal dog adrenal gland. G, Zona glomerulosa; F, zona fasciculata; R, zona reticularis. Micrograph generously provided by I. Doniach.

ulosa and contain abundant lipid droplets which occupy 10–15% of the cell volume (Rhodin 1971). In man the cells of the zona fasciculata are termed clear cells.

The zona reticularis is the innermost zone of the adrenal cortex. Compared with the zona fasciculata, the cells are fewer, smaller and arranged in clusters. The zona reticularis comprises only about 5% of the total cortical volume in the rat, as compared with 25% in the human.

The cells of the three cortical zones have distinct ultrastructural features, particularly in relation to the shape and size of the mitochondria. For a detailed review of the ultrastructure of the mammalian adrenal cortex see Idelman (1978).

The intermediate zone is a feature of the rat adrenal cortex and is located between the zona glomerulosa and the zona fasciculata. The cells of this zone are small and contain few lipid droplets (Nussdorfer 1980). It has been suggested that this is the site of most cell division, but mitosis is seen throughout the adrenal cortex.

2.2.2 *The Adrenal Medulla*

The adrenal medulla comprises approximately 10% of the total volume of the adult human adrenal gland and consists mainly of chromaffin cells, so called because they become dark brown when stained with chromic salts. There are also some ganglion cells present. The chromaffin cells are small and contain numerous secretory vesicles, characteristic of neuroendocrine cells (McNicol 1992). There appears to be a degree of intermingling of medullary and cortical tissues, since islets of chromaffin cells have been identified within the zona glomerulosa of the rat adrenal cortex (Gallo-Payet *et al.* 1987).

2.2.3 *Age-related Changes in Adrenal Morphology*

The morphology of the human foetal adrenal gland has been definitively described by Neville and O'Hare (1982). In the human, the foetal adrenals are very large compared to the adult gland: relative to body weight the foetal adrenal is 35 times larger than in the adult. In the foetus, the adrenal gland consists of an outer definitive zone, comprising 15–30% of the gland volume and containing small basophilic cells arranged in arched cords, and a large inner foetal zone, containing larger eosinophilic cells arranged in tightly-packed cords. The adrenal gland increases in size throughout gestation, and at birth reaches a weight of around 4 g (the same size as the adult gland). After birth, however, the foetal zone rapidly degenerates, resulting in a 50% decrease in adrenal weight within the first month of life. Thereafter, the gland gradually decreases in size, reaching a weight of just 1 g at 12–24 months. During childhood, the gland grows only slowly, attaining a weight of 2 g by 10 years of age. During puberty, however, there is a rapid increase in adrenal growth, and by 15–18 years the gland has reached the adult size. Neville & O'Hare (1982) report that no sex differences have been noted during adrenal development.

No major changes in adrenal morphology occur in adult life, although this area has not been extensively studied. It has been reported that the pattern of cortical zonation becomes irregular with increasing years, although the zona reticularis remains well-defined. The adrenal glands of older people are also likely to bear adrenal nodules, which are 'overgrowths of adrenocortical cells . . . representing a variation of adrenal structure occurring predominantly as part of the aging process' (Neville & O'Hare 1982). These nodules do not appear to be associated with any abnormality of adrenocortical function.

Rat adrenal glands show marked changes with age. There is hypertrophy of the zona fasciculata and reticularis, with increases in both cell size and cell number (Rebuffat *et al.* 1992). It is likely that these changes result from increases in the circulating levels of ACTH seen in these aged rats (Rebuffat *et al.* 1992).

2.2.4 Sex Differences in Adrenal Morphology

In the rat, there is a well-documented sexual dimorphism: the adrenal gland of the female is significantly larger than that of the male, although the relative difference varies between strains. Adult female rats show increased size of all zones of the adrenal cortex, which may be attributable to the effects of oestrogens (Malendowicz 1994). This sex difference is not seen in either the dog or human adrenal.

2.2.5 Vascular Supply to the Adrenal

The adrenal gland is a highly vascular organ. The adrenal gland of the rat comprises approximately 0.02% of the body weight, but it receives about 0.14% of the cardiac output (Sapirstein & Goldman 1959). The arterial blood supply to the adrenal gland is mainly from the dorsal aorta, via several small arteries. In addition, in the human, and some strains of rat, there is an additional supply from branches of the renal and inferior phrenic arteries. Within the connective tissue capsule of the adrenal gland these arterioles anastamose and form a network which distributes blood into cortical sinuses (Figure 2.4). There has been some debate as to whether the blood supply to the cortex and medulla are continuous, or whether the medulla receives a separate supply directly from the subcapsular arteriolar plexus, via medullary arteries. The significance of these medullary arteries appears to vary according to the species. In the rat there are few medullary arteries, around six per gland, and these are unlikely to have a major functional role. In other species, however, most

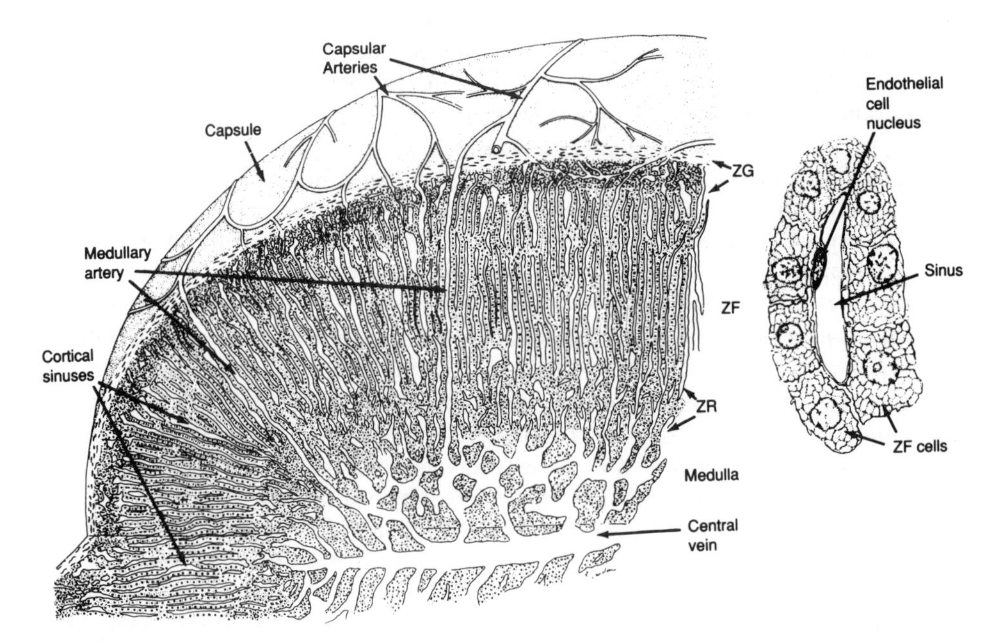

Figure 2.4 The vasculature of the mammalian adrenal gland. ZG, Zona glomerulosa; ZF, zona fasciculata; ZR, zona reticularis. The inset figure shows the relationship between a cortical sinus and zona fasciculata cells. Figure reproduced with permission from Vinson *et al.* (1992).

notably the cat and cow, there are up to 600 medullary arteries per gland, and it is therefore assumed that at least a part of the blood supply to the medulla is independent of the cortical vasculature (Vinson & Hinson 1992).

Blood flows centripetally through the adrenal cortex into the large medullary sinusoids, which drain into a central vein. There is no evidence for retrograde blood flow from the medulla to the cortical sinusoids in the rat. The intraglandular vasculature of the human adrenal is more complex and contrasts markedly with that of the rat, in that blood from the inner adrenocortical zones, and even from the medulla, may reach the outer cortex in two possible ways: first, via the arterio-venous loop; and second, in the region around the central vein, termed the cortical cuff, where there is an involution of the cortex and blood travels from medullary tissue through the cortex to the capsule (Vinson & Hinson 1992).

The arrangement of sinusoids within the gland is such that almost every cell of the adrenal cortex is in direct contact with a blood vessel. This high degree of vascularity has obvious implications both for the function of the gland and for the delivery of toxins to the adrenal. Blood flow through the adrenal is maintained independently of changes in systemic blood pressure and appears to be regulated to some degree by the splanchnic nerve, at least in dogs (Engeland *et al.* 1985). In the rat, dog and human there is evidence that ACTH regulates adrenal blood flow by causing vasodilation in the gland (Neville & O'Hare 1982, Vinson & Hinson 1992).

2.2.6 Innervation of the Adrenal Gland

The adrenal medulla is a component of the sympathetic nervous system, and receives pre-ganglionic innervation via the splanchnic nerve. The chromaffin tissue of the adrenal medulla can be considered to be a modified sympathetic ganglion, and catecholamine secretion by this tissue is regulated by the activity of the splanchnic nerve. In common with other sympathetic pre-ganglionic innervation, the innervation of the adrenal medulla is cholinergic. For many years it was a commonly held view that the nerve supply to the adrenal gland was exclusively to the medulla, and that the nerve bundles passed directly through the cortex without branching. Much evidence to the contrary has accumulated, however, and it is now widely accepted that the adrenal cortex receives a rich innervation, mainly in the region of the zona glomerulosa and the connective tissue capsule. A range of neurotransmitters has been identified in the adrenal cortex, including both catecholamines and neuropeptides (Kondo 1985, Vinson *et al.* 1994). Nerve terminals have been found in close contact with both blood vessels and the adrenocortical cells, and it has been shown that certain neurotransmitters are able to influence blood flow and steroid secretion in the adrenal gland (Vinson *et al.* 1994).

Some of the nerve fibres in the adrenal cortex appear to originate in the medulla and are regulated by the splanchnic nerve, while others appear to be independent of splanchnic nerve activity and are presumed to originate outside the adrenal gland (Holzwarth *et al.* 1987). Although the great majority of these studies have been carried out in rats, there is evidence of adrenocortical innervation in many other species, including the dog and human (for reviews see Edwards & Jones 1993, Malendowicz 1993, Vinson *et al.* 1994).

2.2.7 Effects of ACTH and Other Agents on Adrenocortical Morphology

The morphology of the adrenal gland changes in response to both increased and decreased levels of ACTH, and also in response to changes in electrolyte status. Lack of ACTH results in adrenal atrophy, with a marked decrease in adrenal weight, mostly due to a decrease in cell number and cell volume in the zona fasciculata and zona reticularis. There is only a transitory effect on the zona glomerulosa, however, and several weeks after hypophysectomy the zona glomerulosa has a normal appearance, although the responsiveness of the cells may be blunted (Vinson *et al.* 1992).

High levels of ACTH cause an increase in the width of the zona fasciculata and zona reticularis, with an increase in both cell size and number, resulting in a significant increase in adrenal weight. There is evidence, at least in the rat, that chronic ACTH administration causes a loss of zona glomerulosa function, with transformation of the glomerulosa cells into fasciculata-type cells, accompanied by a decrease in aldosterone biosynthetic capacity (Abayasekara *et al.* 1989). ACTH also has a profound effect on the adrenal vasculature causing a marked vasodilation, which may ultimately result in a breakdown of the delicate internal vasculature, with extravasation of erythrocytes and adrenal infarction (Pudney *et al.* 1984). Similar effects are seen in the adrenals of patients following prolonged stress (Neville & O'Hare 1982). There is also evidence that anticoagulant therapy, with either warfarin or heparin, may result in adrenal haemorrhagic infarction (O'Connell & Aston 1974), which causes acute adrenal insufficiency. The mechanism of this effect is unknown, but the combination of a stress with anticoagulant therapy greatly increases the risk of adrenal haemorrhage (Rao 1995).

The morphology of the zona glomerulosa changes markedly in response to changes in electrolyte status: both sodium deficiency and potassium loading cause an increase in the width of the zona glomerulosa, with increased cell size and number. Sodium loading has the opposite effect (Nussdorfer 1980). Several other agents have been shown to exert a trophic effect on the rat zona glomerulosa, including angiotensin II. The angiotensin-converting enzyme inhibitor, captopril, which is used in the treatment of hypertension, causes atrophy of the rat zona glomerulosa and concomitant loss of aldosterone secretory capacity. This effect is reversed by angiotensin II infusion (Mazzocchi & Nussdorfer 1984).

One agent known to cause distinctive morphological changes in the adrenal cortex is spironolactone, the mineralocorticoid receptor antagonist commonly used to treat hyperaldosteronism. This drug has been reported to induce the formation of 'spironolactone bodies' which are small, round, laminated concentric whorls of membrane in the cytoplasm of zona glomerulosa cells (see Neville & O'Hare 1982). A wide range of drugs and other chemicals have been reported to cause lesions in both the adrenal cortex and medulla. These are reviewed by Ribelin (1984).

There are several features of the morphology of the adrenal gland which render this tissue particularly susceptible to toxins: the blood supply and high lipid content have been described above. In addition, adrenocortical membranes have a high content of unsaturated fatty acids which act as substrates for lipid peroxidation. It has been suggested that this accounts for the particular susceptibility of the gland to peroxidative damage (Colby & Longhurst 1992).

2.3 Hormones of the Adrenal Gland

2.3.1 *Corticosteroids*

The adrenal cortex is capable of producing about 50 different steroids with a wide range of activities. In most species, including the human, the most physiologically important of these corticosteroids are aldosterone, a mineralocorticoid, and cortisol, a glucocorticoid. The most abundant steroid produced by the adrenal cortex, however, is an androgen, dehydroepiandrosterone sulphate (DHAS). The adrenal cortex also produces oestrogens, progesterone, and a wide range of precursors and metabolites of these steroids. In the rat, and a few other rodents which lack the 17α-hydroxylase activity necessary for cortisol and androgen production, the major glucocorticoid is corticosterone and there is negligible androgen production. Recent evidence suggests that the mammalian adrenal cortex may also produce a ouabain-like compound, with Na/K-ATPase-inhibiting activity (Hinson *et al.* 1995).

2.3.2 *Steroid Biosynthesis*

The adrenal steroids are all synthesized from cholesterol, mainly by a series of hydroxylations involving the cytochrome P450 family of enzymes. The major pathways of adrenocortical steroid biosynthesis are shown in Figure 2.5, and the characteristics of the enzymes involved are shown in Table 2.1.

The cholesterol used in steroid biosynthesis is derived from two sources: *de novo* synthesis from acetate in the adrenal, or receptor-mediated uptake of plasma lipoproteins. In the rat and human, around 80% of the cholesterol used in steroidogenesis is derived from circulating lipoproteins, although this varies between species and according to the state of stimulation of the gland. In the human most cholesterol in plasma is associated with low density lipoproteins (LDL), while in rats cholesterol is predominantly associated with high density lipoproteins (HDL). It follows that any lipophilic toxin associated with these plasma lipoproteins will also be taken up and stored within the adrenal gland. There is only a small amount of free cholesterol in adrenocortical cells. Most cholesterol is stored in lipid droplets, in an esterified form, which is rapidly accessible in response to acute stimulation of steroidogenesis and is then replenished (Vinson *et al.* 1992).

The endpoint of conversion of cholesterol to steroid hormones is zone-specific. In all mammalian species the zona glomerulosa is the only site of aldosterone synthesis. In rats the inner adrenocortical zones produce corticosterone as the major secretory product. In the dog and human the inner zones favour the 17α-hydroxy pathway, with the zona fasciculata mainly producing cortisol and the zona reticularis mainly producing androgens and sulphated steroids.

The pathway for aldosterone biosynthesis has, until recently, been something of an enigma. It was recognized that the formation of aldosterone was catalysed by a mitochondrial enzyme, very similar to the cytochrome P450 enzyme converting deoxycorticosterone to corticosterone. This enzyme was therefore termed cytochrome $P450_{11\beta/18}$ as it appeared to carry out both 11β-hydroxylation and 18-hydroxylation. It was usually found, however, that deoxycorticosterone was more readily converted to aldosterone than was corticosterone, and recently a second

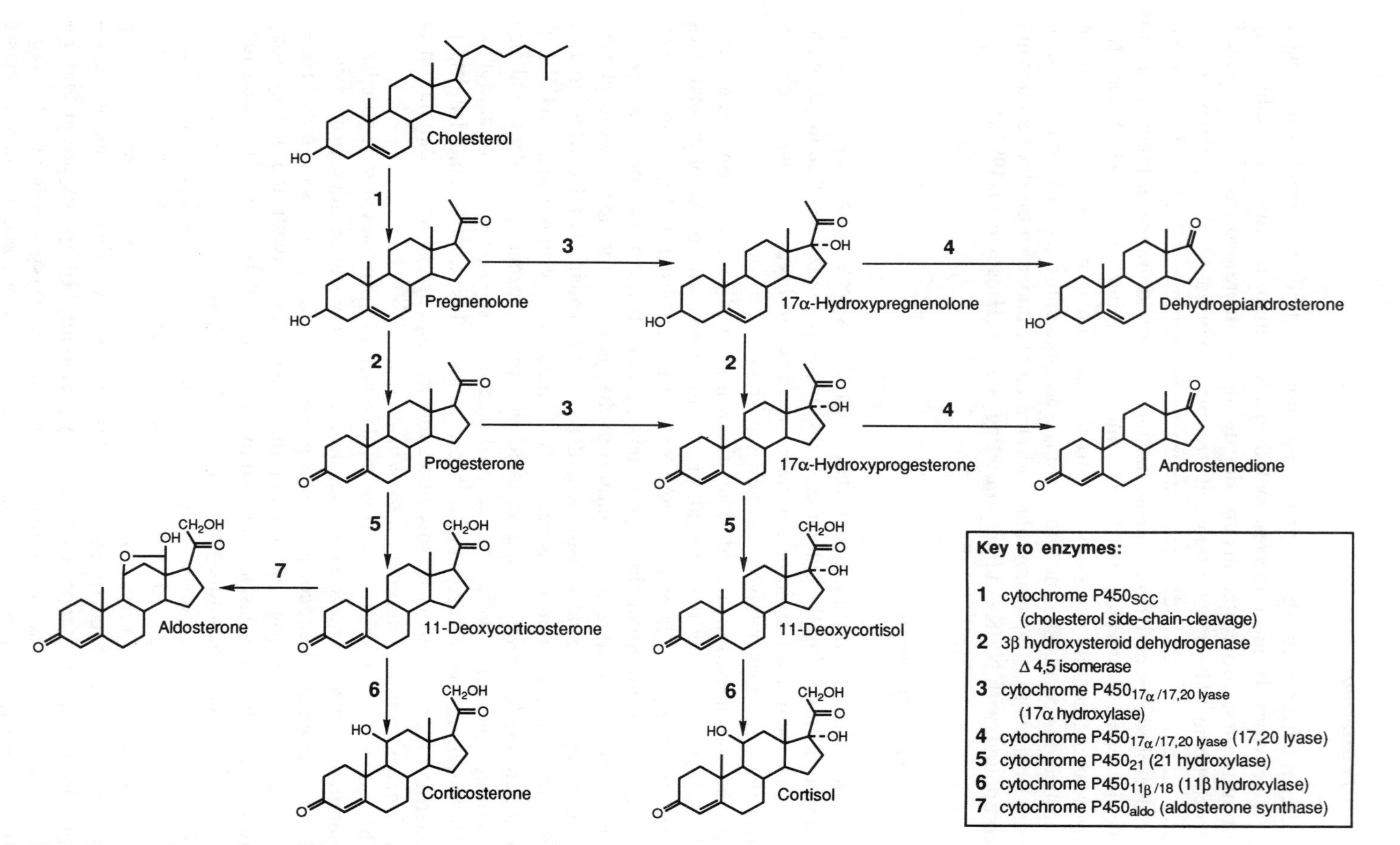

Figure 2.5 Pathways of steroid biosynthesis in the adrenal cortex.

Table 2.1 Characteristics of the enzymes of corticosteroid biosynthesis

Enzyme	Activity	Major biosynthetic step	Site	Cofactor	Inhibitors
$P450_{SCC}$ (P450 XIA family)	20α,22R Hydroxylase; Cholesterol side chain cleavage	Cholesterol → pregnenolone	Mitochondrial	NADPH	[a]Aminoglutethimide (Camacho *et al.* 1967)
3β-Hydroxysteroid dehydrogenase	5-ene-3β-Hydroxysteroid oxoreductase	Pregnenolone → progesterone	Microsomal	$NADP^+$	Cyanoketone (McCarthy *et al.* 1996)
	5-ene-3-Oxosteroid-4,5-isomerase	17α-OH-Pregnenolone → 17α-OH progesterone			[a]Trilostane (Potts *et al.* 1978)
$P450_{17\alpha}$	17α Hydroxylase	Pregnenolone → 17α-OH-pregnenolone	Microsomal	NADPH	Spironolactone (Kossor *et al.* 1991)
(P450 XVII family)	17–20 Lyase	Progesterone → 17α-OH-progesterone			[a]Ketoconazole (Loose *et al.* 1983)
$P450_{21}$	21 Hydroxylase	Progesterone → DOC	Microsomal	NADPH	RU486 (Albertson *et al.* 1994)
(P450 XXI family)		17α-OH-Progesterone → 11-deoxycortisol			
$P450_{11\beta}$	11β Hydroxylase	DOC → corticosterone	Mitochondrial	NADPH	[a]Metyrapone (Liddle *et al.* 1958)
(P450 XIB family)	18 Hydroxylase	11-Deoxycortisol → cortisol			[a]*o,p*′-DDD (mitotane) and DMBA (Hornsby 1989)
Aldosterone synthase	?	DOC → aldosterone	Mitochondrial	NADPH	Guanabenz-related amidinohydrolases (Soll *et al.* 1994)

Note: [a] Used therapeutically due to effects on adrenal enzymes.

form of the enzyme has been identified which catalyses all the reactions involved in this conversion, without apparently releasing any intermediates (Yanagibashi *et al.* 1986, Lauber *et al.* 1987). This enzyme has been termed aldosterone synthase and, in the rat adrenal cortex, is found exclusively in the zona glomerulosa (Yabu *et al.* 1991). A comparable enzyme is present in the human adrenal (Mornet *et al.* 1989).

2.3.3 Adrenocortical Cytochrome P450 Enzymes

The enzymes which are responsible for biosynthetic hydroxylations in the adrenal cortex (see Table 2.1) are from the same family of cytochrome P450 enzymes found in the liver. The hepatic enzymes have been extensively studied due to their involvement in activation of toxins and detoxification processes (Williams 1973). In the adrenal cortex, in addition to their central role in steroidogenesis, cytochrome P450 enzymes are important mediators of toxicological effects: they activate molecular oxygen for oxidation of substrate; they are sensitive to pseudosubstrate toxicity; and they are targets for pharmacological inhibitors and exogenous toxins. As a result of the use of molecular oxygen in corticosteroid biosynthesis, cells are particularly susceptible to the toxic effects of free radicals, such as lipid peroxidation (Hornsby and Crivello 1983a).

Adrenocortical cells contain high concentrations of biological antioxidants including superoxide dismutase (SOD), catalase, α-tocopherol, glutathione and, in particular, ascorbic acid (Hornsby and Crivello 1983b). It has been suggested that these high levels of antioxidants serve to protect cytochrome P450 enzymes from toxic effects of oxygen radicals generated during steroid biosynthesis (Hornsby 1989). It has been observed that adrenocortical levels of α-tocopherol are much higher in female than male rats (Feingold and Colby 1992), although the significance of this observation remains unclear.

In vitro studies have shown that some steroids, whose structure prevents their hydroxylation by a particular cytochrome P450, are capable of acting as pseudosubstrates, resulting in peroxidative damage to the enzyme itself (Hallberg 1990). It has been suggested that the high concentrations of steroids capable of acting as pseudosubstrates, together with the centripetal blood flow of the adrenal gland, provide a mechanism by which regional differentiation of biosynthetic activity may be maintained (Hornsby and Crivello 1983b).

Examples of pharmacological inhibitors and exogenous toxins which target specific adrenal cytochrome P450 enzymes are given in Table 2.1. Although some of these have direct effects on the enzyme, it is increasingly recognized that adrenocortical cytochrome P450 enzymes may play a role in the activation of xenobiotics (Hallberg 1990). For example, metyrapone acts as a competitive inhibitor of $P450_{11\beta}$, resulting in decreased synthesis of corticosterone and cortisol and in increased production of their 11-deoxy precursors, particularly 11-deoxycorticosterone (DOC) and 11-deoxycortisol (Dollery 1991). In contrast, spironolactone requires activation by $P450_{17\alpha}$ to yield 7α-thiol-spironolactone, which causes irreversible damage to $P450_{17\alpha}$ and other adrenal cytochrome P450 enzymes (Kossor *et al.* 1991). The activity of adrenal cytochrome P450 enzymes in metabolizing xenobiotics shows large interspecies differences, and is also age- and sex-dependent (Hallberg 1990). For example, adrenal microsomes from human foetus (Rifkind *et al.* 1987) and rat (Montelius 1982) have higher aryl hydrocarbon hydroxylase (AHH) activities than their hepatic counterparts, while adult human adrenal microsomes appear not to possess any AHH activity (Papadopoulos *et al.* 1984). Similarly, *o,p'*-DDD (mitotane) is a potent inhibitor of steroidogenesis in dogs and humans, but not in rats (Vinson *et al.* 1992).

2.3.4 *Catecholamines*

The major secretory products of the adrenal medulla are the catecholamines, although this tissue also secretes a variety of neuropeptides. Catecholamines are the group of amines containing a 3,4-dihydroxyphenyl (catechol) nucleus. The endogenous catecholamines are adrenaline, noradrenaline and dopamine, which are synthesized in the adrenal medulla from the amino acid, tyrosine (see Figure 2.6). The first stage in catecholamine synthesis is the uptake of tyrosine by the chromaffin cells. This occurs by a passive mechanism of facilitated diffusion, and may be inhibited by structural analogues of tyrosine (Levin 1986). All of the enzymes involved in catecholamine biosynthesis are found in the cytosol of chromaffin cells, except dopamine β-hydroxylase, which is located within the secretory granules. For adrenaline biosynthesis, therefore, the noradrenaline must be transported from the secretory granules into the cytosol, and the adrenaline formed is then repackaged in secretory granules (Cryer 1992). The characteristics of the enzymes involved in catecholamine biosynthesis are shown in Table 2.2. Adrenaline is the major catecholamine produced in the human, dog and rat. In contrast to the other enzymes of catecholamine biosynthesis, phenylethanolamine-N-methyltransferase (PNMT), the enzyme responsible for adrenaline production, is found almost exclusively in the adrenal medulla, although it is not expressed in all chromaffin cells. The expression of PNMT is regulated by glucocorticoids (Ungar & Phillips 1983). Of all the mammalian species studied, the cat is the only species which produces equal amounts of adrenaline and noradrenaline. Noradrenaline comprises approximately 20% of the secreted product in the human and dog, but only 10% in rodents (Ungar & Phillips 1983).

The catecholamines are stored in secretory granules within the cell prior to their release. The granules also contain ATP and proteins, termed chromogranins, which are co-secreted with the catecholamines.

2.4 Regulation of Adrenal Function

In relation to the hormones of the adrenal cortex, the terms 'synthesis' and 'secretion' are used interchangeably as no significant storage of aldosterone or cortisol occurs in adrenocortical cells. Instead, the cells of the adrenal cortex store the precursor of steroid hormone biosynthesis, cholesterol. Exactly the reverse is true of the hormones of the adrenal medulla. The chromaffin cells store preformed catecholamines in vesicles, from which they can be rapidly released in response to stimulation. Thus, in the adrenal medulla it is appropriate to consider synthesis and secretion of catecholamines as distinct events.

2.4.1 *Control of Glucocorticoid Secretion*

Glucocorticoid secretion is regulated almost exclusively by corticotrophin (ACTH), a 39 amino acid peptide hormone released by the anterior pituitary gland. ACTH is

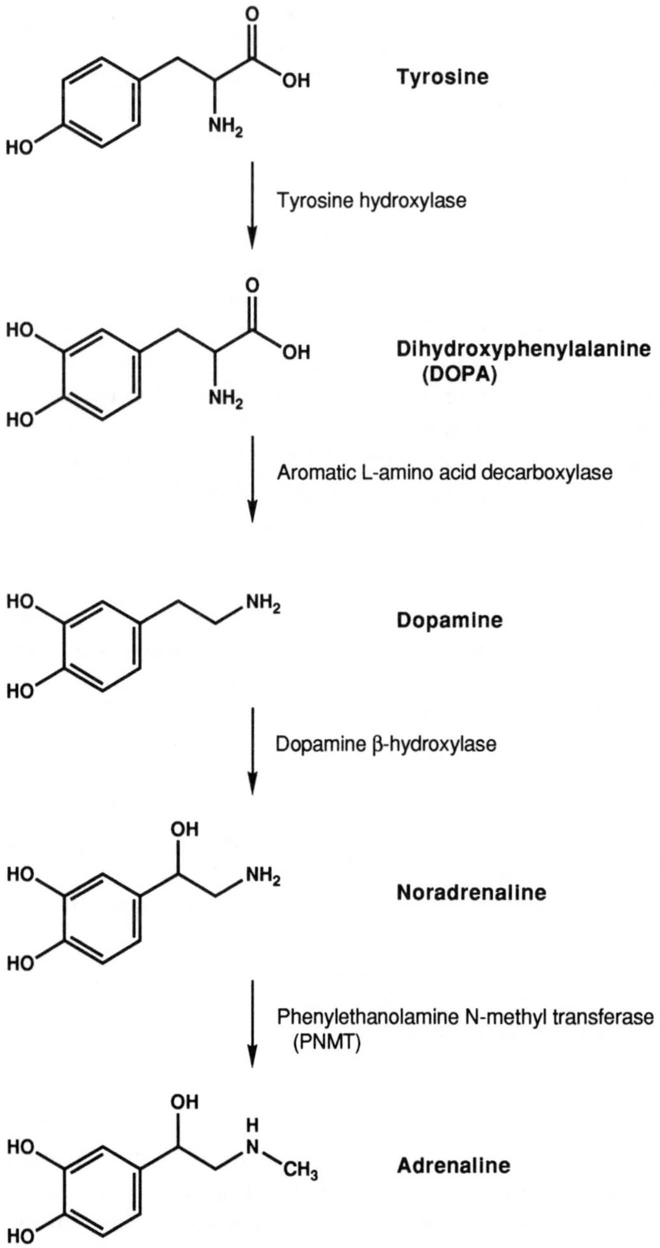

Tyrosine

Tyrosine hydroxylase

Dihydroxyphenylalanine (DOPA)

Aromatic L-amino acid decarboxylase

Dopamine

Dopamine β-hydroxylase

Noradrenaline

Phenylethanolamine N-methyl transferase (PNMT)

Adrenaline

Figure 2.6 Pathway of catecholamine biosynthesis in the adrenal medulla.

produced as a cleavage product of a large precursor molecule, proopiomelanocortin, which also gives rise to opioid peptides and the melanocyte-stimulating hormones. The release of ACTH is controlled by corticotrophin-releasing hormone (CRH) and vasopressin (AVP) from the hypothalamus, and also by negative feedback by glucocorticoids (Figure 2.7). The major stimulus to activation of the hypothalamo-pituitary-adrenal (HPA) axis is stress, which may be a psychological stressor, or a

Table 2.2 Characteristics of the enzymes of catecholamine biosynthesis

Enzyme	Biosynthetic step	Site	Co-factor	Inhibitors
Tyrosine hydroxylase	Tyrosine → DOPA	Cytosol	Tetrahydrobiopterin	α-Methyl-*p*-tyrosine (Levin 1986)
Aromatic-L-amino acid decarboxylase (DOPA decarboxylase)	DOPA → dopamine	Cytosol	Pyridoxal phosphate	[a]Benserazide, [a]carbidopa (Yahr 1973)
Dopamine-β-hydroxylase	Dopamine → noradrenaline	Catecholamine storage granules	Ascorbate	Disulfiram (Musacchio *et al.* 1964)
Phenylethanolamine N-methyl transferase	Noradrenaline → adrenaline	Cytosol	S-Adenosyl methionine (SAM)	13,14-Dihydro-15-keto-prostaglandin $F_{2\alpha}$ (Morimoto *et al.* 1983)

Note: [a] Used therapeutically due to effects on adrenal enzymes.

physical stressor such as exercise, cold exposure, infection, hypoglycaemia or general anaesthesia [for a review of ACTH secretion see Buckingham *et al.* (1992) and Dallman *et al.* (1987)].

Corticotrophin has several distinct effects on the adrenal gland which are evident at different time intervals after stimulation. Acutely, ACTH causes an increase in the rate of blood flow through the adrenal gland and an increase in the rate of steroid secretion, which is immediately preceded by a decrease in the adrenal content of ascorbic acid. The long-term effects of ACTH include stimulation of the growth of the adrenal cortex and increased expression of the enzymes involved in steroid biosynthesis (Simpson & Waterman 1992).

The adrenal requires a certain level of ACTH secretion to maintain its normal structure and function. In the absence of this peptide, glucocorticoid secretion

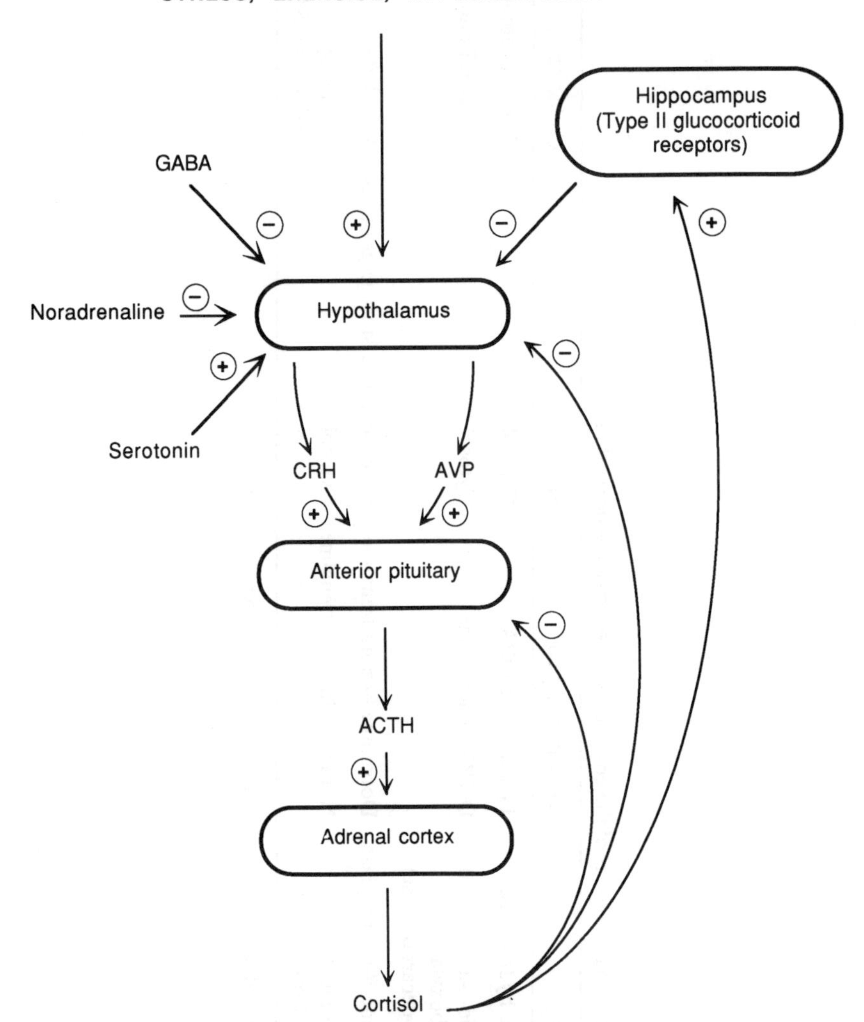

Figure 2.7 Regulation of cortisol secretion. A positive sign indicates a stimulatory effect, a negative sign indicates an inhibitory effect; AVP, arginine vasopressin; CRH, corticotrophin-releasing hormone; ACTH, corticotrophin.

rapidly falls to undetectable levels and the adrenal gland becomes atrophied. One side-effect of the prolonged use of high levels of synthetic glucocorticoids, used to treat inflammatory disorders for example, is the long-term inhibition of ACTH secretion. This results in loss of the HPA response to stress, which is potentially life-threatening. The HPA axis can take up to 2 years to recover fully from the suppressive effects of exogenous glucocorticoids (Graber *et al.* 1965). Other agents which have been reported to inhibit ACTH secretion include cyproheptadine, an antihistamine, and bromocriptine, a dopamine agonist (Lamberts *et al.* 1980, Allolio *et al.* 1987). Excessive secretion of ACTH leads to adrenal hypertrophy with over-production of cortisol resulting in Cushing's syndrome (see Chapter 3), and both cyproheptadine and bromocriptine have been used in the treatment of this disorder (Walker & Edwards 1992).

There is evidence that other factors may play a role in regulating glucocorticoid secretion. In the rat there is a close relationship between the rate of blood flow through the gland and the rate of corticosterone biosynthesis. Endothelin has been shown to stimulate glucocorticoid secretion by both rat and human adrenal cells, and it has been suggested that this peptide may mediate the effect of flow on steroidogenesis (Hinson *et al.* 1991).

There are significant age and sex differences in the control of glucocorticoid secretion. There is evidence of altered HPA function with increased age, certainly in rats, but possibly also in humans (Seeman & Robbins 1994). In response to a given stressor, the cortisol response remains elevated for a longer period with increasing age, suggesting decreased resiliency of the HPA axis. In rats, there is a decrease in the rate of corticosterone secretion per gram of adrenal wet weight, and a diminished response to ACTH (Reaven *et al.* 1988). However, there is disagreement as to the effect of these changes on circulating corticosterone levels. Some authors suggest that there is no change in the circulating level of corticosterone, as ACTH concentrations are increased, and the gland is enlarged, presumably compensating for the diminished capacity of the adrenocortical cells to secrete steroids (Rebuffat *et al.* 1992). Other authors, however, describe increased levels of circulating corticosterone in aged rats, which they attribute to increased activity of the hypothalamus and pituitary (Scaccianoce *et al.* 1990). In rats, there is also flattening of the diurnal rhythm of corticosterone secretion, with attenuation of the ACTH–corticosterone response to increases in CRH (Hauger *et al.* 1994). Thus while the direction of change in corticosterone secretion in aged rats is disputed, there is general agreement that the hypothalamus and pituitary show increased activity. There is also evidence in rats of sex differences in the responsiveness of adrenocortical tissue. Adrenals from female rats show a much greater response to ACTH stimulation than those of male rats (Vinson *et al.* 1978). It is not clear whether this observation is related to the sexual dimorphism seen in rat adrenal morphology (see above).

2.4.2 Regulation of Aldosterone Secretion

The regulation of aldosterone secretion is more complex, involving the interaction of several different systemic factors [for a review see Muller (1988) and Vinson *et al.* (1992)]. Of these factors, the renin–angiotensin system is arguably the most important (Figure 2.8). Renin is an enzyme produced by the juxtaglomerular cells of the kidney in response to a decreased perfusion pressure or a fall in the plasma sodium concentration. Secretion of renin is also altered in response to postural changes in the human: assuming an upright posture stimulates renin and thus aldosterone secretion. Despite being released in an apparently endocrine manner, renin cannot be considered to be a hormone because it does not act via a receptor. In fact, it is an enzyme which cleaves angiotensinogen, a circulating protein synthesized by the liver, to produce angiotensin I. This decapeptide is cleaved by the imaginatively-termed angiotensin converting enzyme (ACE) which is found in vascular epithelial cells, mainly in the lung. The product of this action is an octapeptide, angiotensin II, which is a potent vasoconstrictor in addition to being the major stimulus to aldosterone secretion. The inhibition of ACE activity by captopril is used to treat certain types of hypertension. In addition to preventing the vasoconstrictive effects of angiotensin II, captopril also causes inhibition of aldosterone secretion.

The history of the investigation into the adrenal actions of angiotensin II serves as a salutary lesson in the pitfalls of endocrine research. The effect of angiotensin II infusion on aldosterone secretion in humans was demonstrated in 1960 (Laragh *et al.* 1960), and the effects of this peptide in dogs were reported in 1962 (Davis 1962, Ganong *et al.* 1962). Use of adrenal slices *in vitro* confirmed that this was a direct effect of angiotensin II on the adrenal cortex (Kaplan & Bartter 1962). However, it was not until several years later that the effect of this peptide on aldosterone secretion in the rat was confirmed, since the response of rat adrenals to angiotensin II is highly dependent on tissue preparation. No response is seen in intact anaesthetized rats (Cade & Perenich 1965, Marieb & Mulrow 1965), nor in quartered or sliced rat adrenal preparations *in vitro*. These authors concluded that angiotensin II was

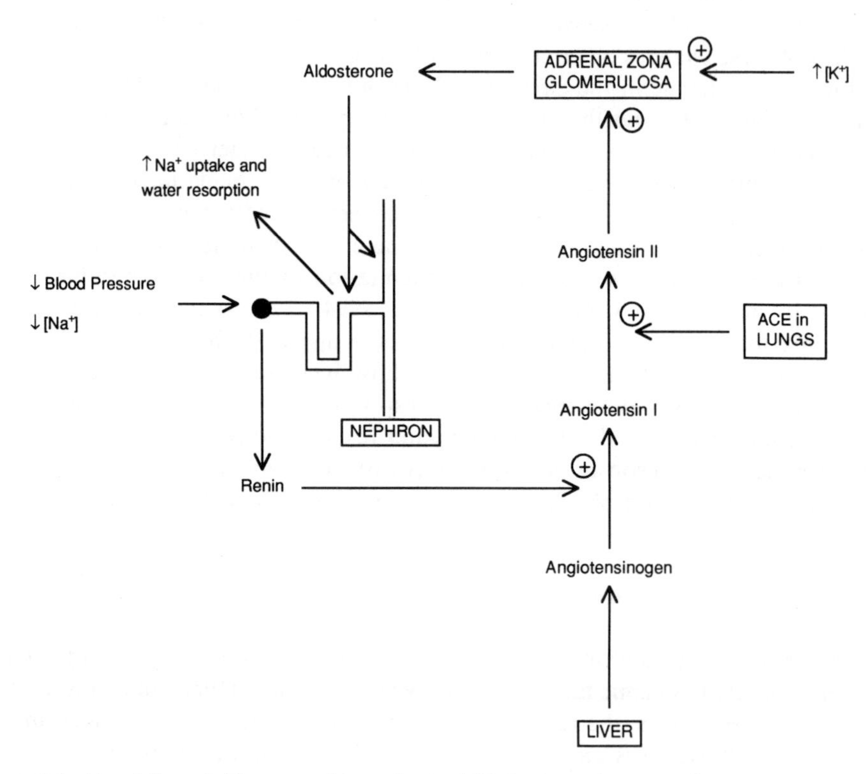

Figure 2.8 Regulation of aldosterone biosynthesis. ACE, Angiotensin-converting enzyme.

unlikely to have a significant role in regulating aldosterone secretion in the rat. It was only when enzymatically-dispersed cell preparations were used that the direct effects of angiotensin II on rat adrenal function were established (Haning *et al.* 1970). Subsequent studies revealed that the effects of angiotensin II in the rat are significantly attenuated by anaesthesia (Coleman *et al.* 1974). Clearly the choice of an appropriate adrenal preparation is crucial when investigating the effects of different agents on adrenocortical function [for a review see Vinson *et al.* (1985)].

Angiotensin II is not the only stimulant of aldosterone secretion, however. It has been known for many years that both electrolyte status and a factor from the pituitary gland have important roles in regulating aldosterone secretion. The pituitary factor was evidently not ACTH as chronic ACTH administration causes loss of aldosterone biosynthetic capacity. It now appears most likely that α-melanocyte-stimulating hormone (α-MSH) has a role in regulating zona glomerulosa function, both in the rat and human (Vinson *et al.* 1981, Henville *et al.* 1989). Plasma potassium concentration is also important. Supraphysiological concentrations of potassium ions act directly on the zona glomerulosa to stimulate aldosterone secretion. There is evidence, however, that very small changes in potassium ion concentration, within the normal physiological range, act to modulate adrenocortical sensitivity to stimulation by angiotensin II, by causing changes in the number, and possibly affinity, of angiotensin II receptors (Douglas & Catt 1976). There are several other factors known to stimulate aldosterone secretion, including endothelin (Hinson *et al.* 1991), a range of neuropeptides (Hinson *et al.* 1994) and serotonin, which is effective in the rat, dog and human, but not in sheep (Muller 1988). The diuretic drug, frusemide, is known to stimulate aldosterone secretion, presumably due to its effects on electrolyte status.

In addition to the factors which cause stimulation of aldosterone secretion, several agents have been shown to inhibit zona glomerulosa function. Atrial natriuretic peptide (ANP) is secreted by the cardiac atrium in response to volume expansion and high plasma sodium concentration (Atlas 1986). This hormone acts on the kidney to cause natriuresis and inhibition of renin release, and also acts on the adrenal zona glomerulosa to inhibit aldosterone secretion (Kudo & Baird 1984, Atlas 1986). In dogs, rats and humans, this inhibitory effect is specific to the zona glomerulosa as glucocorticoid secretion is not affected by ANP (Maack *et al.* 1984, Atarashi *et al.* 1985, Richards *et al.* 1985). Dopamine and its agonists also inhibit aldosterone secretion, and it has been suggested that the zona glomerulosa may be under the tonic inhibitory control of dopamine (Carey *et al.* 1979). Metoclopramide, a dopamine antagonist, has been reported to stimulate aldosterone secretion in the human (Norbiato *et al.* 1977, Lauer *et al.* 1982). In rats, however, there are reports of both stimulatory (Aguilera & Catt 1984) and inhibitory (Campbell *et al.* 1981, Lauer *et al.* 1982) effects of dopamine on aldosterone secretion. Sowers *et al.* (1981) described a species difference in the adrenal effects of metoclopramide which stimulated aldosterone secretion in primates but not in dogs or rabbits. Thus at least some of the differences in results obtained by other workers may be due to species variation.

Sex differences in aldosterone secretion have been reported. The aldosterone secretion rate of women in the follicular phase of the menstrual cycle is comparable to that of men (135–140 μg/day). In the luteal phase, however, there is a significant increase in aldosterone secretion (235 μg/day) (Gray *et al.* 1968). In rats, blood aldosterone concentration is higher in the oestrus phase compared with dioestrus. In the human, rat and dog, aldosterone secretion rates are significantly increased during the later stages of pregnancy. In the dog and rat this appears to be due to an increase in plasma renin activity, while in the human there is an increase in angiotensinogen production (Muller 1988).

Aldosterone secretion also changes with age. There is a significant decrease in the rate of aldosterone secretion with increasing age in the human. This decrease is particularly marked when patients assume an upright posture, or are sodium depleted, and may be explained by decreases in plasma renin activity (see Muller 1988).

2.4.3 Control of Adrenal Androgen Secretion

Although the foetal adrenal secretes large amounts of androgens, production switches to glucocorticoids soon after birth. Adrenal androgen production remains low until puberty when the adrenal cortex starts to secrete significant amounts of androgen, most notably dehydroepiandrosterone sulphate (De Peretti & Forest 1976). This increase in adrenal androgen production is termed *adrenarche*, and coincides with the appearance of pubic and axillary hair. It has been suggested that a specific adrenal androgen stimulating hormone is responsible for adrenarche, but such a hormone has yet to be identified despite considerable research effort in this area (McKenna & Cunningham 1991). There is a decline in the production of adrenal androgens in the fifth decade of life, a phenomenon which has been termed *adrenopause.*

2.4.4 Intracellular Mechanisms in the Regulation of Steroidogenesis

The major regulators of adrenocortical function, ACTH and angiotensin II, are peptide hormones, and therefore act on plasma membrane receptors. The receptors for both ACTH and angiotensin II are members of the superfamily of plasma membrane receptors which possess seven transmembrane domains. The general mechanism for signal transduction is similar for both hormones, in that the hormone-receptor complex activates a G-protein which in turn relays the signal to an enzyme. This enzyme catalyses the formation of a second messenger, thus leading to the cellular response via the activation of protein kinase (see Figure 2.9).

ACTH binds to specific receptors on adrenocortical cells, causing the G-protein-mediated activation of adenylyl cyclase which results in increased cAMP formation. cAMP activates protein kinase A, causing the phosphorylation of its target proteins, which results in the stimulation of steroidogenesis. In the long term ACTH acts to increase expression of the enzymes involved in steroid biosynthesis, but the acute action of ACTH on steroidogenesis is a result of increased delivery of cholesterol to the cytochrome $P450_{SCC}$, the rate-limiting step of glucocorticoid biosynthesis. In part, this is achieved by increasing the concentration of free cholesterol within the cell, as one of the enzymes activated by protein kinase A is cholesterol ester hydrolase. ACTH also activates a 'labile protein', which is closely related to the hepatic sterol carrier protein, SCP2. This protein facilitates the transfer of cholesterol across the mitochondrial membrane and thus increases the rate of delivery of cholesterol to cytochrome $P450_{SCC}$ which is located on the inner mitochondrial membrane (Vinson

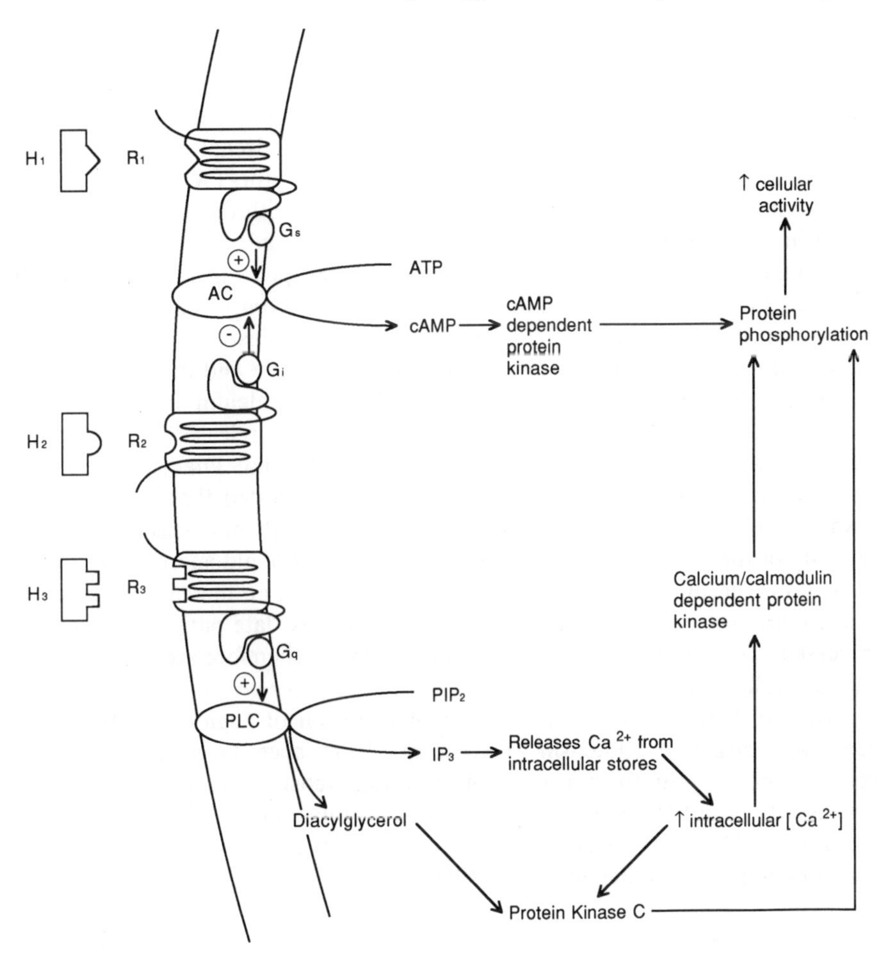

Figure 2.9 Intracellular mechanisms of hormone action. This figure illustrates three different types of hormone (H) interaction with plasma membrane receptor (R) to produce a cellular response. (1) Stimulation of adenylyl cyclase (AC) is mediated by a stimulatory G-protein (G_S), while (2) inhibition of adenylyl cyclase is mediated by an inhibitory G-protein (G_i). In (3) there is G-protein-mediated activation of phospholipase C (PLC), resulting in increased conversion of phosphatidylinositol bisphosphate (PIP_2) to diacylglycerol and inositol trisphosphate (IP_3). Activation of corticotrophin receptors and β-adrenoceptors stimulates adenylyl cyclase activity (mechanism 1). Angiotensin II acts mainly through mechanism 3 to stimulate PLC activity, but also inhibits adenylyl cyclase (mechanism 2). Activation of α_1-adrenoceptors stimulates PLC activity (mechanism 3), while activation of α_2-adrenoceptors inhibits adenylyl cyclase activity (mechanism 2).

1987). It has been suggested that the labile protein may be the peptide which is known as both diazepam-binding inhibitor (DBI) and endozepine (Krueger & Papadopoulos 1992). This peptide binds with high affinity to mitochondrial benzodiazepine receptors, resulting in increased delivery of cholesterol to the inner mitochondrial membrane (Whitehouse 1992).

The mechanism of ACTH signal transduction provides several toxicological targets. For example, cholera toxin is well known to stimulate cAMP production in the gut by covalently modifying the alpha subunit of the G-protein, causing the adenylyl cyclase to be permanently activated, and has been shown to stimulate adrenocortical steroidogenesis *in vitro* (Kowal *et al.* 1977). In general, benzodiazepines which bind with high affinity to central $GABA_A$ receptors have low affinity for mitochondrial benzodiazepine receptors. However, diazepam and several other benzodiazepines have been shown to cause stimulation of adrenal steroidogenesis via their interaction with mitochondrial benzodiazepine receptors (Whitehouse 1992).

The mechanism of action of angiotensin II contrasts with the effects of ACTH, as angiotensin acts via the generation of phospholipid second messengers rather than cAMP. Angiotensin II binds to plasma membrane receptors (AT1 receptors) which act through G-proteins to stimulate phospholipase C activity. This enzyme hydrolyses the membrane phospholipid, phosphatidyl inositol bisphosphate, causing the formation of two second messengers, inositol trisphosphate and diacylglycerol. Inositol trisphosphate (IP_3) acts to cause the release of calcium from intracellular stores, mainly in the endoplasmic reticulum. The calcium released by the action of IP_3 activates a calcium/calmodulin-dependent protein kinase. Diacylglycerol, on the other hand, activates protein kinase C. It has been suggested that the intracellular pathways activated by the actions of angiotensin II result in a specific increase in the activity of the 'late pathway', meaning the final steps of aldosterone biosynthesis. Acutely, however, all the protein kinases appear to effect an increase in the conversion of cholesterol to pregnenolone. It is likely that any 'late pathway' effect is due to increased expression of aldosterone synthase, and is therefore seen as a long-term effect (see Muller 1988).

As with ACTH, the signal transduction mechanism of angiotensin II may be a toxicological target. Angiotensin II receptors have been targeted by the pharmaceutical industry in their search for anti-hypertensive agents. Saralasin is an angiotensin II receptor antagonist which causes inhibition of aldosterone secretion. The binding of IP_3 to its intracellular receptors and its calcium-releasing activity are inhibited by heparin (Guillemette *et al.* 1989).

2.4.5 Regulation of Catecholamine Synthesis and Secretion

Catecholamine secretion is regulated by the activity of the splanchnic nerve, a branch of the sympathetic nervous system. In general, the release of catecholamines is stimulated by similar factors to those stimulating cortisol release: stress, exercise, cold exposure and hypoglycaemia, for example. However, any factor that increases sympathetic activity will also stimulate catecholamine secretion. The splanchnic nerve, in common with other pre-ganglionic sympathetic nerves, uses acetylcholine as its transmitter. This acts on both nicotinic and muscarinic receptors in the adrenal medulla. Most of the studies on catecholamine secretion have been carried out in cats, although it appears that the mechanisms involved are rather different in this species. In cats, muscarinic agonists selectively stimulate adrenaline and nicotinic agonists selectively release noradrenaline. In other species, including the dog, these catecholamines are secreted in a fixed proportion, whether stimulated by nicotinic or muscarinic agonists (Critchley *et al.* 1986). It has been suggested that ACTH (Valenta *et al.* 1986) and angiotensin II may have a role in regulating catecholamine secretion, possibly by potentiating the effect of neural stimulation (Ungar & Phillips 1983).

The actions of acetylcholine cause the generation of an action potential in chromaffin cells, resulting in the secretion of catecholamines. This process has been termed 'stimulus-secretion coupling' (Douglas & Rubin 1963). During the action potential produced in response to acetylcholine, there is a large influx of calcium ions into the chromaffin cells. This influx is blocked by the organophosphates, malathion and methyl parathion (Liu *et al.* 1994). The increase in intracellular calcium is an obligatory requirement for the secretion of catecholamines, although its role is not fully understood. The mechanism of secretion is complex, but is thought to involve the fusion of several secretory granules together. The membrane of the secretory granule then fuses with the plasma membrane of the chromaffin cell and the entire contents of the granule, with the exception of membrane-bound enzyme, are expelled by exocytosis. The plasma membrane then reseals and the granule is recovered and refilled with catecholamines (Levin 1986).

The stimulation of chromaffin cells does not, however, cause a significant decrease in the tissue content of catecholamines, suggesting that the biosynthesis of these hormones occurs rapidly in response to stimulation. Tyrosine hydroxylase is the rate-limiting step of catecholamine biosynthesis. It was originally thought that the activity of this enzyme was principally regulated by negative feedback inhibition, as it is known that adrenaline and noradrenaline inhibit tyrosine hydroxylase activity. The obvious flaw in this scheme is that, while tyrosine hydroxylase is a cytosolic enzyme, adrenaline and noradrenaline are stored within secretory granules and are not present in significant quantities in the cytoplasm. It is unlikely, therefore, that negative feedback inhibition has a major role in regulating catecholamine biosynthesis (Levin 1986). There is evidence that tyrosine hydroxylase undergoes a conformational change following stimulation of the chromaffin cell. This change may involve phosphorylation by a cAMP-dependent kinase, or by a calcium/calmodulin-dependent kinase. The result of the conformational change is that the binding affinity of tyrosine hydroxylase for its substrate and cofactors is increased, while its affinity for noradrenaline is decreased, thus causing a rapid increase in the rate of tyrosine hydroxylase activity (Landsberg & Young 1992). The long-term regulation of catecholamine biosynthesis, in response to chronic stimulation of the adrenal medulla, involves *de novo* synthesis of all the major biosynthetic enzymes (Ungar & Phillips 1983).

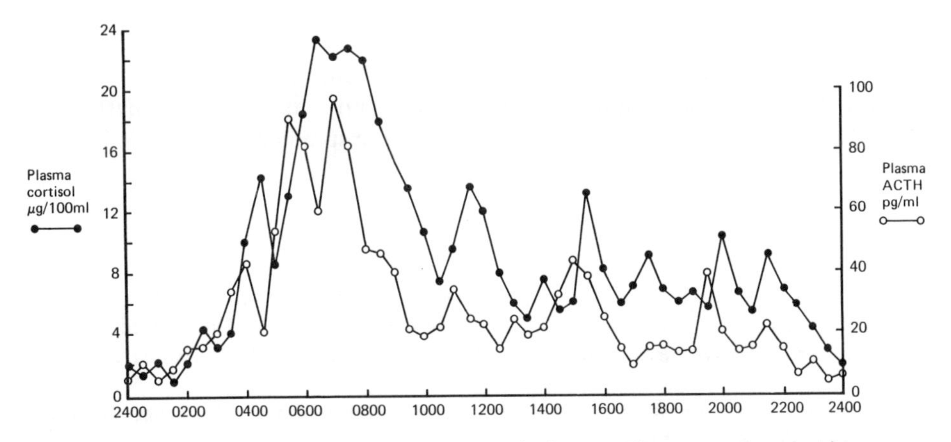

Figure 2.10 Diurnal variation in cortisol secretion in the human. Figure reproduced with permission from Vinson *et al.* (1992).

Table 2.3 Characteristic features of the adrenal gland which render it vulnerable to toxicants

1. Highly vascular gland with high rate of blood flow
2. Mechanisms for uptake and storage of lipoproteins (and associated lipophilic toxins)
3. Free radical generation during steroid biosynthesis
4. Potential for bioactivation of toxicants by cytochrome P450 enzymes
5. High membrane content of unsaturated fatty acids (substrates for lipid peroxidation)
6. Multiple sites at which function may be influenced

There is an increase in plasma catecholamines with increasing age. Although adrenaline is not affected, noradrenaline is significantly increased, with a greater effect seen at night (Ziegler *et al.* 1976, Prinz *et al.* 1979).

2.4.6 Diurnal Variation

In all mammalian species, with the exception of the dog (Dallman *et al.* 1987), there is a distinct circadian pattern of adrenocortical steroid secretion, driven by ACTH, with peak secretion between 6 and 9 am and an evening nadir (see Figure 2.10). In the rat, a nocturnal animal, the rhythm is reversed with peak levels in the evening. The timing of plasma cortisol measurements must be taken into account when adrenal function is investigated. There is evidence suggesting that diurnal variation of plasma corticosterone in the rat may be regulated by splanchnic nerve activity (Jasper & Engeland 1994). In this context it is interesting to note that plasma catecholamines in the human also show diurnal variation, with a peak occurring late in the morning, and a nadir at night (Prinz *et al.* 1979).

Aldosterone shows a diurnal rhythm in its secretion which is often synchronous with that of cortisol. It is clear that the control of the diurnal rhythm of aldosterone is more complex, however, with posture and the renin–angiotensin system influencing the circadian pattern of secretion (Muller 1988).

2.4.7 Adrenal Vulnerability to Toxicants

As we have sought to emphasize in this chapter, there are characteristic morphological and biochemical features of the adrenal gland which render it particularly susceptible to the actions of toxins. These features are summarized in Table 2.3.

References

ABAYASEKARA, D. R. E., VAZIR, H., WHITEHOUSE, B. J., PRICE, G. M., HINSON, J. P. & VINSON, G. P. (1989) Studies on the mechanism of ACTH-induced inhibition of aldosterone biosynthesis in the rat aderenal cortex. *Journal of Endocrinology*, **122**, 625–632.

AGUILERA, G. & CATT, K. J. (1984) Dopaminergic modulation of aldosterone secretion in the rat. *Endocrinology*, **114**, 176–181.

ALBERTSON, B. D., HILL, R. B., SPRAGUE, K. A., WOOD, K. E., NIEMAN, L. K. &

LORIAUX, D. L. (1994) Effect of the antiglucocorticoid RU486 on adrenal steroidogenic enzyme activity and steroidogenesis. *European Journal of Endocrinology*, **130**, 195–200.

ALLOLIO, B., SCHULTE, H. M., DEUSS, U. & WINKELMANN, W. (1987) Cyproheptadine inhibits the corticotropin-releasing hormone (CRH)-induced hormone release in normal subjects. *Hormone and Metabolic Research Supplement*, **16**, 36–38.

ATARASHI, K., MULROW, P. J. & FRANCO-SAENZ, R. (1985) Effect of atrial peptides on aldosterone production. *Journal of Clinical Investigation*, **76**, 1807–1811.

ATLAS, S. A. (1986) Atrial natriuretic factor: a new hormone of cardiac origin. *Recent Progress in Hormone Research*, **42**, 207–249.

BUCKINGHAM, J. C., SMITH, T. & LOXLEY, H. D. (1992) The control of ACTH secretion. In JAMES, V. H. T. (Ed.), *The Adrenal Gland*, 2nd Edn, pp. 131–158, New York: Academic Press.

CADE, R. & PERENICH, T. (1965) Secretion of aldosterone by rats. *American Journal of Physiology*, **208**, 1026–1030.

CAMACHO, A. M., CASH, R., BROUGH, A. J. & WILROY, R. S. (1967) Inhibition of adrenal steroidogenesis by aminoglutethimide and the mechanism of action. *Journal of the American Chemical Society*, **202**, 114–120.

CAMPBELL, D. J., MENDELSOHN, F. A. O., ADAM, W. R. & FUNDER, J. W. (1981) Metoclopramide does not elevate aldosterone in the rat. *Endocrinology*, **109**, 1484–1491.

CAREY, R. M., THORNER, M. O. & ORTT, E. M. (1979) Effects of metoclopramide on the renin–angiotensin–aldosterone system in man: dopaminergic control of aldosterone secretion. *Journal of Clinical Investigation*, **63**, 727–735.

COLBY, H. D. & LONGHURST, P. A. (1992) Toxicology of the adrenal gland. In ATTEMILL, C. K. & FLACK, J. D. (Eds), *Endocrine Toxicology*, pp. 243–281, Cambridge: Cambridge University Press.

COLEMAN, T. G., MCCAA, R. E. & MCCAA, C. S. (1974) Effect of angiotensin II on aldosterone secretion in the conscious rat. *Journal of Endocrinology*, **60**, 421–427.

CRITCHLEY, J. A. J. H., ELLIS, P., HENDERSON, C. G., UNGAR, A. & WEST, C. P. (1986) Muscarinic and nicotinic mechanisms in the responses of the adrenal medulla of the dog and cat to reflex stimuli and to cholinomimetic drugs. *British Journal of Pharmacology*, **89**, 831–835.

CRYER, P. E. (1992) The adrenal medullae. In JAMES, V. H. T. (Ed.), *The Adrenal Gland*, 2nd Edn, pp. 465–489, New York: Academic Press.

DALLMAN, M. F., AKANA, S. F., CASCIO, C. S., DARLINGTON, D. N., JACOBSON, L. & LEVIN, N. (1987) Regulation of ACTH secretion: variations on a theme of B. *Recent Progress in Hormone Research*, **43**, 113–167.

DAVIS, J. O. (1962) The control of aldosterone secretion. *Physiologist*, **5**, 65–86.

DE PERETTI, E. & FOREST, M. G. (1976) Unconjugated dehydroepiandrosterone plasma levels in normal subjects from birth to adolescence in human: the use of a sensitive radioimmunoassay. *Journal of Clinical Endocrinology and Metabolism*, **43**, 982–991.

DOLLERY, C. (1991) *Therapeutic Drugs*, Edinburgh: Churchill Livingstone.

DOUGLAS, J. G. & CATT, K. J. (1976) Regulation of angiotensin II receptors by dietary electrolytes. *Journal of Clinical Investigation*, **58**, 834–843.

DOUGLAS, W. W. & RUBIN, R. P. (1963) The mechanism of catecholamine release from the adrenal medulla and the role of calcium in stimulus–secretion coupling. *Journal of Physiology (London)*, **167**, 288–310.

EDWARDS, A. V. & JONES, C. T. (1993) Autonomic control of adrenal function. *Journal of Anatomy*, **183**, 291–307.

ENGELAND, W. C., LILLY, M. P. & GANN, D. S. (1985) Sympathetic adrenaldenervation decreases adrenal blood flow without altering the response to haemorrhage. *Endocrinology*, **117**, 1000–1010.

FEINGOLD, I. B. & COLBY, H. D. (1992) Sex differences in adrenal and hepatic α-tocopherol concentrations in rats. *Pharmacology*, **44**, 113–116.

GALLO-PAYET, N., POTHIER, P. & ISLER, H. (1987) On the presence of chromaffin cells in the adrenal cortex: their possible role in adrenocortical function. *Biochemistry and Cell Biology*, **65**, 588–592.

GANONG, W. F., MULROW, P. J., BORYCZKA, A. & CERA, G. (1962) Evidence for a direct effect of angiotensin II on adrenal cortex of the dog. *Proceedings of the Society for Experimental Biology and Medicine*, **109**, 381–384.

GRABER, A. L., NEY, R. L., NICHOLSON, W. E., ISLAND, D. P. & LIDDLE, G. W. (1965) Natural history of pituitary–adrenal recovery following long-term suppression with corticosteroids. *Journal of Clinical Endocrinology*, **25**, 11–16.

GRAY, M. J., STRAUSFELD, K. S., WATANABE, M., SIMS, E. A. H. & SOLOMON, S. (1968) Aldosterone secretory rates in the normal menstrual cycle. *Journal of Clinical Endocrinology and Metabolism*, **28**, 1269–1275.

GUILLEMETTE, G., LAMONTAGNE, S., BOULAY, G. & MOUILLAC, B. (1989) Differential effects of heparin on inositol 1,4,5-trisphosphate binding, metabolism and calcium release activity in the bovine adrenal cortex. *Molecular Pharmacology*, **35**, 339–344.

HALLBERG, E. (1990) Metabolism and toxicity of xenobiotics in the adrenal cortex, with particular reference to 7,12-dimethylbenz(a)anthracene. *Journal of Biochemical Toxicology*, **5**, 71–90.

HANING, R., TAIT, S. A. S. & TAIT, J. F. (1970) *In vitro* effects of ACTH, angiotensins, serotonin and potassium on steroid output and conversion of corticosterone to aldosterone by isolated adrenal cells. *Endocrinology*, **87**, 1147–1167.

HAUGER, R. L., THRIVIKRAMAN, K. V. & PLOTSKY, P. M. (1994) Age-related alterations of hypothalamic-pituitary-adrenal axis function in male Fischer-344 rats. *Endocrinology*, **134**, 1528–1536.

HENVILLE, K. L., HINSON, J. P., VINSON, G. P. & LAIRD, S. M. (1989) Actions of desacetyl-α-MSH on human adrenocortical cells. *Journal of Endocrinology*, **121**, 579–583.

HINSON, J. P. (1990) Paracrine control of adrenocortical function: a new role for the medulla? *Journal of Endocrinology*, **124**, 7–9.

HINSON, J. P., KAPAS, S., TEJA, R. & VINSON, G. P. (1991) The role of endothelin in the control of adrenocortical function: stimulation of endothelin release by ACTH and the effects of endothelin-1 and endothelin-3 on steroidogenesis in rat and human adrenocortical cells. *Journal of Endocrinology*, **128**, 275–280.

HINSON, J. P., VINSON, G. P., KAPAS, S. & TEJA, R. (1991) The relationship between adrenal vascular events and steroid secretion: the role of mast cells and endothelin. *Journal of Steroid Biochemistry and Molecular Biology*, **40**, 381–389.

HINSON, J. P., CAMERON, L. A., PURBRICK, A. & KAPAS, S. (1994) The role of neuropeptides in the regulation of adrenal zona glomerulosa function: effects of substance P, neuropeptide Y, neurotensin, Met-enkephalin, Leu-enkephalin and corticotrophin-releasing hormone on aldosterone secretion in the intact perfused rat adrenal. *Journal of Endocrinology*, **140**, 91–96.

HINSON, J. P., DAWNAY, A. B. & RAVEN, P. W. (1995) Why we should give a cautious welcome to ouabain: a whole new family of adrenal steroid hormones? *Journal of Endocrinology*, **146**, 369–372.

HOLZWARTH, M. A., CUNNINGHAM, L. A. & KLEITMAN, N. (1987) The role of adrenal nerves in the regulation of adrenocortical functions. *Annals of the New York Academy of Sciences*, **512**, 449–464.

HORNSBY, P. J. (1989) Steroid and xenobiotic effects on the adrenal cortex: mediation by oxidative and other mechanisms. *Free Radical Biology and Medicine*, **6**, 103–115.

HORNSBY, P. J. & CRIVELLO, J. F. (1983a) The role of lipid peroxidation and biological antioxidants in the function of the adrenal cortex. Part 1: a background review. *Molecular and Cellular Endocrinology*, **30**, 1–20.

HORNSBY, P. J. & CRIVELLO, J. F. (1983b) The role of lipid peroxidation and biological antioxidants in the function of the adrenal cortex. Part 2. *Molecular and Cellular Endocrinology*, **30**, 123–147.

IDELMAN, S. (1978) The structure of the mammalian adrenal cortex. In CHESTER JONES, I. & HENDERSON, I. W. (Eds), *General, Comparative and Clinical Endocrinology of the Adrenal Cortex*, Vol. 2, pp. 1–199, London: Academic Press.

JASPER, M. S. & ENGELAND, W. C. (1994) Splanchnic neural activity modulates ultradian and circadian rhythms in adrenocortical secretion in awake rats. *Neuroendocrinology*, **59**, 97–109.

KAPLAN, N. M. & BARTTER, F. C. (1962) The effect of ACTH, renin, angiotensin II and various precursors on biosynthesis of aldosterone by adrenal slices. *Journal of Clinical Investigation*, **41**, 715–724.

KONDO, H. (1985) Immunochemical analysis of the localisation of neuropeptides in the adrenal gland. *Archives of Histology (Japan)*, **48**, 453–481.

KOSSOR, D. C., KOMINAMI, S., TAKEMORI, S. & COLBY, H. D. (1991) Role of the steroid 17α-hydroxylase in spironolactone-mediated destruction of adrenal cytochrome P-450. *Molecular Pharmacology*, **40**, 321–325.

KOWAL, J., HORST, I., PENSKY, J. & ALFONZO, M. (1977) A comparison of the effects of ACTH, vasoactive intestinal peptide and cholera toxin on adrenal cyclic AMP and steroid synthesis. *Annals of the New York Academy of Sciences*, **297**, 314–328.

KRUEGER, K. E. & PAPADOPOULOS, V. (1992) Mitochondrial benzodiazepine receptors and the regulation of steroid biosynthesis. *Annual Review of Pharmacology and Toxicology*, **32**, 211–237.

KUDO, T. & BAIRD, A. (1984) Inhibition of aldosterone production in the adrenal zona glomerulosa by atrial natriuretic factor. *Nature*, **312**, 756–757.

LAMBERTS, S. W. J., KLIJN, J. G. M., DE QUIJADA, M., TIMMERMANS, H. A. T., UITTERLINDEN, P., DE JONG, F. H. & BIRKENHAGER, J. C. (1980) The mechanism of the suppressive action of bromocriptine on adrenocorticotropin secretion in patients with Cushing's disease and Nelson's syndrome. *Journal of Clinical Endocrinology and Metabolism*, **51**, 307–311.

LANDSBERG, L. & YOUNG, J. B. (1992) Catecholamines and the adrenal medulla. In WILSON, J. D. & FOSTER, D. W. (Eds), *Williams Textbook of Endocrinology*, 8th Edn, pp. 621–705, Philadelphia: W. B. Saunders.

LARAGH, J. H., ANGERS, M., KELLY, W. G. & LIEBERMAN, S. (1960) Hypotensive agents and pressor substances. The effect of epinephrine, norepinephrine, angiotensin II and others on the secretory rate of aldosterone in man. *Journal of the American Medical Association*, **174**, 234–240.

LAUBER, M., SUGANO, S., OHNISHI, T., OKAMOTO, M. & MULLER, J. (1987) Aldosterone biosynthesis and cytochrome P-$450_{11\beta}$. Evidence for two different forms of the enzyme in rats. *Journal of Steroid Biochemistry*, **26**, 693–698.

LAUER, C. G., BRALEY, L. M., MENACHERY, A. I. & WILLIAMS, G. H. (1982) Metoclopramide inhibits aldosterone biosynthesis *in vitro*. *Endocrinology*, **111**, 238–243.

LEVIN, J. A. (1986) Adrenal medulla. In MULROW, P. J. (Eds), *The Adrenal Gland*, Amsterdam: Elsevier.

LIDDLE, G. W., ISLAND, D. P., LANCE, E. M. & HARRIS, A. P. (1958) Alterations of adrenal steroid patterns in man resulting from treatment with a chemical inhibitor of 11β-hydroxylase. *Journal of Clinical Endocrinology*, **18**, 906–912.

LIU, P. S., KAO, L. S. & LIN, M. K. (1994) Organophosphates inhibit catecholamine secretion and calcium influx in bovine adrenal chromaffin cells. *Toxicology*, **90**, 81–91.

LOOSE, D. S., KAN, P. B., HIRST, M. A., MARCUS, R. A. & FELDMAN, D. (1983) Ketoconazole blocks adrenal steroidogenesis by inhibiting cytochrome P-450-dependent enzymes. *Journal of Clinical Investigation*, **71**, 1495–1499.

MAACK, T., MARION, D. N., CAMARGO, M. J. F., KLEINERT, H. D., LARAGH, J. H., VAUGHAN, E. D. & ATLAS, S. A. (1984) Effects of auriculin (atrial natriuretic factor) on blood pressure, renal function and the renin–aldosterone system in dogs. *American Journal of Medicine*, **77**, 1069–1075.

MALENDOWICZ, L. K. (1993) Involvement of neuropeptides in the regulation of growth, structure and function of the adrenal cortex. *Histology and Histopathology*, **8**, 173–186.

MALENDOWICZ, L. K. (1994) *Cytophysiology of the Mammalian Adrenal Cortex*. Poznan: PTPN.

MARIEB, N. J. & MULROW, P. J. (1965) Role of the renin–angiotensin system in the regulation of aldosterone secretion in the rat. *Endocrinology*, **76** 657–664.

MAZZOCCHI, G. & NUSSDORFER, G. G. (1984) Long-term effects of captopril on the morphology of normal rat adrenal zona glomerulosa. A morphometric study. *Experimental and Clinical Endocrinology*, **84**, 148–152.

MCCARTHY, J. L., REITZ, C. W. & WESSON, L. K. (1966) Inhibition of adrenal corticosteroidogenesis in the rat by cyanotrimethylandrostenolone, a synthetic androstane. *Endocrinology*, **79**, 1123–1139.

MCKENNA, T. J. & CUNNINGHAM, S. K. (1991) The control of adrenal androgen secretion. *Journal of Endocrinology*, **129**, 1–3.

MCNICOL, A. M. (1992) The human adrenal gland. In JAMES, V. H. T. (Ed.), *The Adrenal Gland*, 2nd Edn, pp. 1–42, New York: Academic Press.

MONTELIUS, J., PAPADOPOULOS, D., BENGTSSON, M. & RYDSTROM, J. (1982) Metabolism of polycyclic aromatic hydrocarbons and covalent binding of metabolities to protein in rat adrenal gland. *Cancer Research*, **42**, 1479–1486.

MORIMOTO, T., SEKIZAWA, A., HIROSE, K., SUZUKI, A., SAITO, H. & YANAIHARA, T. (1993) Effect of labor and prostaglandins on phenylethanolamine N-methyltransferase in human foetal membranes. *Endocrine Journal*, **40**, 179–183.

MORNET, E., DUPONT, J., VITEK, A. & WHITE, P. C. (1989) Characterisation of two genes encoding human steroid 11 beta-hydroxylase (P-450-11 beta). *Journal of Biological Chemistry*, **264**, 20961–20967.

MULLER, J. (1988) *Regulation of Aldosterone Biosynthesis. Physiological and Clinical Aspects, Monographs on Endocrinology 29*, Berlin: Springer.

MUSACCHIO, J., KOPIN, I. J. & SNYDER, S. (1964) Effects of disulfiram on tissue norepineprine content and subcellular distribution of dopamine, tyramine, and their beta-hydroxylated metabolites. *Life Sciences*, **3**, 769–775.

NEVILLE, A. M. & O'HARE, M. J. (1982) *The Human Adrenal Cortex*, Berlin: Springer.

NORBIATO, G. M., BEVILACQUA, M., RAGGI, D., MICOSSI, P. & MORONI, C. (1977) Metoclopramide increases plasma aldosterone in man. *Journal of Clinical Endocrinology and Metabolism*, **45**, 1313–1316.

NUSSDORFER, G. G. (1980) Cytophysiology of the adrenal zona glomerulosa. *International Review of Cytology*, **64**, 307–368.

O'CONNELL, T. X. & ASTON, S. J. (1974) Acute adrenal haemorrhage complicating anticoagulant therapy. *Surgery, Gynecology and Obstetrics*, **139**, 355.

PAPADOPOULOS, D., SEIDERGARD, J. & RYDSTROM, J. (1984) Metabolism of xenobiotics in the human adrenal gland. *Cancer Letters*, **22**, 23–30.

POTTS, G. O., CREANGE, J. E., HARDING, H. R. & SCHANE, H. P. (1978) Trilostane, an orally active inhibitor of steroid biosynthesis. *Steroids*, **32**, 257–267.

PRINZ, P. N., HALTER, J., BENEDETTI, C. & RASKIND, M. (1979) Circadian variation of plasma catecholamines in young and old men: relation to rapid eye movement and slow wave sleep. *Journal of Clinical Endocrinology and Metabolism*, **49**, 300–304.

PUDNEY, J., PRICE, G. M., WHITEHOUSE, B. J. & VINSON, G. P. (1984) Effects of chronic ACTH stimulation on the morphology of the rat adrenal cortex. *Anatomical Record*, **210**, 603–615.

RAO, R. H. (1995) Bilateral massive adrenal hemorrhage. *Medical Clinics of North America*, **79**, 107–129.

REAVEN, E., KOSTRNA, M., RAMACHANDRAN, J. & AZHAR, S. (1988) Structure and function changes in rat adrenal glands during aging. *American Journal of Physiology*, **255**, E903–E911.

REBUFFAT, P., BELLONI, A. S., ROCCO, S., ANDREIS, P. G., NERI, G., MALENDO-

WICZ, L. K., GOTTARDO, G., MAZZOCCHI, G. & NUSSDORFER, G. G. (1992) The effects of ageing on the morphology and function of the zonae fasciculata/reticularis of the rat adrenal cortex. *Cell and Tissue Research*, **270**, 265–272.

RHODIN, J. A. G. (1971) The ultrastructure of the adrenal cortex of the rat under normal and experimental conditions. *Journal of Ultrastructural Research*, **34**, 23–71.

RIBELIN, W. E. (1984) The effects of drugs and chemicals upon the structure of the adrenal gland. *Fundamental and Applied Toxicology*, **4**, 105–119.

RICHARDS, A. M., NICHOLLS, M. G., ESPINER, E. A., IKRAM, H., YANDLE, T. G., JOYCE, S. L. & CULLENS, M. M. (1985) Effects of alpha-human atrial natriuretic peptide in essential hypertension. *Hypertension*, **7**, 812–817.

RIFKIND, A. B., TSENG, L., HIRSCH, M. B. & LAUERSEN, N. H. (1987) Aryl hydrocarbon hydroxylase activity and microsomal cytochrome content of human fetal tissues. *Cancer Research*, **38**, 1572–1577.

SAPIRSTEIN, L. A. & GOLDMAN, H. (1959) Adrenal blood flow in the albino rat. *American Journal of Physiology*, **196**, 159–162.

SCACCIANOCE, S., DI SCULLO, A. & ANGELUCCI, L. (1990) Age-related changes in hypothalamo-pituitary-adrenocortical axis activity in the rat. *Neuroendocrinology*, **52**, 150–155.

SEEMAN, T. E. & ROBBINS, R. J. (1994) Aging and hypothalamic-pituitary-adrenal response to challenge in humans. *Endocrine Reviews*, **15**, 233–260.

SIMPSON, E. R. & WATERMAN, M. R. (1992) Regulation of expression of adrenocortical enzymes. In JAMES, V. H. T. (Ed.), *The Adrenal Gland*, 2nd Edn, pp. 191–207, New York: Academic Press.

SOLL, R. M., DOLLINGS, P. J., MITCHELL, R. D. & HAFNER, D. A. (1994) Guanabenz-related amidinohydrazones: potent non-azole inhibitors of aldosterone biosynthesis. *European Journal of Medical Chemistry*, **29**, 223–232.

SOWERS, J. R., SHARP, B., LEVIN, E. R., GOLUB, M. S. & EGGENA, P. (1981) Metoclopramide, a dopamine antagonist, stimulates aldosterone secretion in rhesus monkeys but not in dogs or rabbits. *Life Sciences*, **29**, 2171–2175.

UNGAR, A. & PHILLIPS, J. H. (1983) Regulation of the adrenal medulla. *Physiological Reviews*, **63**, 787–843.

VALENTA, L. J., ELIAS, A. N. & EISENBERG, H. (1986) ACTH stimulation of adrenal epinephrine and norepinephrine release. *Hormone Research*, **23**, 16–20.

VINSON, G. P. (1987) The stimulation of steroidogenesis by corticotrophin: the role of intracellular regulatory peptides and proteins. *Journal of Endocrinology*, **114**, 163–165.

VINSON, G. P. & HINSON, J. P. (1992) Blood flow and hormone secretion in the adrenal gland. In JAMES, V. H. T. (Ed.), *The Adrenal Gland*, 2nd Edn, pp. 71–86, New York: Academic Press.

VINSON, G. P., WHITEHOUSE, B. J. & GODDARD, C. (1978) The effect of sex and strain of rats on the *in vitro* response of adrenocortical tissue to ACTH stimulation. *Journal of Steroid Biochemistry*, **9**, 553–560.

VINSON, G. P., WHITEHOUSE, B. J., DELL, A., ETIENNE, A. & MORRIS, H. R. (1981) Characterisation of an adrenal zona glomerulosa-stimulating component of posterior pituitary extracts as α-MSH. *Nature*, **284**, 464–467.

VINSON, G. P., HINSON, J. P. & RAVEN, P. W. (1985) The relationship between tissue preparation and function: methods for the study of aldosterone secretion: a review. *Cell Biochemistry and Function*, **3**, 235–253.

VINSON, G. P., WHITEHOUSE, B. J. & HINSON, J. P. (1992) *The Adrenal Cortex*, New Jersey: Prentice Hall.

VINSON, G. P., HINSON, J. P. & TOTH, I. (1994) The neuroendocrinology of the adrenal cortex. *Journal of Neuroendocrinology* **6**, 235–246.

WALKER, B. R. & EDWARDS, C. R. W. (1992) Cushing's syndrome. In JAMES, V. H. T. (Ed.), *The Adrenal Gland*, 2nd Edn, pp. 289–318, New York: Academic Press.

WHITEHOUSE, B. J. (1992) Benzodiazepines and steroidogenesis. *Journal of Endocrinology*, **134**, 1–3.

WILLIAMS, R. T. (1973) *Detoxication Mechanisms*, London: Chapman & Hall.

YABU, M., SENDA, T., NONAKA, Y., MATSUKAWA, N., OKAMOTO, M. & FUJITA, H. (1991) Localization of the gene transcripts of 11β-hydroxylase and aldosterone synthase in the rat adrenal cortex by *in situ* hybridization. *Histochemistry*, **96**, 391–394.

YAHR, M. D. (1973) *Advances in Neurology*, Vol. 2, *Treatment of Parkinsonism: The Role of Dopa Decarboxylase Inhibitors*, New York: Raven Press.

YANAGIBASHI, K., HANUI, M., SHIVELY, J. E., SHEN, W. H. & HALL, P. (1986) The synthesis of aldosterone by the adrenal cortex. *Journal of Biological Chemistry*, **261**, 3556–3562.

ZIEGLER, M. G., LAKE, C. R. & KOPIN, I. J. (1976) Plasma noradrenalin increases with age. *Nature*, **261**, 333.

3

Transport, Actions and Metabolism of Adrenal Hormones and Pathology and Pharmacology of the Adrenal Gland

P. W. RAVEN

Institute of Psychiatry, London

J. P. HINSON

Queen Mary and Westfield College, London

3.1 Introduction

In the previous chapter various aspects of adrenal morphology and physiological function, and the way in which these contribute to the gland's susceptibility to toxins, were discussed. In this chapter the actions and metabolism of adrenal hormones, together with the pathology and pharmacology of the gland, will be described.

3.2 Transport of Hormones in Blood

The steroids secreted by the adrenal cortex, being hydrophobic in nature, are carried in blood mostly bound to plasma proteins. Aldosterone is mostly carried by plasma albumin, to which it binds weakly, while the glucocorticoids have a specific carrier protein, termed corticosteroid binding globulin (CBG). This difference in transport is reflected in the biological half-life of these steroids: the half-life of cortisol is around 90 min, while that of aldosterone is about 15 min. There is a dynamic equilibrium between the free and bound glucocorticoid in plasma, and it has been argued that only the free steroid is available to be biologically active (for a review see Mendel 1989). It is certainly possible that CBG may function as a large reservoir of cortisol in plasma: under certain conditions of stress CBG levels in plasma fall rapidly, presumably making more free cortisol available. CBG may also act as a local delivery mechanism for cortisol. CBG is structurally similar to the serpine (serine protease inhibitor) superfamily of proteins, which are cleaved by their target enzymes. It has been demonstrated that CBG is cleaved by the elastase released by neutrophils at sites of inflammation, resulting in a 10-fold decrease in its affinity for cortisol. It has also been demonstrated that CBG, bound to cortisol, binds to specific receptor sites in a variety of tissues, and induces a cellular response. For a review of the functions of CBG see Rosner (1990).

Whether CBG acts as a simple reservoir of cortisol, or a specific targeting/delivery mechanism, it is clearly an important factor when considering the effects of glucocorticoid hormones. There is evidence that certain non-steroidal anti-inflammatory drugs (NSAIDs) may cause a decrease in the binding capacity of CBG, apparently by a mechanism other than competitive displacement of the bound glucocorticoid (Engelhardt 1978).

Catecholamines have a very short half-life in plasma of around 1–2 min. They are rapidly taken up by sympathochromaffin and other tissues. They are also inactivated by conjugation in the circulation, with the result that $<5\%$ of circulating catecholamines are excreted unchanged in the urine (Cryer 1992).

3.3 Actions of Adrenal Hormones

3.3.1 *Corticosteroids*

The major classes of corticosteroids, glucocorticoids and mineralocorticoids, exert their effects via binding to intracellular receptors resulting in induction of RNA and protein synthesis. The mechanism by which this is achieved is similar for all classes of steroid hormones with intracellular receptors (glucocorticoid, mineralocorticoid, oestrogen, progesterone and androgen) as well as for 1,25-dihydroxy vitamin D_3, retinoic acid and thyroid hormones (Parker 1988). Steroids are mostly able to enter cells by passive diffusion and, in their target cells, bind to specific receptor proteins found in both cytosol and nucleus [Figure 3.1(a)]. Steroid binding induces conformational changes in the receptor which activate it, allowing the steroid–receptor complex to bind to a variety of nuclear binding sites with high affinity. In general, this results in activation of specific genes and induction of mRNA and protein synthesis, with consequent effects on cell function, growth or differentiation. Some of the actions of steroid hormones appear to be mediated through decreased gene activity, although these receptor–gene interactions are less well understood (Orth *et al.* 1992).

Glucocorticoids were named for their effects on carbohydrate metabolism, but they have such a wide range of actions in so many tissues that it is more appropriate to consider a glucocorticoid effect as being a result of the activation of specific glucocorticoid receptors. Glucocorticoid receptors mediate negative feedback of the hypothalamo–pituitary–adrenocortical (HPA) axis (see Chapter 2), in addition to having effects on intermediary metabolism, immune function, fluid and electrolyte balance, bone and connective tissue, mood and behaviour and developmental processes. Glucocorticoid receptor antagonists include progesterone, RU486, dexamethasone 21-mesylate and dexamethasone oxetanone, although *in vivo* these compounds may have mixed agonist/antagonist effects (Clark *et al.* 1992).

Glucocorticoid effects on intermediary metabolism are essentially antagonistic to the actions of insulin. Glucocorticoids increase hepatic glycogenesis by activating glycogen synthase and inactivating glycogen phosphorylase (Stalmans & Laloux 1979). They also increase hepatic gluconeogenesis by a variety of mechanisms, including direct activation of glucose-6-phosphatase and pyruvate kinase and mobilization of glucogenic substrates from peripheral tissues as a result of increased protein catabolism and lipolysis (Exton 1979). In addition, glucocorticoids inhibit glucose uptake and utilization by peripheral tissues (Exton 1979, Fain 1979) and

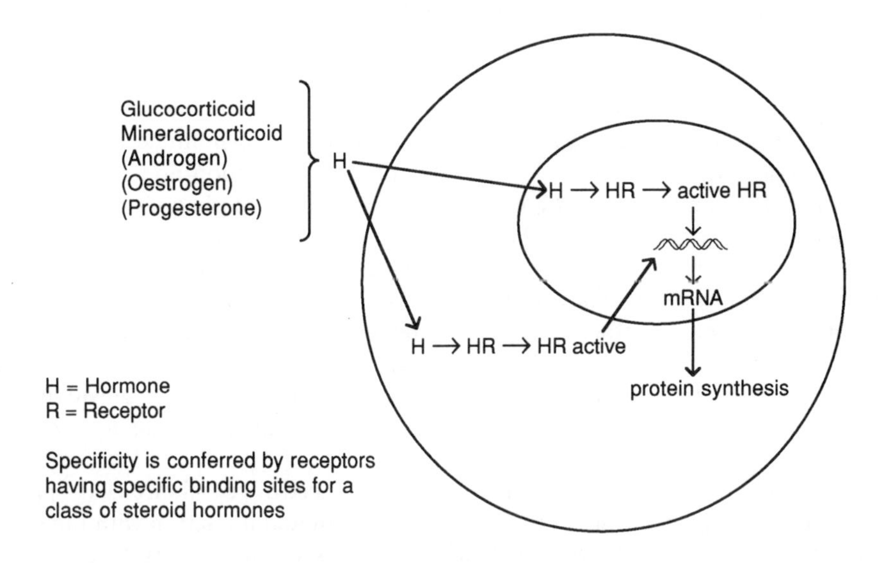

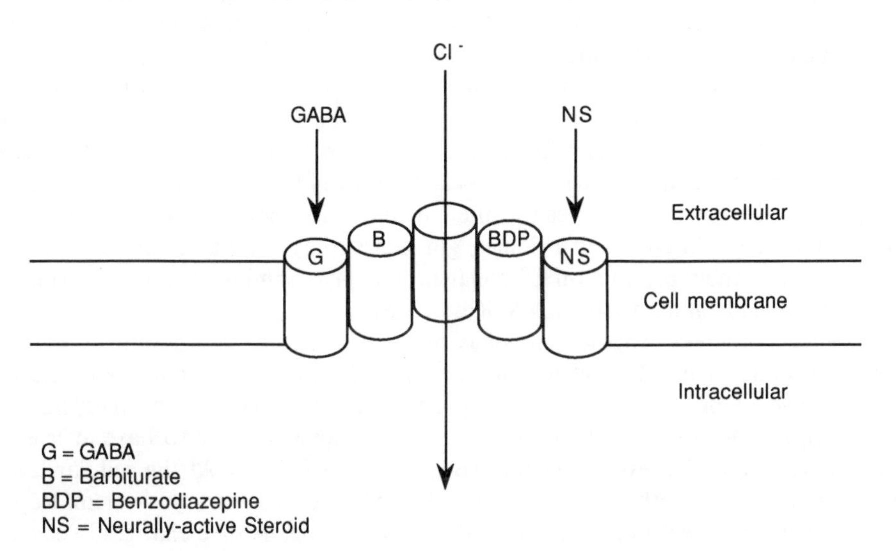

Figure 3.1 Mechanism of action of steroid hormones: (a) intracellular receptor and (b) surface receptor. The example shown is for GABAergic agonist steroids such as tetrahydrodeoxycorticosterone, whose binding to surface receptors increases GABA binding at $GABA_A$ receptors.

have permissive effects on the actions of other gluconeogenic hormones such as glucagon and adrenaline (Malbon *et al.* 1988).

The anti-inflammatory and immunosuppressive actions of glucocorticoids are prominent when glucocorticoid levels are high and are therefore of considerable importance in modulation and suppression of the acute stress response (Munck *et al.* 1984) and for pharmacological uses of steroids (see later chapters). However, the role which glucocorticoids play in the normal modulation of immune processes remains unclear. Glucocorticoids decrease numbers of circulating lymphocytes and monocytes and increase circulating granulocytes by effects on the distribution of these cells between body compartments. The functioning of immune system cells may also be affected, with inhibition of B-lymphocyte activation and proliferation (Cupps *et al.* 1985), monocyte differentiation and macrophage phagocytic and cytotoxic functions (Orth *et al.* 1992). The anti-inflammatory effects of glucocorticoids are partly due to inhibition of synthesis and release of many chemical mediators of inflammation (Fauci 1979). These effects include antagonism of histamine and inhibition of prostaglandin synthesis, which may be mediated by induction of a lipocortin (Flower 1986). It should be noted that there are marked species differences in these effects, with animals such as rats and mice showing far greater sensitivity in the response of their immune systems to glucocorticoids than humans and guinea pigs (Vinson *et al.* 1992).

Glucocorticoids have certain effects on fluid and electrolyte balance that appear to be mediated via glucocorticoid receptors rather than an interaction with mineralocorticoid receptors. Glucocorticoid deficiency results in an inability to excrete a water load and is associated with increased levels of plasma vasopressin (AVP; Raff 1987). The role of glucocorticoids in maintaining the ability to excrete a salt load was thought to be due to glucocorticoid effects on increasing glomerular filtration rate, but may be due to the role of glucocorticoids in inducing synthesis of atrial natriuretic peptides (ANP; Gardner *et al.* 1986).

Glucocorticoids have many effects on bone and mineral metabolism, including decreased calcium absorption from the gut with consequent increased serum levels of parathyroid hormone, inhibition of osteoblast function and therefore decreased new bone formation, and increased osteoclast numbers (Hahn *et al.* 1979). The consequences of these actions are seen in the osteoporosis associated with chronic glucocorticoid excess. In connective tissues, glucocorticoids inhibit the proliferation of fibroblasts and their production of collagen and glycosaminoglycans, resulting in impaired wound healing (Leibovich & Ross 1975).

The effects of glucocorticoids on behaviour and mood are well-described, but the mechanisms underlying these effects are poorly understood. Glucocorticoids affect a diverse range of processes such as sleep pattern, cognition and the reception of sensory input (McEwen 1979), and glucocorticoid excess appears to have at least a maintaining role in depressive disorder (Bearn & Raven 1993). At the cellular level, glucocorticoids have both rapid and long-term effects on the nervous system. Rapid effects, which occur within 2 min of exposure to hormone, involve changes in electrical activity such as hyperpolarization of the cell membrane (Hua & Chen 1989). It is not known how these effects are mediated as they occur too rapidly to be explained by a genomic mechanism and glucocorticoids do not appear to have activity at ligand-gated ion channels (see below). More long-term effects of glucocorticoids in the nervous system include induction of enzyme activity in glial cells (Kumar *et al.* 1986).

Developmental effects include the inhibition of linear growth in children by glucocorticoid levels as little as two or three times higher than normal (Loeb 1976). The mechanism is unclear, as glucocorticoids increase the growth hormone response to growth hormone-releasing hormone (GHRH) and do not suppress levels of insulin-like growth factor (IGF-1) which mediates the effects of growth hormone (Wehrenberg *et al.* 1983). Glucocorticoids also stimulate the differentiation of many cell types. Of particular interest are lung cells, where glucocorticoids are responsible for the normal development of surfactant secretion by foetal lung pneumocytes, and the nervous system, where glucocorticoids regulate the differentiation of neural crest epithelial cells into chromaffin cells. During foetal development of the adrenal gland, neural crest precursor cells which migrate into the gland are exposed to glucocorticoids. As a result they cease to express neuron-specific gene products, acquire the characteristic morphology of medullary chromaffin cells, and begin to produce the enzymes of catecholamine biosynthesis (Ballard 1979).

Mineralocorticoid effects are those which are associated with regulation of water and electrolyte balance and are mediated by specific mineralocorticoid receptors. The principal effect of aldosterone and other mineralocorticoids is to increase the reabsorption of sodium in the kidney and in secretory epithelia, thereby reducing the sodium content of urine, saliva, sweat, gastric juice and faeces. In the kidney tubules the overall effect of mineralocorticoids is to cause sodium ions to be reabsorbed in exchange for either potassium or hydrogen ions, leading to decreased sodium ion excretion, increased potassium ion excretion and increased urine acidity (Vinson *et al.* 1992). As with the glucocorticoid effects described above, mineralocorticoid effects are mediated by a genomic mechanism resulting in synthesis of specific mRNA and protein. Mineralocorticoid receptors are also present in high concentrations in the brain and, in particular, the hippocampus, where they operate in conjunction with glucocorticoid receptors as part of the HPA negative feedback system (Sapolsky *et al.* 1986). Mineralocorticoid effects are antagonized by spironolactone and sodium canrenoate, which both act as competitive inhibitors at the mineralocorticoid receptor (Vinson *et al.* 1992).

Finally, there are groups of steroid hormones which do not act through the classical intracellular, genomic mechanism described above. Although it has been known for 50 years that some steroids possess anaesthetic and sedative properties (Selye 1942), it is only recently that the mechanism of action of these compounds, known as neurally-active steroids, has begun to be understood (Majewska *et al.* 1986). The neurally-active steroids which have been most extensively studied are those which exert their rapid effects by binding with high affinity to a surface receptor associated with the $GABA_A$ receptor in brain [Figure 3.1(b)], and include the corticosteroid metabolites allopreganolone and 3α-tetrahydrodeoxycorticosterone. These steroids augment GABAergic neurotransmission by increasing GABA-activated chloride ion currents in a similar manner to barbiturates, but do not appear to bind to either the barbiturate or benzodiazepine receptors present on the $GABA_A$ receptor (Paul & Purdy 1992). These effects may be seen with picomolar concentrations of neurally-active steroid, suggesting a physiological relevance (Hiemke *et al.* 1991). Other corticosteroid metabolites, such as pregnenolone sulphate and dehydroepiandrosterone sulphate (DHEAS), act as excitatory neurally-active steroids and are both potent GABA antagonists and positive allosteric modulators at the N-methyl-D-aspartate (NMDA) receptor (Mellon 1994).

3.3.2 *Catecholamines*

Unlike the adrenal cortex, the adrenal medulla is not essential for life, due to the fact that it forms only a part of the sympathetic nervous system. While most of the

Table 3.1 Characteristics of adrenoreceptors

	Alpha receptor		Beta receptor		
Antagonists	Phentolamine	Phenoxybenzamine	Propranolol, Alprenolol, Nadolol, Timolol		
Subtypes	α_1	α_2	β_1	β_2	β_3
Agonist potency	NA = A	A ≥ NA	NA ≥ NA	A > NA	NA > A
Selective agonists	Phenylephrine, methoxamine	Clonidine, methylnoradrenaline	Na[a] Xamoterol[a]	Procaterol	BRL 37344
Second messenger	↑Ca^{++} ↑IP_3 + DAG	↓cAMP	↑cAMP	↑cAMP	↑cAMP
Selective antagonists	Prazosin, corynanthine	Yohimbine, rauwolscine	Atenolol	α-Methyl + propranolol	—

Note:[a] Selective relative to β_2 receptor. A, Adrenaline; NA, Noradrenaline.

Table 3.2 Tissue distribution and effects of activation of adrenoreceptors

Receptor type	Tissue	Actions
α_1	Liver	Glycogenolysis
	Vascular smooth muscle	Contraction
	Heart	Increased force of contraction
	Pupillary dilator muscle	Pupil dilation (muscle contraction)
α_2	Adipocytes	Inhibition of lipolysis
	Some vascular smooth muscle	Contraction
	Gastrointestinal tract	Smooth muscle relaxation
	Pancreas	Inhibition of insulin release
	Platelets	Aggregation
β_1	Heart	Increased rate and increased force of contraction
β_2	Smooth muscle of respiratory tract, vasculature gastrointestinal tract and genito-urinary tract	Relaxation
	Pancreas	Stimulation of insulin release
	Liver	Increased glycogenolysis and gluconeogenesis
	Kidney	Increased renin release
β_3	Adipocytes	Increased lipolysis

effects of the sympathetic nervous system are ultimately due to the release of noradrenaline from nerve terminals, the adrenal medulla releases mostly adrenaline into the circulation. Although in the physiological state these processes are inseparable, consideration of the specific effects of medullary hormones is crucial to an understanding of the pharmacology and toxicology of this gland.

Catecholamines act via plasma membrane receptors, termed adrenoceptors, which are members of the seven-transmembrane domain superfamily of hormone receptors. Several distinct subtypes of adrenoceptor have been identified. The characteristics of the major subtypes of adrenoceptor are shown in Table 3.1. Details of their intracellular actions are given in Chapter 2, Figure 2.9. There is species variation in the pharmacology of the α_2 adrenoceptor: in the rat, rauwolscine and yohimbine have 20-fold lower affinity for the α_2 adrenoceptor. The existence of a β_3 adrenoceptor has been demonstrated, but its physiological significance is unclear (Emorine *et al.* 1994).

Adrenoceptors mediate the wide range of effects which the catecholamines exert over physiological and biochemical processes in the body. The tissue distribution of the adrenoceptor subtypes is given in Table 3.2, with a list of the effects of receptor activation. In human pregnancy there is a decrease in myometrial β-adrenoceptors near term. Thyroid hormones cause potentiation of β-adrenergic effects and inhibition of α-adrenergic effects (Landsberg & Young 1992).

3.4 Metabolism of Adrenal Hormones

3.4.1 *Corticosteroids*

The peripheral metabolism of corticosteroids has two functions. First, and most extensively studied, metabolism of steroids may result in structural changes to the molecule which reduce its potency or render it less hydrophobic in preparation for urinary excretion. Second, there is increasing recognition that peripheral metabolism provides a mechanism for fine tuning of corticosteroid activity, in contrast to regulation of gross output of steroid at the level of the adrenal cortex. For example, the corticosteroid 11-deoxycorticosterone (DOC) has weak mineralocorticoid activity while its peripheral metabolite, tetrahydro-DOC, is a potent GABAergic agonist (Mellon 1994). For this reason we prefer to use the term extra-adrenal metabolism, rather than 'catabolism', to describe the peripheral metabolism of corticosteroids.

Most steroid metabolism occurs in the liver, where phase I oxidation or reduction reactions (mostly mediated by cytochrome P450 enzymes) can result in increased or decreased potency. Phase II reactions involve conjugation, and usually result in decreased potency. Corticosteroids undergo reduction of the A-ring and 3-oxo group, leading to the formation of tetrahydro derivatives. Adrenal androgens are similarly reduced in the A-ring and at C-3, resulting in greatly decreased potency. Both groups of steroids can undergo a series of other reactions including, for the C21 steroids, oxidation of the 17-hydroxyl group and reduction of the 20-oxo group. The major pathways of phase I corticosteroid metabolism are outlined in Figure 3.2. Over 90% of steroid excretion is by the urinary route, mostly as conjugates with glucuronide or sulphate (Gower 1984).

While most steroid metabolism occurs in the liver, there is also significant metabolism of cortisol in other tissues, notably the kidney, which possesses significant 11β-hydroxysteroid dehydrogenase (11β-HSD) activity. This enzyme exists in two major isoforms: 11β-HSD I is predominantly reductive and found mainly in the liver, and 11β-HSD II is predominantly oxidative, and found mainly in the kidney (Stewart *et al.* 1994). In the liver, 11β-HSD performs a role in maintaining levels of cortisol by back-conversion from cortisone. In the kidney, 11β-HSD is important in the actions of aldosterone. The mineralocorticoid receptor in the kidney has similar affinities for both aldosterone and cortisol, but as cortisol circulates in concentrations approximately 1000 times greater than aldosterone, it would prevent aldosterone from binding to any significant extent. By converting cortisol to the inactive cortisone, 11β-HSD II serves to protect the mineralocorticoid receptor from the high circulating levels of cortisol. The isoforms of 11β-HSD are recently-identified toxicological targets: glycyrrhetinic acid, one of the active principles of liquorice, is a potent inhibitor of the oxidative actions of 11β-HSD, causing the syndrome of apparent mineralocorticoid excess (Stewart *et al.* 1988). Metyrapone has been identified as an inhibitor of the reductase activity of 11β-HSD, mimicking an inborn error of cortisol metabolism (Raven *et al.* 1995a), and carbenoxolone appears to inhibit both activities (Stewart *et al.* 1990).

The concept that peripheral metabolism of corticosteroids is not simply catabolic is especially relevant to toxicological considerations, since particular pathways of corticosteroid metabolism are influenced by many factors, including disease, hormonal status and drug administration. There are two different ways in which the pathway of corticosteroid metabolism can be modulated, as shown in Figure 3.3.

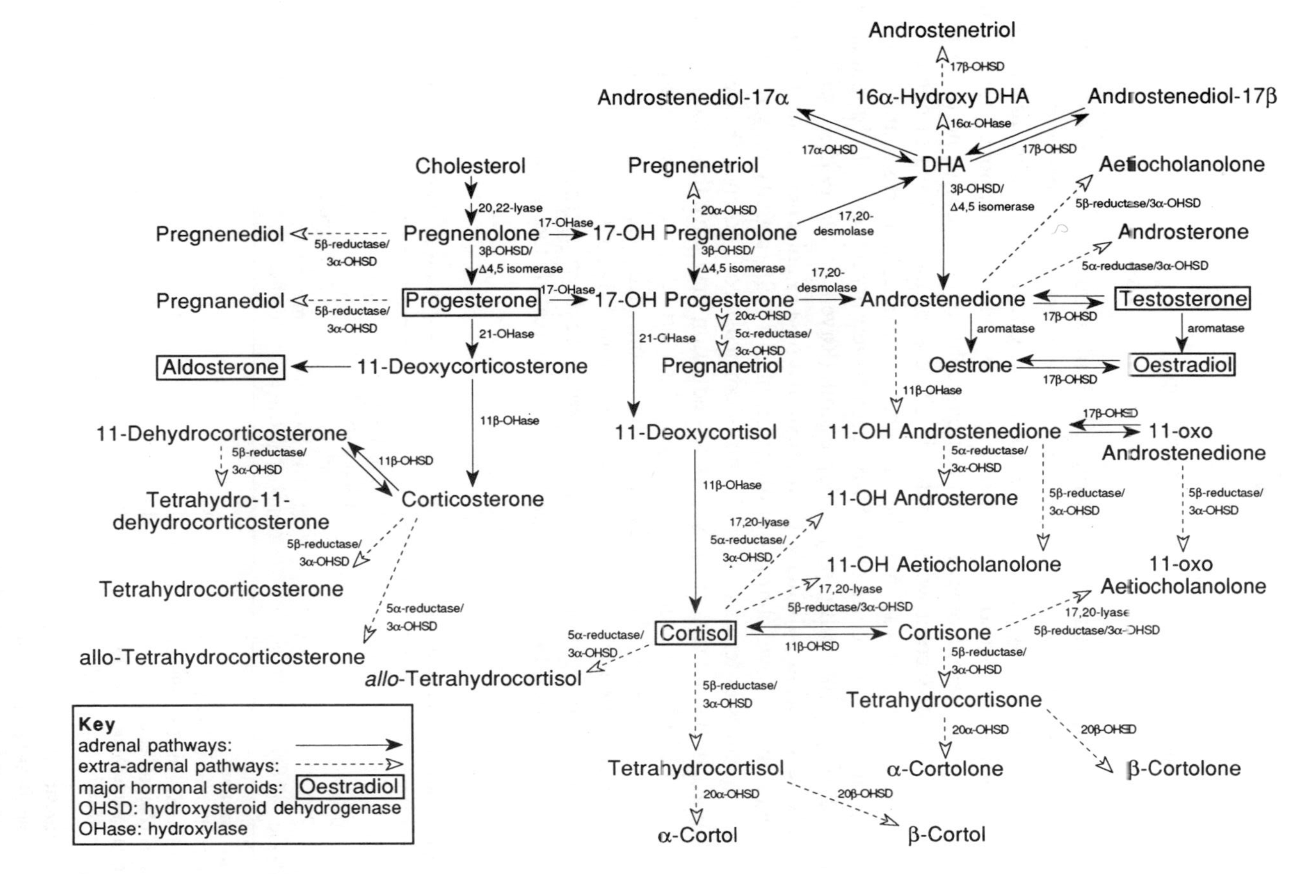

Figure 3.2 Pathways of steroid metabolism.

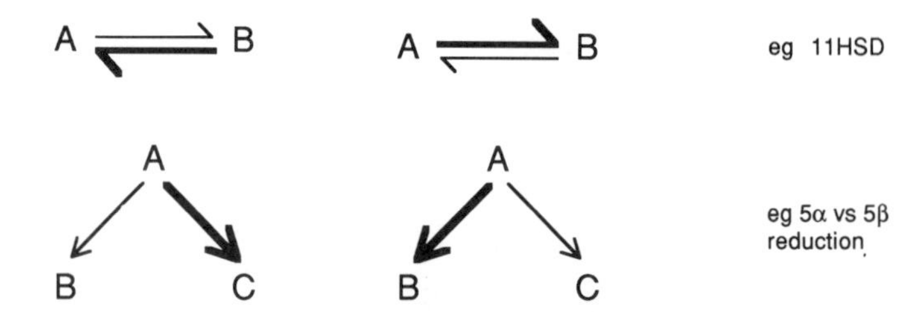

Figure 3.3 Modulation of pathways of steroid metabolism.

First, the direction of activity of an enzyme which interconverts two steroids may be altered. For example, 11β-HSD interconverts the potent glucocorticoid, cortisol, and its less potent 11-dehydro derivative, cortisone: glycyrrhetinic acid inhibits conversion of cortisol to cortisone (Stewart *et al.* 1988), while metyrapone (used as an inhibitor of corticosteroid biosynthesis) inhibits conversion of cortisone to cortisol (Raven *et al.* 1995a). Second, where a steroid site can be metabolized in one of two ways, the favoured pathway may be altered. For example, cortisol may be 5α-reduced to allotetrahydrocortisol or 5β-reduced to tetrahydrocortisol: 5α-reduction is favoured in patients with major depression (Raven *et al.* 1995b), while 5β-reduction is favoured in patients with anorexia nervosa (Vanluchene *et al.* 1979). In the examples given above, changes in direction of 11β-HSD activity may have profound consequences in terms of modulation of corticosteroid potency (Fraser 1990), while changes in the activity of 5α- versus 5β-reduction may have the effect of altering levels of GABA-agonistic neurosteroids, which are all 5α-reduced compounds (Mellon 1994). Some of the clinical conditions and drug treatments which are known to affect corticosteroid metabolism are listed in Table 3.3.

Sex differences in corticosteroid metabolism have been extensively studied in animals (Malendowicz 1994), principally due to interest in the mechanism of neo-

Table 3.3 Effects of clinical conditions and drug treatments on pathways of cortisol metabolism

References	11-Oxo/11-OH metabolites of cortisol (index of 11-HSD direction of activity)	5α/5β Reduced metabolites of cortisol	20-OH/20-Oxo metabolites of cortisol
Clinical conditions			
Cushings (Phillipou 1983)	↓	↓	?
Depression (Raven *et al.* 1995b)	↓	↑	→
Cirrhosis (Zumoff *et al.* 1967)	↓	↓	↑
Anorexia (Vanluchene *et al.* 1979)	↓	↓	?
Hirsutism (Rodin *et al.* 1994)	↑	→	?
Drug treatment			
Metyrapone (Raven *et al.* 1995a)	↑	↑	↑
Glycyrrhetinic acid (Stewart *et al.* 1987)	↓	↑	?
Hydrocortisone (Taylor *et al.* 1994)	↓	→	→
GH replacement in hypopituitarism (Weaver *et al.* 1994)	↑	↓	→

natal imprinting of the metabolic enzymes by sex hormones. Thus, female rats show increased hepatic 5α-reduction (Malendowicz 1994) and decreased hepatic 11β-HSD activity (Low *et al.* 1993), and this sexual dimorphism appears to be related, at least in part, to sex-specific patterns of growth hormone secretion (Mode *et al.* 1989). As an added complication, there are also significant species differences in the pattern of sex-specific metabolism. For example, in the hamster, hepatic 5α-reduction is considerably higher in males than in females (Colby *et al.* 1973). Surprisingly, the sex differences in corticosteroid metabolism in the human have received little attention. It is known that men excrete significantly higher amounts of free cortisol (Lamb *et al.* 1994), and have higher plasma levels of DHEAS but lower levels of DHEA (Zumoff *et al.* 1980). More significantly, there appears to be sexual dimorphism in human 11β-HSD activity, as the ratio of urinary cortisone metabolites to cortisol metabolites is higher in women (Raven *et al.* 1995b). Typical urinary steroid profiles from healthy men and women are shown in Figure 3.4.

While there are important sex and species differences in corticosteroid metabolism, age appears to be a less significant factor. There is no change in the half-life in blood of corticosterone with increasing age in rats (Reaven *et al.* 1988).

3.4.2 *Catecholamines*

There are two specific enzymes responsible for catecholamine degradation. These are monoamine oxidase (MAO) and catechol-*o*-methyltransferase (COMT). The major pathways of catecholamine degradation are shown in Figure 3.5. MAO exists in two forms: MAO-A and MAO-B and is present in most tissues, but is found in particularly high concentrations in the liver, kidney, stomach and intestine. COMT is mainly found in the liver and kidney, but is also present in many different tissues (Karhunen *et al.* 1994). Many inhibitors of these enzymes have been developed and are routinely used therapeutically. Non-specific MAO inhibitors are used as antidepressants, while MAO-B inhibitors are used to treat Parkinson's disease (Tipton 1994). Styrene epoxide, an industrial solvent, has been reported to cause decreased MAO-B activity in peripheral blood cells in man (Pahwa & Kalra 1993). The use of antidepressants with MAO-A inhibitory activity is associated with a significant toxicological interaction (Lavin *et al.* 1993). As MAO-A is the isoform found predominantly in the gut, it is this form of the enzyme which metabolizes ingested amines such as tyramine. Some foodstuffs, such as cheese, are particularly rich in tyramine, and the combination of an MAO-A inhibitor with ingestion of these foodstuffs can lead to severe hypertensive reactions, termed the 'cheese reaction' (Tipton 1994). Inhibitors of COMT activity have been used to treat neurodegenerative disorders. There are reports of COMT inhibition by L1,CP20, an orally active iron chelator (Waldmeier *et al.* 1993). There are significant ethnic differences in COMT activity (Klemetsdal *et al.* 1994, McLeod *et al.* 1994).

3.4.3 *The Foetal Adrenal Cortex*

The morphology of the foetal adrenal has been described in Chapter 2 (see section on 'age-related changes in adrenal morphology'). The foetal adrenal cortex achieves

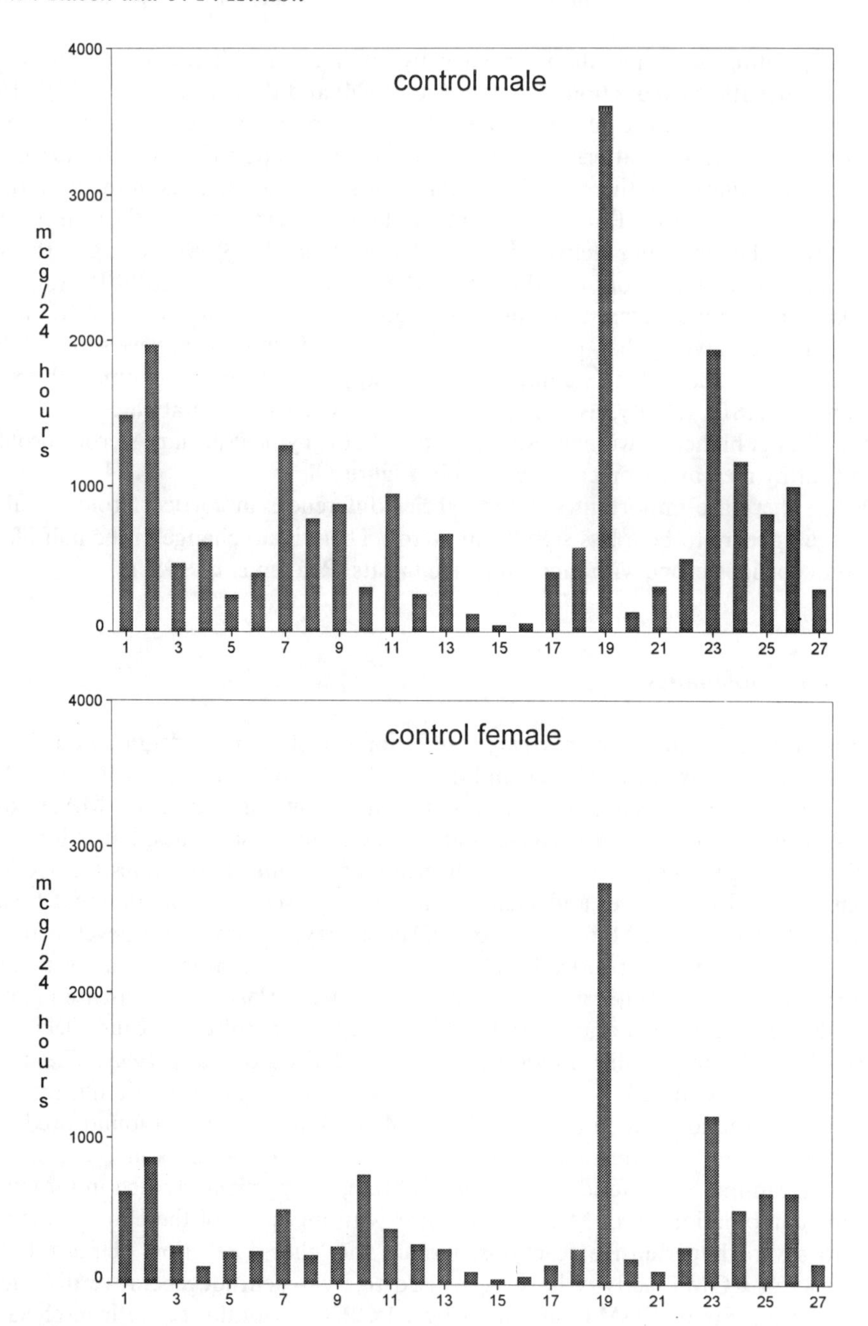

Figure 3.4 Typical chromatograms of urinary steroids. Key to peak numbers: 1, androsterone; 2, aetiocholanolone; 3, androstenediol 17α; 4, DHA, 5, androstenediol 17β; 6, 11-oxoaetiocholanolone; 7, 11β-hydroxyandrosterone; 8, 11β-hydroxyaetiocholanolone; 9, 16α-hydroxyDHA; 10, pregnanediol; 11, pregnanetriol; 12, pregnenediol; 13, androstanetriol; 14, tetrahydro-11-deoxycortisol; 15, tetrahydro-11-deoxycorticosterone; 16, allo-tetrahydro-11-deoxycortisol; 17, pregnenetriol; 18, hexahydro-11-deoxycortisol; 19, tetrahydrocortisone; 20, tetrahydro-11-dehydrocorticosterone; 21, tetrahydrocorticosterone; 22, allo-tetrahydrocorticosterone; 23, tetrahydrocortisol; 24, allo-tetrahydrocortisol; 25, α-cortolone; 26, β-cortolone and β-cortol; 27, α-cortol.

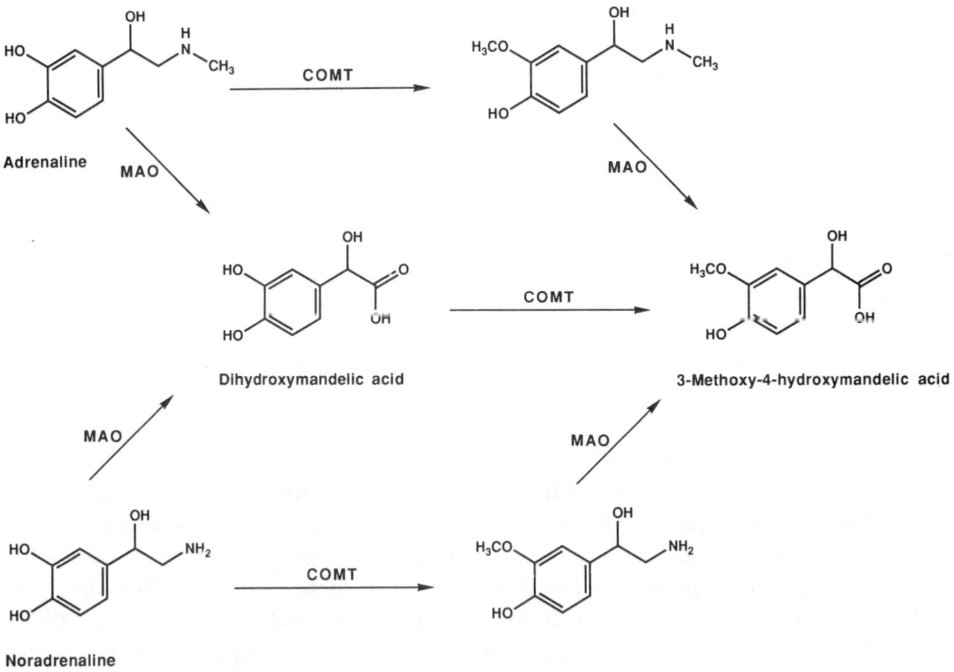

Figure 3.5 Major pathways of catecholamine metabolism.

the enzymic capability to synthesize cortisol and aldosterone during the second trimester (Villee 1969). Compared to the adult, however, the foetal adrenal has relatively low activity of the 3β-hydroxysteroid dehydrogenase system, high sulphurylating activity and low sulphate hydrolysing activity (Shackleton 1984). As a result, the major product of *de novo* synthesis is DHEAS. About 25% of foetally circulating cortisol is of maternal origin, as it freely crosses the placenta (Beitins *et al.* 1970), with most of the remainder being synthesized from progesterone of placental origin. Peripheral metabolism by 11β-HSD in both the foetus and placenta favours the oxidation of cortisol to cortisone, so that cortisone is the major circulating glucocorticoid in the foetus (Shackleton 1984).

The human foetus develops a high capacity for peripheral phase I metabolism of corticosteroids from early in gestation, in contrast to most experimental animals such as the rat (Pelkonen 1978). As a consequence there is extensive metabolism of xenobiotics in the human foetus. In addition, the foetal adrenal cortex has a higher xenobiotic metabolizing activity than in the adult (Juchau & Pedersen 1973). This suggests that metabolic activation of toxins may be particularly important in the foetal adrenal.

The profile of urinary steroids excreted in the neonate is very different from that seen in adults and is made up of metabolites of progesterone and oestrogen derived from the placenta *in utero*, metabolites of DHEA and pregnenolone derived from foetal adrenal cortex and metabolites of corticosteroids. Analysis of neonatal urinary steroid profiles by methods such as high resolution gas chromatography plays an important role in the diagnosis of inborn errors of steroid biosynthesis and metabolism (Shackleton 1986).

3.4.4 The Adrenal Cortex and Pregnancy

Most of the changes in levels of circulating and excreted corticosteroids during pregnancy are the result of altered metabolism rather than synthesis. In addition, levels of CBG rise progressively through pregnancy. There are many changes in cortisol metabolism (Fotherby 1984) as a result of which plasma cortisol levels and urinary free cortisol excretion are both increased. Near term there is also a decrease in the metabolic clearance rate of cortisol. Most of these changes are believed to be due to the effects of increased circulating oestrogens on the peripheral metabolism of corticosteroids. There are similar increases in both plasma aldosterone levels and plasma renin activity (Ledoux *et al.* 1975).

Most of the greatly increased maternal excretion of oestriol in pregnancy is derived from placental metabolism of foetal adrenal DHEAS and 16α-hydroxy DHEAS (Shackleton 1984), an example of the complex interplay between foetal, placental and maternal synthesis and metabolism.

There are many species differences in the endocrinology of pregnancy (Shackleton & Mitchell 1975), particularly in the hepatic metabolism of corticosteroids in late pregnancy. For example, only humans and great apes produce such a large range of progesterone metabolites in such large quantities (Sjovall 1970). Hepatic 5α-reduction appears to be particularly prominent (Anderson *et al.* 1990), and it may be that the physiological relevance of these changes lies in the recent recognition (Mellon 1994) that 5α-reduced progesterone derivatives belong to the class of neurally-active steroids which are potent GABAergic agonists.

3.5 Pathology of the Adrenal Gland

3.5.1 Cushing's Syndrome

Excessively high circulating concentrations of glucocorticoids result in Cushing's syndrome. This disorder has a range of characteristic features, including obesity, facial rounding, skin striae, plethora, poor wound healing, hypertension and osteoporosis (Orth *et al.* 1992). Cushing's syndrome has several possible causes, including pituitary adenoma, adrenal adenoma, and ectopic ACTH production, but is most often iatrogenic, caused by the exogenous administration of glucocorticoids (Walker & Edwards 1992). The features of iatrogenic Cushing's syndrome may differ from those associated with other causes since it is the result of administration of one specific glucocorticoid, while in Cushing's syndrome resulting from excessive endogenous production a wide range of active steroids is produced. Thus hypertension, which is a very common feature of Cushing's syndrome, is rarely seen in iatrogenic Cushing's (Gomez-Sanchez 1986). Similarly, endogenous glucocorticoid excess is associated primarily with depressive disorder, whereas iatrogenic Cushing's syndrome is associated primarily with mania. Chronic alcoholism is also associated with Cushing's syndrome: in cirrhosis, while cortisol biosynthesis is normal, there is inhibition of cortisol metabolism, resulting in hypercortisolaemia (Zumoff *et al.* 1967). While Cushing's syndrome is a serious disorder with a poor prognosis, due to the likelihood of severe infection and cardiovascular complications, several of the inhibitors of steroidogenic enzymes (Chapter 2, Table 2.1) may be exploited therapeutically in the treatment of Cushing's syndrome (see below).

3.5.2 Addison's Disease (Primary Adrenal Insufficiency)

There are few adrenal toxins which cause direct pharmacologically-mediated hyperfunction of the adrenal cortex: most of the toxins affecting adrenocortical function result in impaired steroid hormone production, which may, in extreme cases, lead to primary adrenocortical insufficiency (Addison's disease).

The adrenal cortex and its regulatory mechanisms are able to compensate for a considerable loss of functional adrenal tissue, and in unstressed conditions the symptoms of Addison's disease do not appear until more than 90% of the adrenal tissue in the body is destroyed (Vinson *et al.* 1992). When the adrenal tissue is destroyed slowly, increases in the rate of ACTH secretion can usually induce the secretion of sufficient amounts of glucocorticoid to maintain homeostatis. The features of Addison's disease of slow onset include postural hypotension, fatigue, muscle weakness, and pigmentation. The pigmentation (Nelson's syndrome) is thought to be due to the melanocyte-stimulating activity of ACTH, and the peptides co-secreted with ACTH, which are usually present in very high concentrations in primary adrenal insufficiency. In the case of adrenal haemorrhage, however, or when stressful conditions are also present, the symptoms of Addison's disease may appear suddenly, presenting as an acute adrenal crisis. The features of an Addisonian crisis, which are life-threatening, include severe hypotension leading to circulatory collapse, hypoglycaemia, nausea and vomiting, followed by coma and death.

Major surgery is generally acknowledged to be a particularly stressful condition: patients receiving steroid replacement therapy are given higher doses to cope with the stress of surgery. Clearly it is important that the adrenal glands should be fully functional at such times. Etomidate is a sedative agent, which was originally licensed for use in the induction and maintenance of anaesthesia. In 1982, a report was published advancing the use of etomidate for the sedation of patients in intensive care (Edbrooke *et al.* 1982), as an alternative to benzodiazepines. In June 1983, Ledingham & Watt published a letter in the *Lancet*, reporting that patients in intensive care had a significantly worse prognosis if they received etomidate rather than benzodiazepines. In fact, of the patients with the most severe injuries, nine patients were treated with etomidate from which group none survived, while 13 out of 20 patients treated with benzodiazepines survived (Ledingham & Watt 1983). They followed this initial report with a suggestion that adrenocortical function may be compromised in patients receiving etomidate (Ledingham *et al.* 1983). There followed a flurry of letters to the *Lancet* over the following weeks: some reporting similar findings (Fellows *et al.* 1983), while others strongly defended the use of etomidate (Doenicke 1983). Soon afterwards, evidence was published suggesting that etomidate inhibited the adrenal response to ACTH (Fellows *et al.* 1983), and the use of etomidate and the clinical complications of its use were reviewed in a *Lancet* editorial (1983) which concluded that there was 'no strong case for continued use of etomidate'. Subsequent studies have revealed that etomidate is a potent inhibitor of 11β-hydroxylase activity in the adrenal cortex, and thus prevents the normal adrenocortical response to stress (Wagner *et al.* 1984; for a review see Preziosi & Vacca 1988). It appears to be safe as an anaesthetic induction agent, the adrenocortical effects only becoming evident with prolonged use. Etomidate represents the most extreme case of an adrenocortical toxin: a potent inhibitor of glucocorticoid biosynthesis, administered at a time of severe physiological stress, when adequate adrenocortical function is most vital.

3.5.3 Phaeochromocytoma

Phaeochromocytomas are tumours of chromaffin tissue which secrete catecholamines. About 90% of phaeochromocytomas occur in the adrenal medulla, although they may also arise in extra-adrenal sites (Cryer 1992). The symptoms of this condition include palpitations, headache, anxiety and chest pain. These symptoms occur episodically, with episodes usually lasting several minutes, and are due to increased release of noradrenaline. Hypertension is a major feature of phaeochromocytoma, and cardiovascular effects of this condition are life-threatening. In patients with tumours secreting adrenaline, hypotension and tachycardia may occur. Several agents have been reported to cause adrenal medullary hypertrophy (Ribelin 1984), and MAO inhibitors have been reported to mimic the effects of phaeochromocytoma (Landsberg & Young 1992).

Other tumours of the adrenal medulla which may be associated with excessive catecholamine secretion are more rare and include neuroblastoma, ganglioneuroblastoma and ganglioneuroma (Landsberg & Young 1992). The toxicology and pathology of the medulla is covered in detail in Chapter 6.

3.6 The Adrenal Gland as a Pharmacological Target

The principal use of drugs targeting the adrenal gland is in the treatment of cortical overactivity. Underactivity of corticosteroid biosynthesis is treated by administering replacement hormone, including both glucocorticoid (typically hydrocortisone 30 mg/day) and mineralocorticoid if necessary (fludrocortisone 0.1–0.2 mg/day). Increased doses of glucocorticoid are required at times of stress. Underactivity of medullary hormone secretion does not require treatment, while overactivity of the adrenal medulla, as in phaeochromocytoma, is treated by blocking the effects of excess catecholamines and surgical removal of the tumour.

3.6.1 Inhibitors of Corticosteroid Synthesis (Figure 3.6)

3.6.1.1. Aminoglutethimide

This derivative of the hypnotic glutethimide was originally developed as an anticonvulsant, but was found to be a potent inhibitor of corticosteroid biosynthesis (Hughes & Burley 1970). It reversibly inhibits several of the cytochrome P450 enzymes in the steroid biosynthetic pathway, including 20α,22R-hydroxylase required for conversion of cholesterol to pregnenolone, 21-hydroxylase and 11β-hydroxylase (see Chapter 2). In addition, aminoglutethimide inhibits aromatase, an enzyme which is required for oestrogen synthesis, and also inhibits biosynthesis of thyroid hormones. Tissue concentrations of aminoglutethimide needed to inhibit these different enzymes vary, with inhibition of cholesterol side-chain cleavage and aromatization at low concentrations and of 11β-hydroxylase only at much higher concentrations (Dollery 1991).

Aminoglutethimide is also an inducer of liver microsomal enzymes, including those involved in corticosteroid metabolism such as 6β-hydroxylase. Thus, treatment with aminoglutethimide reduces the half-life of dexamethasone, a synthetic

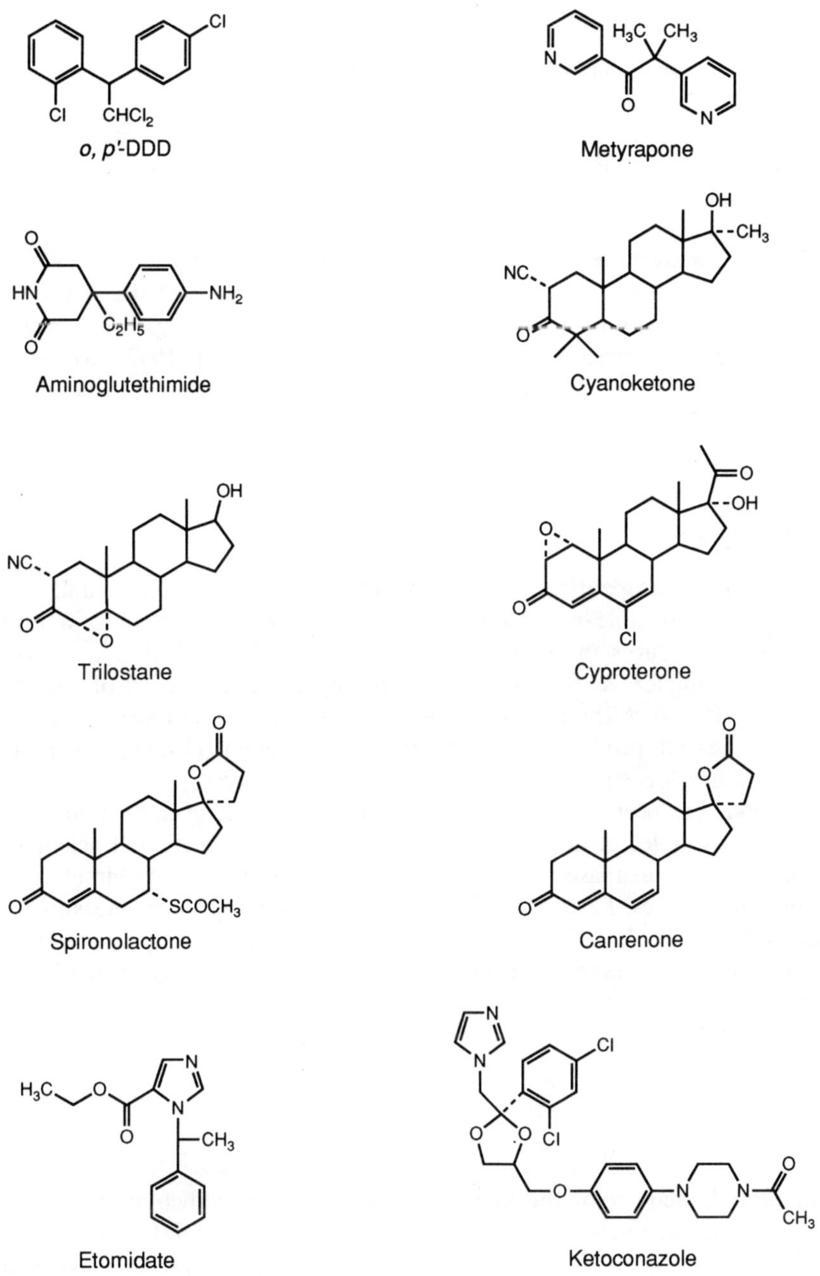

o, p'-DDD

Metyrapone

Aminoglutethimide

Cyanoketone

Trilostane

Cyproterone

Spironolactone

Canrenone

Etomidate

Ketoconazole

Figure 3.6 Inhibitors of corticosteroid biosynthesis.

glucocorticoid, by almost half, but has a less pronounced effect on the half-life of administered hydrocortisone. It has been used both alone, and in conjunction with *o,p'*-DDD (see below) in the treatment of Cushing's syndrome, particularly where cortisol levels are very high as in ectopic ACTH syndrome and adrenocortical carcinoma.

3.6.1.2 *Ketoconazole*

Ketoconazole is an imidazole which has broad-spectrum antifungal activity due to its inhibition of the synthesis of ergosterol, the major sterol component of fungal cell membranes. At higher concentrations, ketoconazole also has widespread inhibitory effects on cytochrome P450 enzymes involved in steroidogenesis, including 17α-hydroxylase and 11β-hydroxylase (Loose *et al.* 1983). Since it does not inhibit 21-hydroxylase, it results in decreased plasma cortisol levels and increased 11-deoxycorticosterone levels. Ketoconazole is now recommended as the agent of choice to decrease excessive cortisol production, due to the fact that its use is not associated with the rise in ACTH and consequent 'cortisol escape' seen with other inhibitors of steroidogenesis (American Medical Association 1993). Ketoconazole has also been used in the treatment of major depression, a disorder which is associated with hypercortisolaemia (Wolkowitz *et al.* 1993).

3.6.1.3 *Metyrapone*

Metyrapone is a competitive inhibitor of adrenal 11β-hydroxylase and of most phenobarbitone-inducible microsomal cytochrome P450 enzymes. As a result of these actions, it reduces the bioavailability of cortisol by both decreasing cortisol synthesis (Dominguez & Samuels 1963) and by increasing cortisol metabolism (Raven *et al.* 1995a). Although metyrapone treatment also suppresses aldosterone synthesis, increased production of the mineralocorticoid 11-deoxycorticosterone means that mineralocorticoid deficiency does not usually occur.

Metyrapone has been used as a test for pituitary reserve, since its administration normally results in decreased plasma cortisol and therefore decreased negative feedback, leading to increased ACTH and therefore increased circulating 11-deoxycortisol. This test has now largely been superceded by direct measurement of circulating ACTH levels. Like aminoglutethimide, metyrapone may be used alone or in combination to suppress cortisol production in Cushing's syndrome and, like ketoconazole, metyrapone has also been used in the treatment of major depression (O'Dwyer *et al.* 1995).

3.6.1.4 *o,p'-DDD (Mitotane)*

The adrenolytic activity of the insecticide dichlorodiphenyldichloroethane (DDD) was found to be due to the presence of small amounts of *o,p'*-DDD (Cueto & Brown 1958). In contrast to other inhibitors of steroidogenesis, *o,p'*-DDD is cytotoxic for cells of the zonae fasciculata and reticularis (but not zona glomerulosa), resulting in adrenal atrophy and therefore decreased corticosteroid synthesis. In addition, cholesterol side-chain cleavage and 11β-hydroxylase are inhibited in surviving cells, and induction of hepatic microsomal enzymes results in altered peripheral metabolism of corticosteroids. There is a marked species difference in the response to *o,p'*-DDD, with the dog and human showing greater sensitivity to its effects than the rat (Vinson & Whitehouse 1987). Clinically, mitotane has been used in the treatment of Cushing's syndrome and, in particular, adrenocortical carcinoma.

3.6.1.5 *Trilostane*

This synthetic steroid which has no hormonal activity is a competitive inhibitor of 3β-hydroxysteroid dehydrogenase (Potts *et al.* 1978). Unlike most other inhibitors of steroidogenesis it has no effect on other steroid biosynthetic enzymes and therefore does not significantly suppress testicular production of testosterone. In spite of this advantage, the clinical use of trilostane in Cushing's syndrome is limited by the considerable variability in dose-response.

3.6.1.6 *Other Therapeutic Agents which Affect Corticosteroid Biosynthesis*

The principal action of spironolactone is competitive antagonism of aldosterone in the collecting duct of the renal tubule. However, aldosterone synthesis may be directly inhibited by high doses (>100 mg/day) of spironolactone, possibly as a result of the formation of its metabolite, canrenone (Cheng *et al.* 1976).

Adrenal insufficiency has been noted in patients treated with suramin, and this antiparasitic drug has subsequently been shown to inhibit several of the enzymes of corticosteroid biosynthesis (Ashby *et al.* 1989).

The inhibition of corticosteroid biosynthesis by the anaesthetic etomidate has been described earlier, and is due to inhibition of cytochrome P450 enzymes, particularly 11β-hydroxylase and cholesterol side-chain cleavage.

High doses of itraconazole, used to treat severe fungal infection, have been noted to result in a blunted cortisol response to ACTH, and effects consistent with inhibition of 11β-hydroxylase and, to a lesser extent, 17α-hydroxylase (Sharkey *et al.* 1991).

Steroid analogues which inhibit steroid biosynthesis include danazol, used as an inhibitor of pituitary gonadotrophin secretion, which inhibits 11β- and 21-hydroxylation, and cyproterone acetate, an anti-androgen, which inhibits 11β- and 21-hydroxylation and 3β-hydroxysteroid dehydrogenase (Vinson *et al.* 1992).

While most therapeutic agents have inhibitory effects, benzodiazepines stimulate corticosteroid synthesis *in vitro* via an action on mitochondrial benzodiazepine receptors which facilitate cholesterol side-chain cleavage (Krueger & Papadopoulos 1992; see Chapter 2). In fact, this interaction has little practical effect due to the predominant central effects of benzodiazepines in increasing ACTH secretion (Schuckit *et al.* 1992).

3.6.2 *Inhibitors of Catecholamine Biosynthesis and Metabolism (Figure 3.7)*

The only inhibitors of medullary catecholamine synthesis used therapeutically are L-aromatic amino acid decarboxylase (LAAD) inhibitors, used to improve the efficacy of administered levodopa by preventing its peripheral metabolism, and MAO inhibitors, used for the treatment of depression, in which their principal action is in the brain. In neither case is the action on the adrenal medulla central to the therapeutic effect (Dollery 1991).

The two commonly used inhibitors of LAAD are benserazide and carbidopa. Neither readily penetrates the blood–brain barrier and their effects are due to the

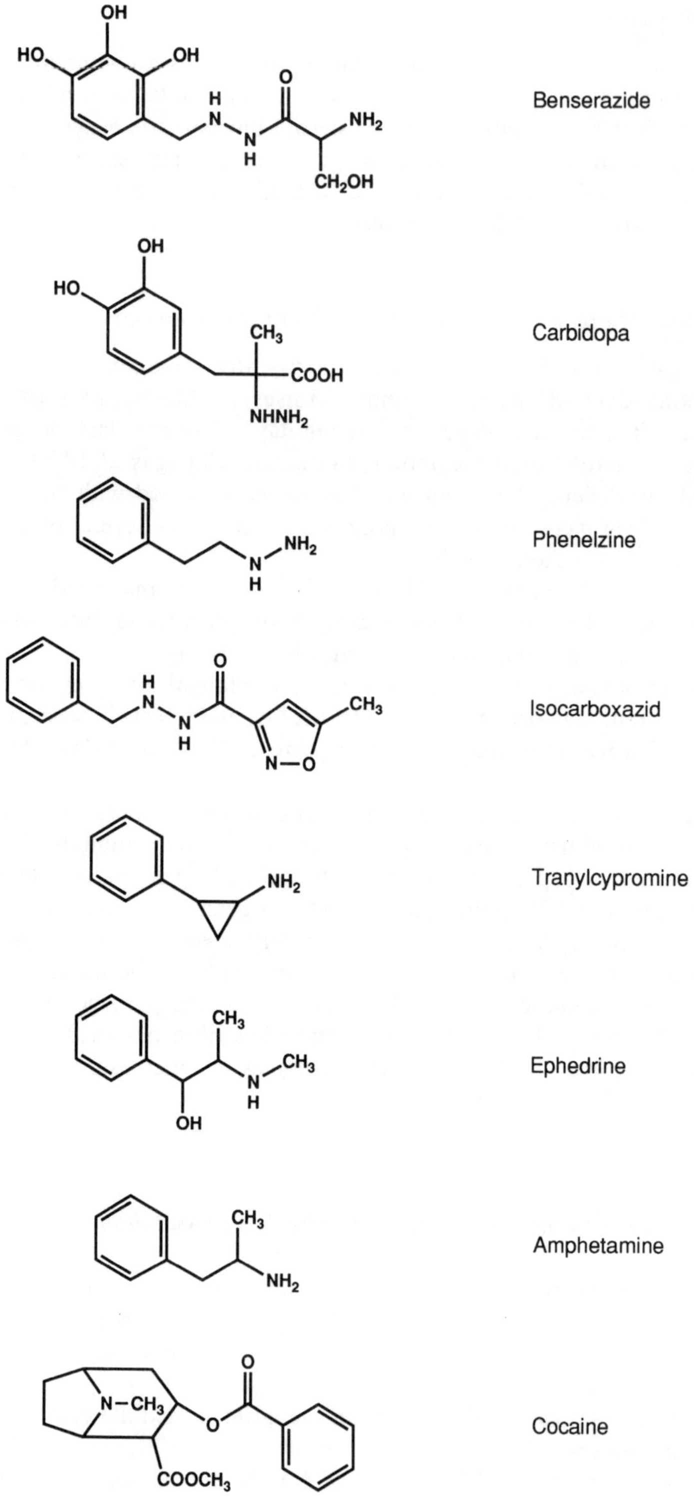

Benserazide

Carbidopa

Phenelzine

Isocarboxazid

Tranylcypromine

Ephedrine

Amphetamine

Cocaine

Figure 3.7 Drugs which modulate adrenal medullary function.

inhibition of peripheral metabolism of levodopa by LAAD in a variety of tissues including adrenal medulla, gut, liver, kidney, salivary glands and pancreas.

The most commonly used MAO inhibitors are phenelzine, isocarboxiazid and tranylcypromine. All are irreversible and non-selective inhibitors of MAO types A and B. Although MAO inhibitors result in increased circulating levels of catecholamines, this effect is peripheral to their mode of action as antidepressants which depends on increased synaptic concentrations of catecholamine neurotransmitters.

3.6.2.1 *Other Therapeutic Agents Which Affect Catecholamine Synthesis and Secretion*

Ephedrine is a naturally-occurring non-catechol sympathomimetic which has some direct activity on adrenoceptors, but acts primarily through inducing the release of stored catecholamines (Hoffman 1995). Other sympathomimetics which have significant peripheral actions due to their effects on releasing medullary catecholamines are amphetamine (Seidon & Sabol 1993) and cocaine (Wilkins 1992).

Radioiodine-labelled meta-iodobenzylguanidine (mIBG) is specifically taken up by the cells of the adrenal medulla. This uptake was originally exploited in the diagnosis of phaeochromocytoma by radioimaging, but is now used for targeted radiotherapy of malignant phaeochromocytoma (Ackery & Lewington 1992).

3.7 Summary

In this and in the preceding chapter we have described the multiplicity of chemical sites at which adrenal function may be influenced, both in health and disease. Table 3.4 lists some of these sites in relation to adrenal cortical function, ranging from the release of hormones which stimulate the cortex, through to the interaction of

Table 3.4 Sites at which exobiotics may influence adrenocortical function, with examples. Details are given in this and previous chapter

Actions on the formation of adrenocortical stimulants	
ACTH release	Cyproheptadine
	Bromocriptine
	Synthetic steroids
Angiotensin II formation	Captopril
Direct action on the adrenal	
Adrenal vasculature	Anticoagulants: heparin, warfarin
Angiotensin II receptors	Saralasin
Second messenger formation	Cholera toxin
Second messenger action	Heparin
Enzymes of steroid biosynthesis	Etomidate, etc.
Effects on secreted steroids	
Corticosteroid binding globulin	NSAIDS
Steroid metabolism	Glycyrrhetinic acid
Steroid/receptor interaction	RU486
	Spironolactone

corticosteroids with their receptors. We have listed an example of an exobiotic which influences each identified step, but it should be remembered that this is far from an exhaustive list of the exobiotics which affect adrenal function. Many of these are discussed in greater detail in later chapters.

References

ACKERY, D. & LEWINGTON, V. (1992) Treatment of malignant phaeochromocytoma with mIBG. In GROSSMAN, A. B. (Ed.) *Clinical Endocrinology*, pp. 486–489. London: Blackwell Scientific.

AMERICAN MEDICAL ASSOCIATION (1993) Drug evaluations annual. Chicago: American Medical Association.

ANDERSON, R. A., BAILLIE, T. A., AXELSON, M., CRONHOLM, T., SJOVALL, K. & SJOVALL, J. (1990) Stable isotope studies on steroid metabolism and kinetics: sulfates of 3α-hydroxy-5α-pregnane derivatives in human pregnancy. *Steroids*, **55**, 443–457.

ASHBY, H., DIMATTINA, M. LINEHAN, W. N., ROBERTSON, C. N., QUEENAN, J. J. & ALBERTSON, B. D. (1989) The inhibition of human adrenal steroidogenic enzyme activities by Suramin. *Journal of Clinical Endocrinology*, **68**, 505–508.

BALLARD, P. L. (1979) Glucocorticoids and differentiation. In BAXTER, J. D. & ROUSSEAU, G. G. (Eds), *Glucocorticoid Hormone Action*, New York: Springer.

BEARN, J. A. & RAVEN, P. W. (1993) Neuroendocrine developments in psychiatric research. In KERWIN, R. (Ed.), *Cambridge Medical Reviews: Neurobiology and Psychiatry*, pp. 71–95, Cambridge: Cambridge University Press.

BEITINS, I. Z., KOWARSKI, A., SHERMETA, D. W., DE LEMOS, & MIGEON, C. J. (1970) Fetal and maternal secretion rate of cortisol in sheep: diffusion resistance of the placenta. *Pediatric Research*, **4**, 129–134.

CHENG, S. C., SUZUKI, K., SADEE, W. & HARDING, B. W. (1976) Effects of spironolactone, canrenone and canrenoate-K on cytochrome P450 and 11β- and 18-hydroxylation in bovine and human adrenal cortical mitochondria. *Endocrinology*, **99**, 1097–2006.

CLARK, J. H., SCHRADER, W. T. & O'MALLEY, B. W. (1992) Mechanisms of action of steroid hormones. In WILSON, J. D. & FOSTER, D. W. (Eds), *Williams Textbook of Endocrinology*, 8th Ed. pp. 35–89, Philadelphia: W. B. Saunders.

COLBY, H. D., GASKIN, J. H. & KITAY, J. I. (1973) Requirement of the pituitary gland for gonadal hormone effect on hepatic corticosteroid metabolism in rat and hamster. *Endocrinology*, **92** 769–774.

CRYER, P. E. (1992) The adrenal medullae. In JAMES, V. H. T. (Ed.), *The Adrenal Gland*, 2nd Edn, pp. 465–489, New York: Academic Press.

CUETO, C. & BROWN, J. H. U. (1958) Biological studies of an adrenocorticolytic agent and isolation of the active components. *Endocrinology*, **62**, 334–339.

CUPPS, T. R., GERRARD, T. L., FALKOFF, R. J., WHALEN, G. & FAUCI, A. S. (1985) Effects of in vitro corticosteroids on B-cell activation, proliferation and differentiation. *Journal of Clinical Investigation*, **75**, 754–761.

DOENICKE, A. (1983) Etomidate. *The Lancet*, **2**, 168.

DOLLERY, C. (1991) *Therapeutic Drugs*, Edinburgh: Churchill Livingstone.

DOMINGUEZ, O. V. & SAMUELS, L. T. (1963) Mechanism of inhibition of adrenal steroid 11β-hydroxylase. *Endocrinology*, **73**, 304–309.

EDBROOKE, D. L., NEWBY, D. M., MATHER, S. J., DIXON , A. M. HEBRON, B. S. (1982) Safer sedation for ventilated patients: a new application for etomidate. *Anaesthesia*, **37**, 765–771.

EDITORIAL (1983) Etomidate. *The Lancet*, **2**, 24–25.

EMORINE, L., BLIN, N. & STROSBERG, A. D. (1994) The human β_3-adrenoceptor: the search for a physiological function. *Trends in Pharmacological Sciences*, **15**, 3–7.

ENGELHARDT, G. (1978) Wirkung von nichtsteroidischen antiphlogistika auf die plasmseiweissbindung des corticosterons. *Arzneimittelforschung*, **20**, 1714–1723.

EXTON, J. H. (1979) Regulation of gluconeogenesis by glucocorticoids. In BAXTER, J. D. & ROUSSEAU, G. G. (Eds), *Glucocorticoid Hormone Action*, New York: Springer.

FAIN, J. H. (1979) Inhibition of glucose transport in fat cells and activation of lipolysis by glucocorticoids. In BAXTER, J. D. & ROUSSEAU, G. G. (Eds), *Glucocorticoid Hormone Action*, pp. 547–560, New York: Springer.

FAUCI, A. S. (1979) Immunosuppressive and anti-inflammatory effects of glucocorticoids. In BAXTER, J. D. & ROUSSEAU, G. G. (Eds), *Glucocorticoid Hormone Action*, New York: Springer.

FELLOWS, I. W., BYRNE, A. J. & ALLISON, S. P. (1983) Adrenocorticol suppression with etomidate. *The Lancet*, **2**, 54–55.

FLOWER, R. J. (1986) The mediators of steroid action. *Nature*, **320**, 20–21.

FOTHERBY, K. (1984) Endocrinology of the menstrual cycle and pregnancy. In MAKIN, H. L. J. (Ed.), *Biochemistry of Steroid Hormones*, 2nd Edn, pp. 409–440, Oxford: Blackwell Scientific.

FRASER, R. (1990) Peripheral metabolism of corticosteroids as a mechanism of potency modulation. *Journal of Endocrinology*, **125**, 1–2.

GARDNER, D. G., HANE, S., TRACHEWSKY, D., SCHENK, D. & BAXTER J. D. (1986) Atrial natriuretic peptide mRNA is regulated by glucocorticoids in vivo. *Biochemical and Biophysical Research Communications*, **139**, 1047–1054.

GOMEZ-SANCHEZ, C. E. (1986) Cushing's syndrome and hypertension. *Hypertension*, **8**, 258–264.

GOWER, D. B. (1984) Steroid catabolism and urinary excretion. In MAKIN, H. L. J. (Ed.), *Biochemistry of Steroid Hormones*, pp. 349–408, Oxford: Blackwell Scientific.

HAHN, T. J., HALSTEAD, L. R., TEITELBAUM, S. L. & HAHN, B. H. (1979) Altered mineral metabolism in glucocorticoid-induced osteopenia. Effect of 25-hydroxy vitamin D administration. *Journal of Clinical Investigation*, **64**, 655–665.

HEIMKE, C., JUSSOFIE, A. & JUPTNER, M. (1991) Evidence that 3α-hydroxy-5α-pregnan-20-one is a physiologically relevent modulator of GABA-ergic neurotransmission. *Psychoneuroendocrinology*, **16**, 517–523.

HOFFMAN, B. B. (1995) Adrenoceptor-activating and other sympathomimetic drugs. In KATZUNG, B. C. (Ed.), *Basic and Clinical Pharmacology*, 6th Edn, pp. 115–131, Norwalk, CT: Appleton & Lange.

HUA, S. Y. & CHEN, Y. Z. (1989) Membrane receptor-mediated electrophysiological effects of glucocorticoid on mammalian neurons. *Endocrinology*, **124**, 687–691.

HUGHES, S. W. M. & BURLEY, D. M. (1970) Aminoglutethimide: a side-effect turned to therapeutic advantage. *Postgraduate Medicine*, **46**, 409–416.

JUCHAU, M. R. & PEDERSEN, M. G. (1973) Drug biotransformation reactions in the human foetal adrenal gland. *Life Sciences*, **12**, 193–204.

KARHUNEN, T., TILGMANN, C., ULMANEN, I., JULKENEN, I. & PANULA, P. (1994) Distribution of catechol-*o*-methyltransferase enzyme in rat tissues. *Journal of Histochemistry and Cytochemistry*, **42**, 1079–1090.

KLEMETSDAL, B., STRAUME, B., GIVERHAUG, T. & AARBAKKE, J. (1994) Low catechol-*o*-methyltransferase activity in a Saami population. *European Journal of Clinical Pharmacology*, **46**, 231–235.

KRUEGER, K. E. & PAPADOPOULOS, V. (1992) Mitochondrial benzodiazepine receptors and the regulation of steroid biosynthesis. *Annual Review of Pharmacology and Toxicology*, **32**, 211–237.

KUMAR, S., HOLMES, E., SCULLY, S., BIRREN, B. W., WILSON, R. H. & DE VELLIS, J. (1986) The hormonal regulation of gene-expression of glial markers: glutamine synthetase and glycerol phosphate dehydrogenase in primary cultures of rat brain and in C6 cell line. *Journal of Neuroscience Research*, **16**, 251–264.

LAMB, E. J., NOONAN, K. A. & BURRIN, J. M. (1994) Urine-free cortisol excretion: evidence of sex-dependence. *Annals of Clinical Biochemistry*, **31**, 455–458.

LANDSBERG, L. & YOUNG, J. B. (1992) Catecholamines and the adrenal medulla. In WILSON, J. D. & FOSTER, D. W. (Eds), *Williams Textbook of Endocrinology*, 8th Edn, pp. 621–705, Philadelphia: W. B. Saunders.

LAVIN, M. R., MENDELOWITZ, A. & KRONIG, M. H. (1993) Spontaneous hypertensive reactions with monoamine oxidase inhibitors. *Biological Psychiatry*, **34**, 146–151.

LEDINGHAM, I. M. & WATT, I. (1983) Influence of sedation on mortality in critically ill multiple trauma patients. *Lancet*, **1**, 1270.

LEDINGHAM, I. M., FINLAY, W. E. I., WATT. I. & MCKEE, J. I. (1983) Etomidate and adrenocortical function. *Lancet*, **1** 1434.

LEDOUX, F., GENEST, J., NOWACZYNSKI, W., KUCHEL, O. & LEBEL, M. (1975) Plasma progesterone and aldosterone in pregnancy. *Canadian Medical Association Journal*, **112**, 943–947.

LEIBOVICH, S. J. & ROSS, R. (1975) The role of the macrophage in wound repair. A study with hydrocortisone and anti-macrophage serum. *American Journal of Pathology*, **78**, 71–100.

LOEB, J. N. (1976) Corticosteroids and growth. *New England Journal of Medicine*, **295**, 547–552.

LOOSE, D. S., KAN, P. B., HIRST, M. A., MARCUS, R. A. & FELDMAN, D. (1983) Ketoconazole blocks adrenal steroidogenesis by inhibiting cytochrome P-450-dependent enzymes. *Journal of Clinical Investigation*, **71**, 1495–1499.

LOW, S. C., ASSAAD, S. N., RAJAN, V., CHAPMAN, K. E., EDWARDS C. R. W. & SECKL, J. R. (1993) Regulation of 11β-hydroxysteroid dehydrogenase by sex steroids in vivo: further evidence for the existence of a second dehydrogenase in rat kidney. *Journal of Endocrinology*, **139**, 27–35.

MAJEWSKA, M. D., HARRISON, N. L., SCHWARTZ, R. D., BARKER, J. L. & PAUL, S. M. (1986) Steroid hormone metabolites are barbiturate-like modulators of the GABA receptor. *Science*, **232**, 1004–1007.

MALBON, C. C., RAPIEJKO, P. J. & WATKINS, D. C. (1988) Permissive hormone regulation of hormone sensitive effector systems. *Trends in Pharmacological Sciences*, **9**, 33–36.

MALENDOWICZ, L. K. (1994) *Cytophysiology of the Mammalian Adrenal Cortex*, Poznan: PTPN.

MCEWEN, B. S. (1979) Influences of adrenocortical hormones on pituitary and brain function. In BAXTER, J. D. & ROUSSEAU, G. G. (Eds), *Glucocorticoid Hormone Action*, New York: Springer.

MCLEOD, H. L., FANG, L., LUO, X., SCOTT, E. P. & EVANS, W. E. (1994) Ethnic differences in erythrocyte catechol-*o*-methyltransferase activity in black and white Americans. *Journal of Pharmacology and Experimental Therapeutics*, **270**, 26–29.

MELLON, S. H. (1994) Neurosteroids: biochemistry, modes of action, and clinical relevance. *Journal of Clinical Endocrinology and Metabolism*, **78**, 1003–1008.

MENDEL, C. M. (1989) The free hormone hypothesis: a physiologically based mathematical model. *Endocrine Reviews*, **10**, 232–274.

MODE, A., WIERSMA-LARSSON, E., STROM, A., ZAPHIROPOULOS, P. G. & GUSTAFSSON, J.-A. (1989) A dual role of growth hormone as a feminising and masculinising factor in the control of sex-specific cytochrome P-450 isozymes in rat liver. *Journal of Endocrinology*, **120**, 311–317.

MUNCK, A., GUYRE, P. M. & HOLBROOK, N. J. (1984) Physiological functions of glucocorticoids in stress and their relation to pharmacological actions. *Endocrine Reviews*, **5** 25–44.

O'DWYER, A.-M., LIGHTMAN, S. L., MARKS, M. N. & CHECKLEY, S. A. (1995) Treatment of major depression with metyrapone and hydrocortisone. *Journal of Affective Disorders*, **33**, 123–128.

ORTH, D. N., KOVACS, W. J. & DEBOLD, C. R. (1992) The adrenal cortex. In WILSON J. D. & FOSTER, D. W. (Eds), *Williams Textbook of Endocrinology*, 8th Edn, pp. 489–619, Philadelphia: W. B. Saunders.

PAHWA, R. & KALRA, J. (1993) A critical review of the neurotoxicity of styrene in humans. *Veterinary and Human Toxicology*, **35**, 516–520.

PARKER, M. G. (1988) The expanding family of hormone receptors. *Journal of Endocrinology*, **119**, 175–177.

PAUL, S. M. & PURDY, R. H. (1992) Neuroactive steroids. *FASEB Journal*, **6**, 2311–2322.

PELKONEN, O. (1978) In BRIDGES, J. W. & CHASSEAUD, L. F. (Eds), *Progress in Drug Metabolism*, Vol. 2, Chichester: John Wiley.

PHILLIPON, G. (1983) Investigation of urinary steroid profiles as a diagnostic method in Cushing's syndrome. *Clinical Endocrinology*, **16**, 433–439.

POTTS, G. O., CREANGE, J. E., HARDING, H. R. & SCHANE, H. P. (1978) Trilostane, an orally active inhibitor of steroid biosynthesis. *Steroids*, **32**, 257–267.

PREZIOSI, P. & VACCA, M. (1988) Minireview: adrenocortical suppression and other endocrine effects of etomidate. *Life Sciences*, **42**, 477–489.

RAFF, H. (1987) Glucocorticoid inhibition of neurohypophyseal vasopressin secretion. *American Journal of Physiology*, **21**, R635–R644.

RAVEN, P. W., CHECKLEY, S. A. & TAYLOR, N. F. (1995a) Extra-adrenal effects of metyrapone include inhibition of the 11-oxoreductase activity of 11β-hydroxysteroid dehydrogenase: a model for 11-HSD I deficiency. *Clinical Endocrinology*, **43**, (in press).

RAVEN, P. W., TAYLOR, N. F. & CHECKLEY, S. A. (1995b) Sex differences in cortisol metabolism in health and depression: C21 steroids. Submitted for publication. In major depression. *Journal of Endocrinology*, **147**, (Suppl.), 32.

REAVEN, E. KOSTRNA, M., RAMACHANDRAN, J. & AZHAR, S. (1988) Structure and function changes in rat adrenal glands during aging. *American Journal of Physiology*, **255**, E903—E911.

RIBELIN, W. E. (1984) The effects of drugs and chemicals upon the structure of the adrenal gland. *Fundamental and Applied Toxicology*, **4**, 105–119.

RODIN, A., THAKKAR, H., TAYLOR, N. & CLAYTON, R. (1994) Hyperandrogenism in polycystic ovary syndrome: evidence of dysregulation of 11β-hydroxysteroid dehydrogenase. *New England Journal of Medicine*, **330**, 460–465.

ROSNER, W. (1990) The functions of corticosteroid-binding globulin and sex hormone-binding globulin: recent advances. *Endocrine Reviews*, **11**, 80–91.

SAPOLSKY, R. M., KREY, L. C. & MCEWAN, B. S. (1986) The neuroendocrinology of stress and aging: the glucocorticoid cascade hypothesis. *Endocrine Reviews*, **7**, 284–301.

SCHUCKIT, M. A., HAUGER, R. & KLEIN, J. L. (1992) Adrenocorticotropin hormone response to diazepam in healthy young men. *Biological Psychiatry*, **31**, 661–669.

SEIDON, L. S. & SABOL, K. E. (1993) Amphetamine: effects on catecholamine systems and behavior. *Annual Review of Pharmacology and Toxicology*, **32**, 639–677.

SELYE, H. (1942) Correlations between the chemical structure and the pharmacological actions of the steroids. *Endocrinology*, **30**, 437–453.

SHACKLETON, C. H. L. (1984) Steroid synthesis and catabolism in the fetus and neonate. In MAKIN, H. L. J. (Ed.), *Biochemistry of Steroid Hormones*, pp. 441–447, Oxford: Blackwell Scientific.

SHACKLETON, C. H. L. (1986) Profiling steroid hormones and urinary steroids. *Journal of Chromatography*, **379**, 91–156.

SHACKLETON, C. H. L. & MITCHELL, F. L. (1975) The comparison of perinatal steroid endocrinology in simians with a view to finding a suitable animal model to study human problems. *Laboratory Animal Handbooks*, **6**, 159–181.

SHARKEY, P. K., RINALDI, M. G., DUNN, J. F., HARDIN, T. C., FETCHIK, R. J. & GRAYBILL, J. R. (1991) High-dose itraconazole in the treatment of severe mycosis. *Antimicrobial Agents and Chemotherapy*, **35**, 707–713.

SJOVALL, K. (1970) Gas chromatographic determination of steroid sulphates in plasma during pregnancy. *Annals of Clinical Research*, **2**, 393–408.

STALMANS, W. & LALOUX, M. (1979) Glucocorticoids and hepatic glycogen metabolism. In BAXTER, J. D. & ROUSSEAU, G. G. (Eds), *Glucocorticoid Hormone Action*, New York: Springer.

STEWART, P. M., WALLACE, A. M., VALENTINO, R., BURT, D., SHACKLETON, C. M. & EDWARDS, C. R. W. (1987) Mineralocorticoid activity of liquorice: 11β-hydroxysteroid dehydrogenase activity comes of age. *Lancet*, ii, 821–824.

STEWART, P. M., CORRIE, J. E., SHACKLETON, C. H. & EDWARDS, C. R. W. (1988) Syndrome of apparent mineralocorticoid excess. A defect in the cortisol–cortisone shuttle. *Journal of Clinical Investigation*, **82**, 340–349.

STEWART, P. M., WALLACE, A. M., ATHERDEN, S. M., SHEARING, C. H. & EDWARDS, C. R. W. (1990) Mineralocorticoid activity of carbenoxolone: contrasting effects of carbenoxolone and licorice on 11β-hydroxysteroid dehydrogenase activity in man. *Clinical Science*, **78**, 49–54.

STEWART, P. M., MURRAY, B. A. & MASON, J. I. (1994) Human kidney 11β-hydroxysteroid dehydrogenase is a high affinity nicotinamide adenine dinucleotide-dependent enzyme, and differs from the cloned type I isoform. *Journal of Clinical Endocrinology and Metabolism*, **79**, 480–484.

TAYLOR, N. F., FISHER, J., LLEWELLYN, D. E. M. & RAVEN, P. W. (1994) Origin of impaired 11-dehydrogenation of corticosteroids in conditions of corticosteroid excess. *Journal of Endocrinology*, **143**, (Suppl.), p56.

TIPTON, K. F. (1994) Monoamine oxidase inhibition. *Biochemical Society Transactions*, **22**, 764–768.

VANLUCHENE, E., AERTSENS, W. & VANDEKERCKHOVE, D. (1979) Steroid excretion in anorexia nervosa patients. *Acta Endocrinologica*, **90**, 133–138.

VILLEE, D. B. (1969) Development of endocrine function in the human placenta and fetus. *New England Journal of Medicine*, **281**, 473–484.

VINSON, G. P. & WHITEHOUSE, B. J. (1987) Inhibitors of corticosteroid biosynthesis: mitotane, metyrapone, aminoglutethimide. In FURR, B. J. A. & WAKELING, A. E. (Eds), *Pharmacology and Clinical Uses of Inhibitors of Hormone Secretion and Action*, pp. 346–364, London: Bailliere-Tindall.

VINSON, G. P., WHITEHOUSE, B. J. & HINSON, J. P. (1992) *The Adrenal Cortex*, New Jersey: Prentice Hall.

WAGNER, R. L., WHITE, P. F., KAN, P. B., ROSENTHAL, M. H. & FELDMAN, D. (1984) Inhibition of adrenal steroidogenesis by the anaesthetic etomidate. *New England Journal of Medicine*, **310**, 1415–1421.

WALDMEIER, P. C., BUCHLE, A. M. & STEULET, A. F. (1993) Inhibition of catechol-*o*-methyltransferase (COMT) as well as tyrosine and tryptophan hydroxylase by the orally active iron chelator, 1,2-dimethyl-3-hydroxypyridin-4-one (L1,CP20), in rat brain *in vivo*. *Biochemical Pharmacology*, **45**, 2417–2424.

WALKER, B. R. & EDWARDS, C. R. W. (1992) Cushing's syndrome. In JAMES, V. H. T. (Ed.), *The Adrenal Gland*, 2nd Edn, pp. 289–318, New York: Academic Press.

WEAVER, J. V., THAVENTHIRAN, L., NOONAN, K., BURRIN, J. M., TAYLOR, N. F., NORMAN, M. R. & MONSON, J. P. (1994) The effect of growth hormone replacement on cortisol metabolism and glucocorticoid sensitivity in hypopituitary adults. *Clinical Endocrinology*, **41**, 639–648.

WEHRENBERG, W. B., BAIRD, A. & LING, N. (1983) Potent interaction between glucocorticoids and growth hormone releasing factor *in vivo*. *Science*, **221**, 556–558.

WILKINS, J. M. (1992) Brain, lung and cardiovascular interactions with cocaine and cocaine-induced catecholamine effects. *Journal of Addictive Disorders*, **11**, 9–19.

WOLKOWITZ, O. M., REUS, V. I., MANFREDI, F., INGBAR, J., BRIZENDINE, L. & WEINGARTNER, H. (1993) Ketoconazole administration in hypercortisolemic depression. *American Journal of Psychiatry*, **150**, 810–812.

ZUMOFF, B., BRADLOW, H. L., GALLAGHER, T. F. & HELLMAN, L. (1967) Cortisol metabolism in cirrhosis. *Journal of Clinical Investigation*, **46**, 1735–1743.

ZUMOFF, B., ROSENFELD, R. S., STRAIN, G. W., LEVIN, J. & FUKUSHIMA, D. K. (1980) Sex differences in the twenty-four-hour mean plasma concentrations of dehydroisoandrosterone (DHA) and dehydroisoandrosterone sulfate (DHAS) and the DHA to DHAS ratio in normal adults. *Journal of Clinical Endocrinology and Metabolism*, **51**, 330–333.

4

Molecular and Systems Pharmacology of Glucocorticosteroids

MARK GUMBLETON and PAUL J. NICHOLLS

Welsh School of Pharmacy, University of Wales, Cardiff

4.1 Structure and Biosyntheses

The two classes of steroids synthesized within the adrenal cortex, the corticosteroids (glucorticoids and mineralocorticoids) based on a 21-carbon skeleton [Figure 4.1(a)], and the androgens (dehydroepiandrosterone, androstenedione, testosterone)

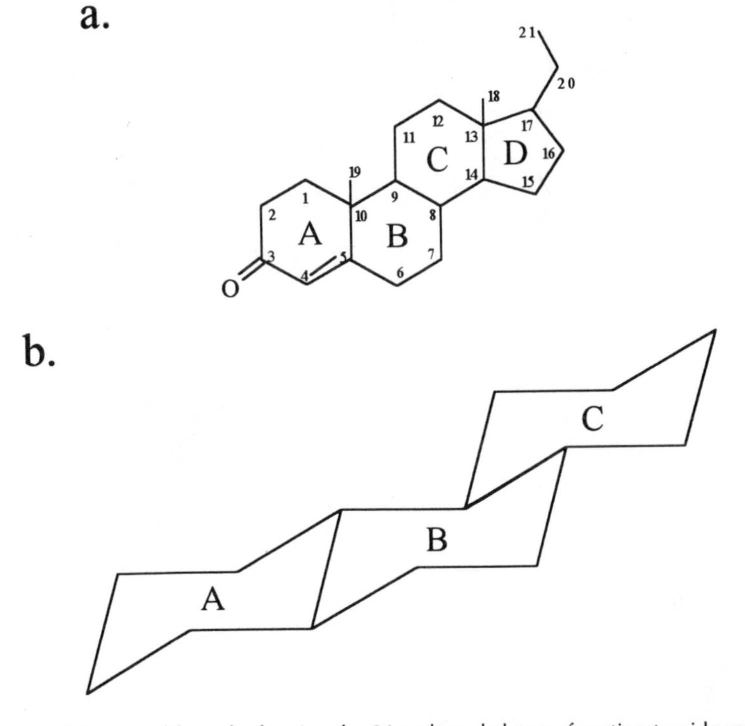

Figure 4.1 (a) Structural formula showing the 21-carbon skeleton of corticosteroids and (b) an illustration showing the corticosteroid cyclohexane rings A, B and C to be orientated in the chair conformation

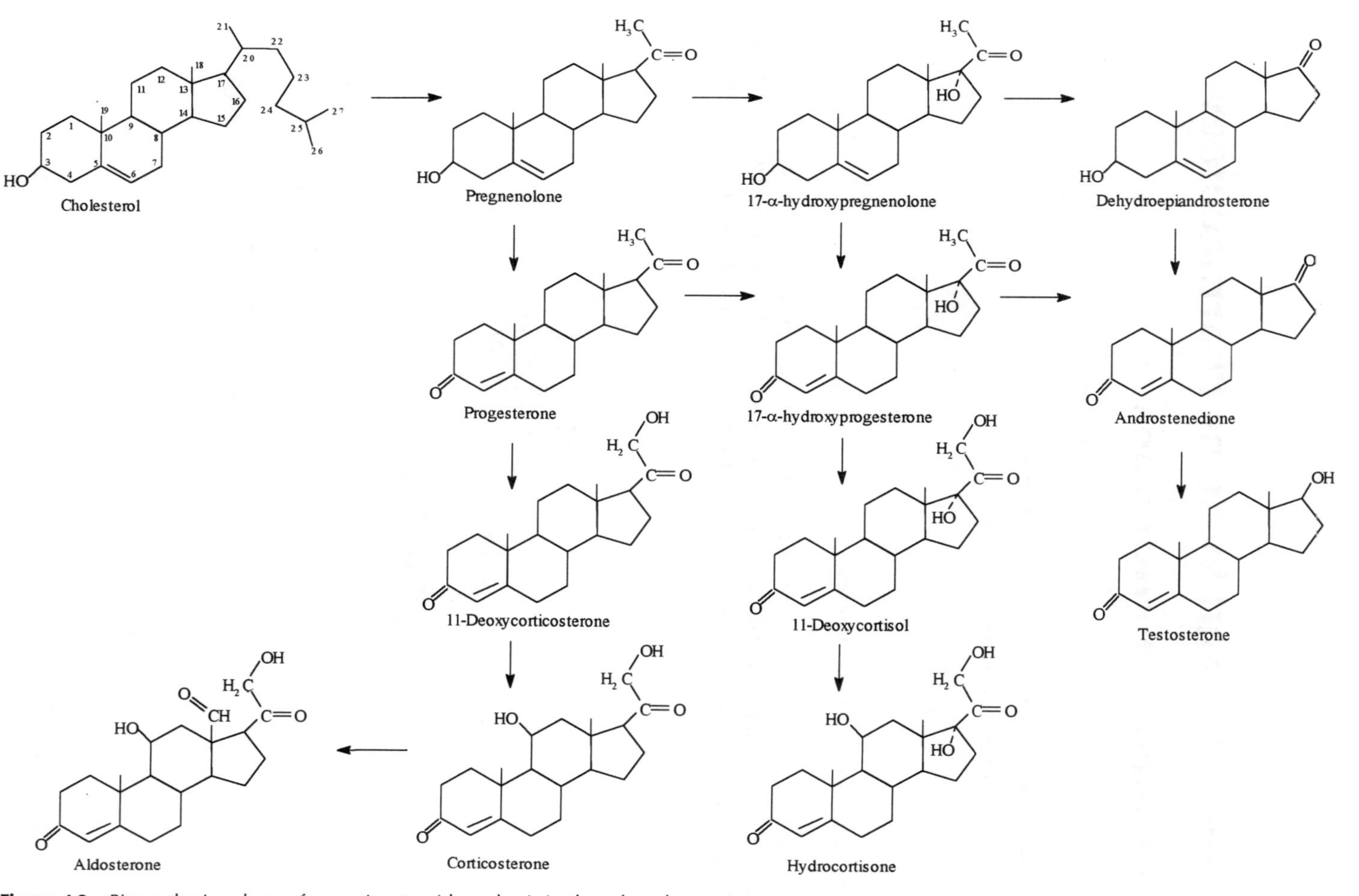

Figure 4.2 Biosynthetic scheme for corticosteroid synthesis in the adrenal cortex

based on a 19-carbon skeleton, possess variable chemical modifications to the basic steroid nucleus of three cyclohexane rings and one cyclopentane ring. Figure 4.1(b) shows the respective cyclohexane rings, A, B and C, orientated in the 'chair' conformation, with rings A and B joined either *cis* or *trans*. In all the steroid hormones the ring junctions B : C and C : D are in the *trans* configuration; the alternative 'boat' structure for the cyclohexane rings is less stable. The corticosteroids are classified into two groups: 17-hydroxylated corticosteroids (ie in Figure 4.2 17α-hydroxy pregnenolone → hydrocortisone) and 17-deoxycorticosteroids (ie in Figure 4.2 pregnenolone → corticosterone → aldosterone) depending on the presence or absence of a hydroxyl group at the 17α position. The former is recognized as the glucocorticoid pathway in corticosteroid biosynthesis and the latter as the mineralocorticoid pathway. However, this classification is frequently misleading, first because glucocorticoids may profoundly affect electrolyte and water balance (although in a different manner to that of aldosterone), and second because the classification is not mutually exclusive. Indeed, hydrocortisone is known to activate both glucocorticoid and mineralocorticoid receptors. Likewise if the hydrocortisone (predominant glucocorticoid in man) pathway is specifically inhibited then increased synthesis of corticosterone can provide the necessary glucocorticoid activity to compensate. To study glucocorticoid action in isolation it is necessary to consider the pharmacology of the synthetic steroids such as dexamethasone, which is essentially devoid of mineralocorticoid activity (see section on quantitative structure–activity relationships).

The adrenal cortex comprises three histologically defined zones constituted (from adrenal capsule to boundary of adrenal medulla) by the zona glomerulosa (ZG), the zona fasciculata (ZF) (comprising the bulk of cortical cell mass, with regular columns of large cells laden with cholesterol-rich droplets) and the zona reticularis (ZR). The zones are functionally distinct in that corticosteroid syntheses are zone-specific: glucocorticoids are mainly produced in the ZF and ZR, the latter zone contributing only a limited amount, while the mineralocorticoid, aldosterone, is formed in the ZG. The sequence of biosynthetic reactions from cholesterol through the 17-deoxycorticosteroid pathway to corticosterone, and through the 17-hydroxylated corticosteroid pathway to hydrocortisone is shown in Figure 4.3. This figure additionally identifies the subcellular location of the steroidogenic transformations; for example, cholesterol → pregnenolone, 11-deoxycorticosterone → corticosterone, and 11-deoxycortisol → hydrocortisone are reactions catalysed within mitochondria, while other reactions depicted in Figure 4.3 are catalysed within endoplasmic reticulum (ER). The initial reaction in steroid hormone biosynthesis is the mitochondrial metabolism of cholesterol to pregnenolone, and it is this initial reaction that is the site of regulation and rate limitation in steroidogenesis (Orme-Johnson 1990). Although the adrenal cortex synthesizes cholesterol from acetyl CoA, the greater part of cholesterol (60–80%) utilized for corticosteroidogenesis comes from exogenous sources. Under basal conditions, cholesterol esterified to long-chain fatty acids is extracted from blood via receptor-mediated endocytosis of low-density lipoprotein complexes and as such is the major source for corticosteroid biosynthesis. However, when production of corticosteroids is stimulated, the adrenally-stored cholesterol becomes an increasingly important precursor. The reaction, cholesterol → pregnenolone is catalysed by an enzyme system comprising NADPH-adrenodoxin reductase, adrenodoxin, and a cytochrome P450 that mediates cholesterol side chain cleavage ($P450_{scc}$ also known as 20,22-desmolase); cDNA

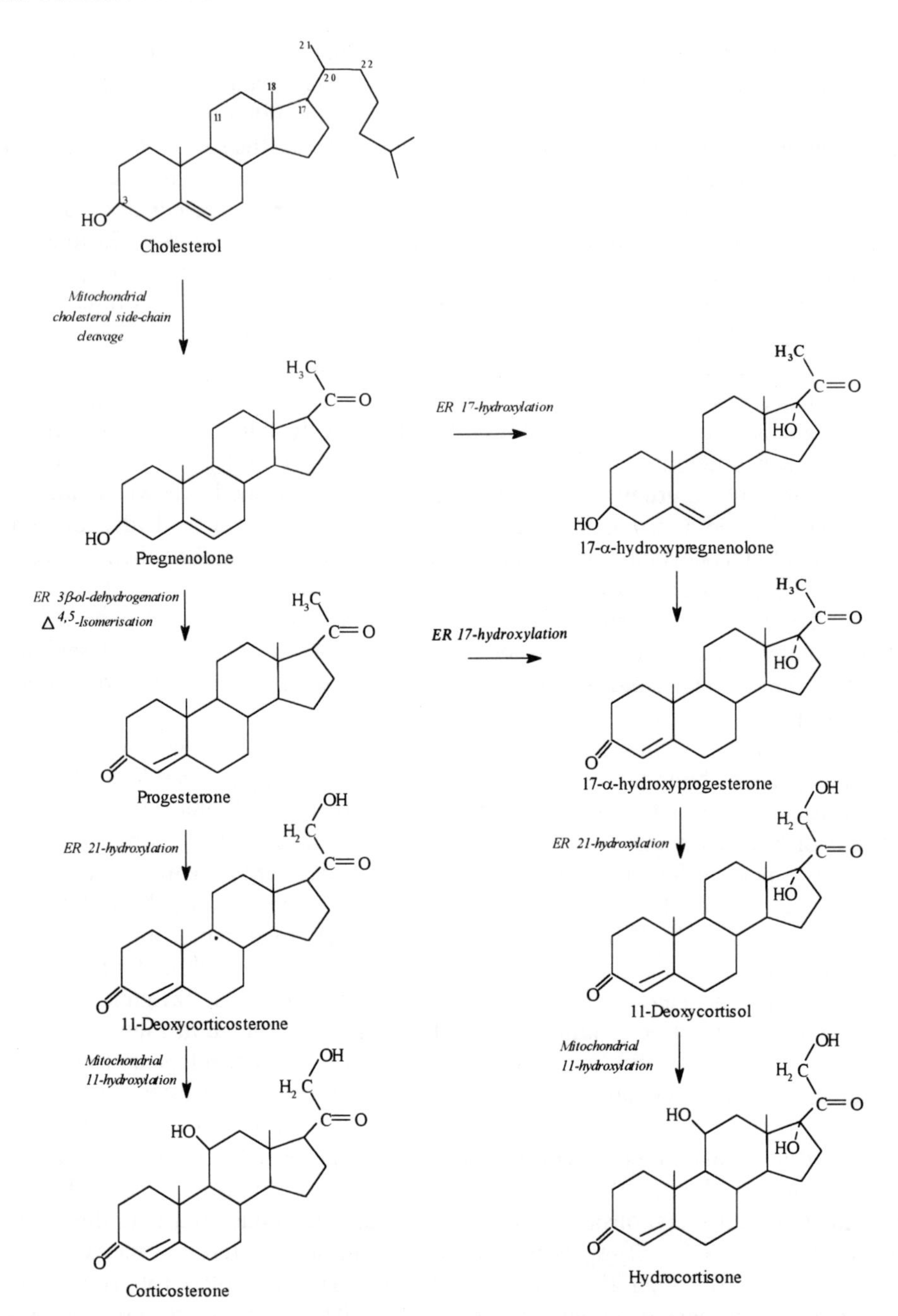

Figure 4.3 Biosynthetic scheme for the glucocorticoid and mineralocorticoid pathways including identification of the subcellular location of the steroidogenic transformations

sequence information affords assignment of $P450_{scc}$ to the P450XIA gene subfamily (Chung *et al.* 1986, Nebert & Gonzalez 1987). The adrenodoxin accepts electrons from the reductase which in turn reduces the cytochrome P450 – the site of cholesterol and O_2 binding and of catalysis. This enzyme system is localized to the inner mitochondrial membrane (Farkash *et al.* 1986) requiring that substrate penetrates to the inner membrane of the mitochondrion for catalysis to proceed. Cytochrome $P450_{scc}$ catalyses a sequence of reactions resulting in oxidative cleavage of a carbon–carbon single bond in the side chain of cholesterol to yield pregnenolone and isocaproaldehyde (Figure 4.4). The reaction proceeds via two sequential monohydroxylations, first at C-22 position then further at C-20, followed by cleavage of the resulting diol. The first steroid formed is 22R-hydroxycholesterol, the

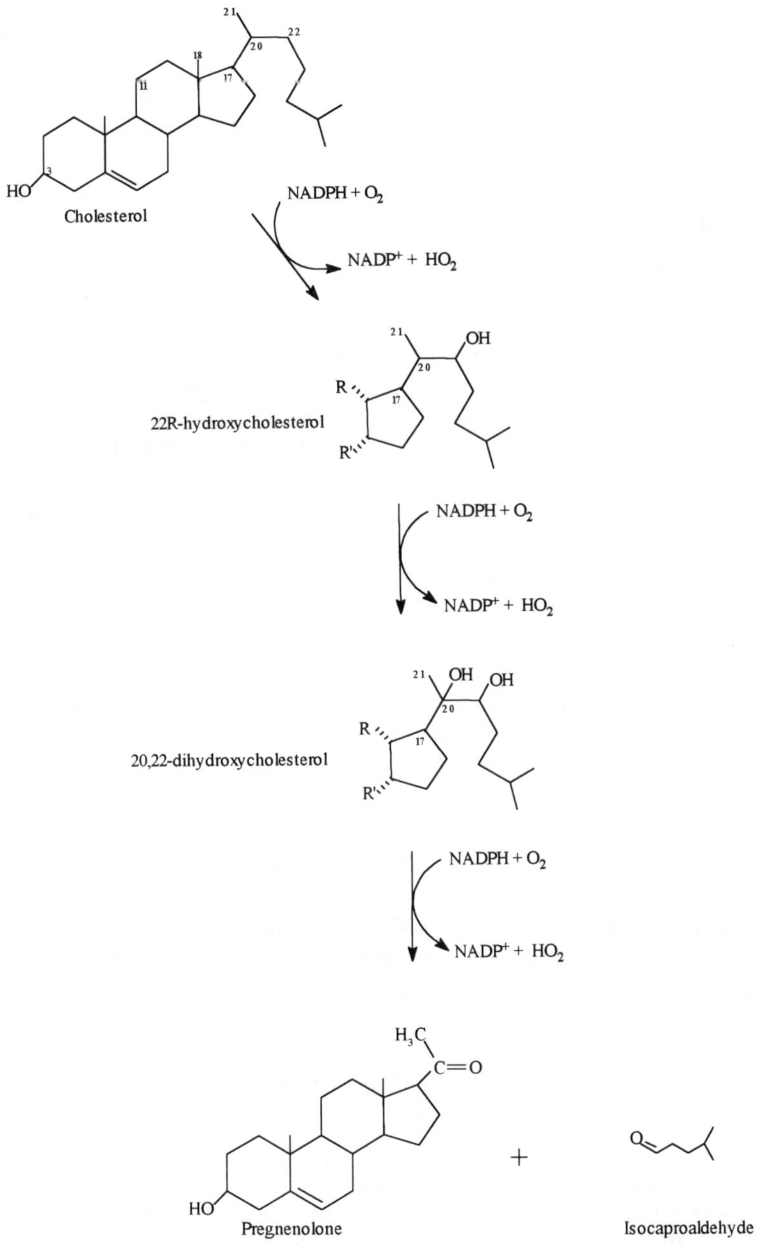

Figure 4.4 Sequence of reactions involved in the oxidative cleavage of the side-chain of cholesterol to yield pregnenolone

second is 20,22-dihydroxycholesterol, and the final steroid produced is the 21-carbon ketosteroid, pregnenolone (Burstein *et al.* 1974, Hume and Boyd 1978, Larroque *et al.* 1981). The side-chain fragment produced is isocaproic aldehyde (4-methyl pentanal) but this is rapidly oxidized to isocaproic acid (4-methyl-pentanoic acid) in many tissues. The synthesized pregnenolone is then exported back to the ER where it is predominantly routed, via P450-dependent catalysis, through the 17-hydroxylated corticosteroid synthetic pathway to produce 11-deoxycortisol. The final step in the synthetic pathway, 11β-hydroxylation of 11-deoxycortisol to yield hydrocortisone, is very efficient ($\sim$95% of 11-deoxycortisol formed is converted to hydrocortisone) and occurs within the inner mitochondrial membrane with catalysis via a P450 enzyme complex involving the P450 isozyme P450XIB1 (Mornet *et al.* 1989, Curnow *et al.* 1991, Mukai *et al.* 1993). Although no analogous human data exists, recent data in the rat (Neri *et al.* 1993) has provided evidence that the 11β-hydroxylation of 11-deoxycorticosterone to generate corticosterone (17-deoxycorticosteroid pathway is the major pathway in the rat) is selectively inhibited by thyrotrophin-releasing hormone (TRH) through a decrease in ACTH-stimulated, but not basal, corticosterone secretion. When hydrocortisone and corticosterone (in man the 17-deoxycorticosteroid pathway is of minor significance under normal conditions) pass back into the ER, the enzyme 11β-hydroxysteroid dehydrogenase can catalyse the reversible conversion to the inactive cortisone and 11-dehydrocorticosterone, respectively. The expression of 11β-hydroxysteroid dehydrogenase, which transforms glucocorticoids to their respective 'biologically inert' 11-dehydro derivatives (Brem & Morris 1993, Maser 1994), clearly has implications in regulating the actions of glucocorticoids. For example, in hepatic tissue 11β-hydroxysteroid dehydrogenase activity is relatively high and can clearly influence steroid exposure to the glucocorticoid receptor (Type II adrenocorticoid receptor). Similarly, localization of 11β-hydroxysteroid dehydrogenase to vascular and cardiac smooth muscle cells (Walker *et al.* 1991) may modulate glucocorticoid effects upon vascular resistance and cardiac output. In Leydig cells of the rat testes 11β-hydroxysteroid dehydrogenase regulates glucocorticoid mediated inhibition of testosterone biosynthesis (Monder *et al.* 1994), and *in vivo* protects the testis from the effects of high levels of circulating glucocorticoids, as may occur with stress and in Cushing's syndrome. Different glucocorticoids exhibit different EC_{50} values for inhibition of testosterone synthesis (Monder *et al.* 1994), reflecting in part differences in their binding affinity to the catalytic site of 11β-hydroxysteroid dehydrogenase, eg dexamethasone is a poor substrate for this enzyme compared to corticosterone. In mineralocorticoid target tissue, eg kidney among others, the activity of 11β-hydroxysteroid dehydrogenase is considered an important mechanism in aldosterone selectivity of the mineralocorticoid receptor (Type I adrenocorticoid receptor). Indeed, 11β-hydroxysteroid dehydrogenase serves a protective function toward the mineralocorticoid receptor, for either decreased 11β-hydroxysteroid dehydrogenase activity or excessive hydrocortisone levels will lead to significant hydrocortisone binding to the mineralocorticoid receptor and to spironolactone-inhibitable Na^+ retention, hypokalemia and hypertension – the syndrome of apparent mineralocorticoid excess (AME) (Laplace *et al.* 1992). Nevertheless, the intrinsic selection by mineralocorticoid receptor itself for aldosterone should not be negated (Lombes *et al.* 1994).

For many investigators a major obstacle in the study of biochemical mechanisms controlling adrenal steroidogenesis, ie isolation and primary culture of cells from each of the distinct adrenocortical zones, may now have been overcome with a recent report on the isolation of an adrenocortical cell line, H295. Under specified conditions this continuous pluripotent cell line maintains the ability to produce all the adrenocortical steroids, ie mineralocorticoids, glucocorticoids and adrenal androgens (Rainey *et al.* 1994).

4.2 ACTH Regulation of Corticosteroid Biosyntheses

Although the adrenal concentrations of corticosteroids are 100–1000-fold higher than plasma levels (Dickerman *et al.* 1984), such stores fail to provide a significant reservoir readily available for release. The amount of corticosteroids present in the adrenal cortex is sufficient to maintain normal secretion for only a few minutes. The fasciculata cells of the adrenal cortex synthesize glucocorticoids in response to adrenocorticotrophic hormone (ACTH), a 4.5 kDa opiomelanocortin polypeptide, secreted from the anterior pituitary. ACTH is secreted upon stimulation with corticotrophin releasing hormone (CRH), arginine vasopressin, and several other neuropeptides (King & Baertshci 1990). CRH, released from the hypothalamus, is the most potent ACTH secretagogue and mediates its effects of stimulating ACTH secretion and biosynthesis through increasing intracellular cyclic adenosine 3′,5′-monophosphate (cAMP) levels. Arginine vasopressin is a weak secretagogue of ACTH but strongly potentiates CRH-stimulated ACTH secretion through the phosphatidylinositol transduction pathway. The regulation of ACTH secretion is among the most complex of all the pituitary hormones (Figure 4.5), exhibiting circadian rhythms, cyclic bursts, feedback control and responding to a wide variety of stimuli. For example, ACTH secretion (at the level of the hypothalamus) is modulated by

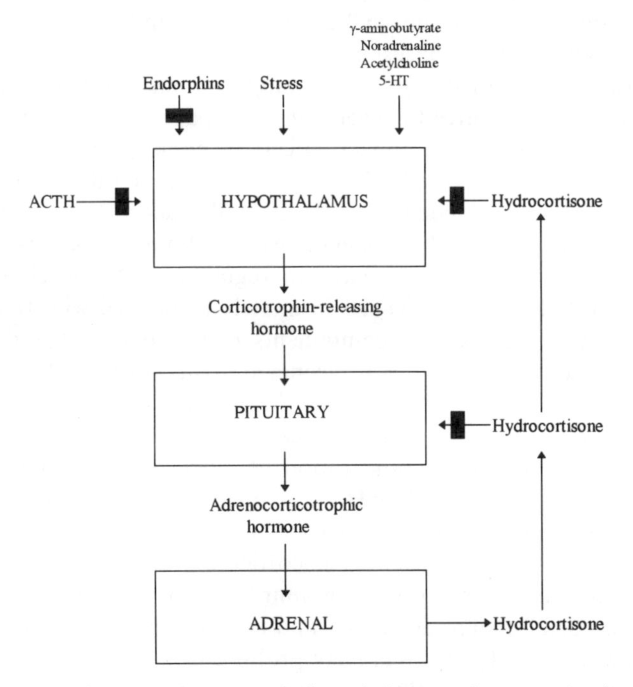

Figure 4.5 Regulation of ACTH secretion at the hypothalamic and pituitary levels

sleep–wake transitions, stress – such as hypoglycaemia, trauma, infection, psychiatric disturbance – such as depression, and is stimulated by chemical mediators such as α-adrenergic agonists, β-adrenergic antagonists, 5HT, γ-aminobutyric and acetylcholine. The encephalins, the opioids and somatostatin are chemical mediators recognized to inhibit ACTH secretion at the level of CRH release. The cytokines – interleukin 1, tumor necrosis factor, interleukin 2 and interleukin 6 – activate the hypothalamic-pituitary-adrenal (HPA) axis leading to elevated levels of glucocorticoids (Hermus & Sweep 1990).

ACTH secretion has a markedly diurnal pattern with a large peak occurring 2–4 h prior to awakening. Thereafter the average level decreases to a nadir just before or after falling asleep. The clocktime of this diurnal pattern can be shifted by systematically altering the sleep–wake cycle for a number of days. The rhythm is diminished or abolished by loss of consciousness, blindness or constant exposure to dark or light. The diurnal pattern is composed of pulses of ACTH release, with major ACTH peaks caused by an increased amplitude rather than an increased frequency of secretory activity. Like ACTH secretion, hydrocortisone secretion exhibits a distinct diurnal variation, with plasma levels of the latter generally following those of the former but with a 15–30 min delay. Feedback inhibition of ACTH secretion is affected by plasma levels of hydrocortisone and by that of synthetic analogues, indeed glucocorticoids exert negative feedback action at multiple levels of the HPA axis. ACTH may also inhibit its own secretion by decreasing CRH release through a short-loop negative feedback. Glucocorticoids inhibit ACTH synthesis by suppressing transcription of the proopiomelanocortin gene (proopiomelanocortin is a 31 kDa precursor protein of ACTH) and attenuate ACTH release by decreasing cAMP accumulation in response to CRH. Indeed, evidence is accumulating that environmental events in the early postnatal period can permanently alter glucocorticoid receptor gene expression in the hippocampus, providing a neural mechanism for the feedback regulation and the development of individual differences in HPA responses. Specifically, it is proposed (Meaney *et al.* 1991) that glucocorticoid receptor density in the hippocampus and frontal cortex can be permanently increased by 'stressful' events in early postnatal life, thereby enhancing the sensitivity to the negative feedback effects of circulating glucocorticoids, and increasing neural inhibition over ACTH secretion. Although there are clear reciprocal relationships between glucocorticoid levels and glucocorticoid receptor expression (Gustafsson *et al.* 1987), evidence exists (Vamvakopoulos *et al.* 1992) demonstrating that glucocorticoids may not regulate mRNA levels of heat shock proteins of the 90 K family – molecules intimately associated with the inactive glucocorticoid receptor and essential components of the glucocorticoid signal transduction pathway (see later). Chronic deficiency of hydrocortisone leads to persistant elevations of plasma ACTH, although the diurnal and pulsatile patterns are preserved, indicating their basic non-feedback control. Chronic hypersecretion of hydrocortisone or long-term administration of glucocorticoid analogues leads to functional atrophy of the CRH–ACTH axis, which may require several months for recovery after the suppressive influence has been removed.

ACTH regulates adrenal cortex glucocorticoid synthesis through both (i) a trophic stimulation (ie achieving or maintaining appropriate levels of enzymes and other proteins needed for steroidogenesis, and invoked through exposure of the cells to increased levels of ACTH for relatively prolonged periods, eg minutes to hours) and (ii) an acute stimulation, enabling an immediate increased glucocorticoid secretory response. The effects of ACTH are mediated through the adenylate cyclase–cAMP pathway acting as the second messenger system (Kimura 1981, Simpson & Waterman 1983); phospholipase C-inositol phosphate may also fulfil an adjunct role in ACTH signal transduction (Farese 1987). Acute stimulation can result in 10–15-fold increases in the rate of steroid biosynthesis and is accomplished without a change in the levels of the enzymes catalysing steroid biosynthesis, including that mediating cholesterol side-chain cleavage (Conneely *et al.* 1984), but rather through an increase in the amount of cholesterol substrate made available to $P450_{scc}$. The increase in cholesterol availability to $P450_{scc}$ may, somewhat arbitrarily, be divided into two processes: The first is the transport of unesterified cholesterol from the lipid vacuoles contained in the cytoplasm of the ZF cell to the outer mitochondrial membrane. The second is the intra-mitochondrial transport of cholesterol from the outer to inner membrane, affording substrate access to $P450_{scc}$. The enzyme cholesterol ester hydrolase is one of the molecules acutely activated by ACTH, specifically via cAMP-dependent kinase mediated phosphorylation (Rae *et al.* 1979, Bisgaier *et al.* 1985). This hydrolase catalyses the de-esterification of cholesterol stored in lipid droplets of the steroidogenic tissue, thereby making available unesterified cholesterol for transport to mitochondria. The accumulation of cholesterol in the mitochondrion in response to ACTH has been found not to be affected by inhibitors of translation (Privalle *et al.* 1983), although microtubule/microfilament cytoskeletal elements and sterol binding proteins have been implicated in the process (Hall 1984). The second process, transport of cholesterol from the outer to inner mitochondrial membrane, is the site of rate limitation in steroidogenesis and is blocked when translation is inhibited (Garren *et al.* 1971, Privalle *et al.* 1983). In the acute response, cAMP also mediates the synthesis of a regulatory protein(s) whose function is to transfer cholesterol from the outer to the inner mitochondrial membrane (Orme-Johnson 1990). Sterol carrier $protein_2$ (SCP_2), a protein stimulated by the trophic actions of ACTH and shown to facilitate cytoplasmic to mitochondrial and intra-mitochondrial transport is not one of the acutely synthesized transfer proteins. Several peptides that may fulfil the above intra-mitochondrial cholesterol transport function have been isolated (Alberta *et al.* 1989, Mertz & Pederson, 1989, Besman *et al.* 1989), although the control and precise nature of the regulatory molecule(s) still remains to be defined. Trophic stimulation of steroidogenesis results in increased amounts of steroid-catalysing enzymes through induction of the transcription of the relevant nuclear genes (John *et al.* 1986). This effect is mediated through cAMP and is postulated to involve increased synthesis of steroid hydroxylase inducing proteins (SHIPs) whose function is considered to activate the transcription of the genes coding for the steroid hydroxylase enzymes (Waterman and Simpson 1985). The rate of synthesis of sterol carrier $protein_2$ (SCP_2), which facilitates the transport of cholesterol to the mitochondrion and within that organelle, is also regulated by trophic stimulation of ACTH (Conneely *et al.* 1984). Recently proto-oncogene expression (c-fos and jun-B) has been implicated in the mediation of the long-term stimulatory effects of ACTH (Viard *et al.* 1992).

4.3 Glucocorticoid Molecular Mechanism of Action

Steroids interact with intracellular receptors rather than plasma membrane-bound receptors. There is a family of steroid receptors with members recognizing the different steroids such as glucocorticoids, mineralocorticoids, androgens and oestrogens.

Indeed, steroid receptors belong to a gene superfamily that also includes thyroid hormone and vitamin D receptors (Evans 1988). All steroid receptors interact with nuclear DNA where they act as modulators of the transcription of specific genes, eg glucocorticoids induce the synthesis of lipocortin proteins which act as 'second messenger' molecules of glucocorticoid action through their inhibition of phospholipase A_2, thereby regulating the synthesis of biologically active pro-inflammatory lipids such as PAF and metabolites of arachidonic acid (Flower 1988, Peers *et al.* 1993). The general mechanism by which glucocorticoids activate, or occasionally suppress, gene transcription is shown schematically in Figure 4.6, and basically involves (i) the reversible binding of the steroid to an intracellular glucocorticoid receptor (GR) complex leading to dissociation of components of the complex to leave a steroid occupied glucocorticoid receptor protein; (ii) the occupied glucocorticoid receptor undergoing homodimer formation with a conformational change allowing it to bind to specific DNA sequences – glucocorticoid response elements (GRE) – in the cell nucleus; and (iii) translocation of the activated glucocorticoid receptor to the nucleus and subsequent binding to GRE leading to transactivation of gene expression.

Upon secretion from the adrenal the glucocorticoids bind to the blood transport protein – corticosteroid-binding globulin or transcortin – and are circulated

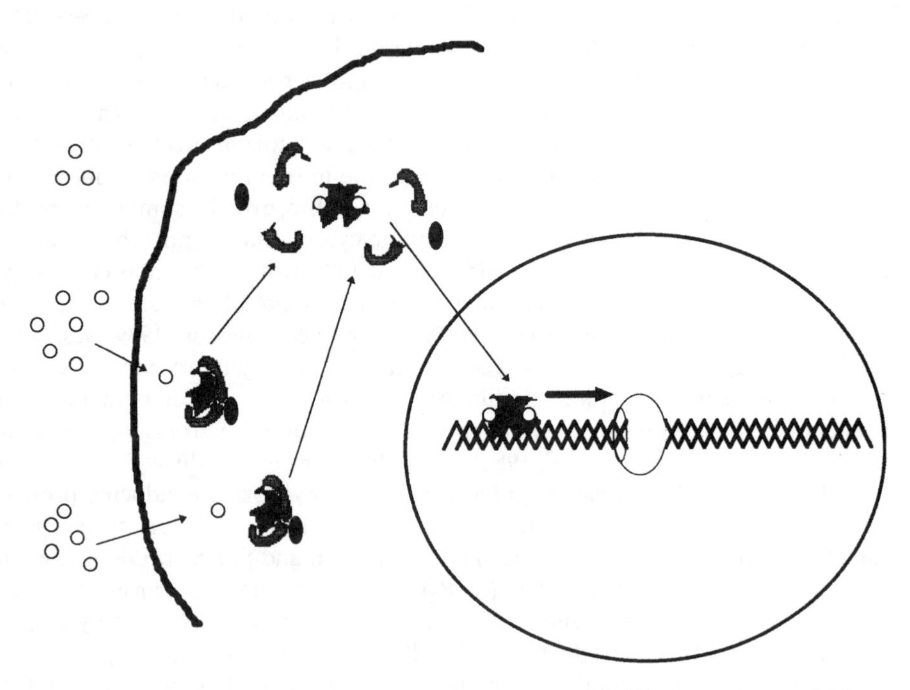

Figure 4.6 Scheme of signal transduction pathway for glucocorticoid hormones. Steroid diffuses into cell cytoplasm where it binds to the steroid-binding site on the glucocorticoid receptor complex. Binding leads to dissociation of components of the complex, eg homodimers of hsp90, and a conformational change in the structure of the occupied glucocorticoid receptor such that homodimers of the activated receptors can form. The glucocorticoid receptors then translocate to the nucleus where they bind, through their DNA-binding domains, to specific sites on the DNA of responsive genes – glucocorticoid response elements – and mediate effects upon the transcriptional process

throughout the blood stream for delivery to 'target cells'. Although glucocorticoid molecules are lipophillic, their intracellular delivery to some target cells may in part be mediated by an active transporter, ie Spindler *et al.* (1991) have shown the existence of an active glucocorticoid uptake system coupled to a Na^+/K^+-ATPase in rat hepatocytes. The first step in the cascade of events leading to gene regulation by glucocorticoids is the binding of the hormone to a glucocorticoid receptor complex located in the cytoplasm of a target cell. Although the unoccupied oestrogen, progesterone and androgen receptors are located in the cell nucleus, the unoccupied glucocorticoid and mineralocorticoid receptors are cytoplasmically located (Picard *et al.* 1990). Consistent with the actions of glucocorticoids, virtually every mammalian cell type contains glucocorticoid receptors (a typical cell containing ~10 000 receptors), although Norman and Litwack (1987) have determined that hepatocytes possess 65 000 receptors per cell. Only a small number of genes in any target cell are directly influenced by glucocorticoid, eg Ivarie and O'Farrell (1978) reported that 30 min after cultured rat hepatocytes were exposed to hydrocortisone only six of 1000 proteins that could be distinguished by two-dimensional gel electrophoresis were found to have increased in amount (~10% of the cell's proteins being detectable on such gels). In the absence of glucocorticoid binding, the glucocorticoid receptor is sequestered in an inactive form by association with other proteins, the best characterized of which is a 90 kDa heat shock protein (hsp90) (Mendel & Orti 1988, Denis & Gustafsson 1989, Bresnick *et al.* 1990, Pratt & Welsh 1994). Indeed, tissue-specific expression of heat shock proteins may confer tissue specificity and gene sensitivity to glucocorticoids (Vamvakopoulos 1993). The unactivated receptor complex has a molecular weight of ~310 kDa (Vedeckis 1983) and is currently thought to comprise one hormone binding glucocorticoid receptor and two hsp90 molecules. Additional components of the complex may also include a 100–200 nucleotide RNA (Sablonniere *et al.* 1988, Unger *et al.* 1988) and p59 protein (Tai *et al.* 1986, Renoir *et al.* 1990). The complex maintains the glucocorticoid receptor in an inactive form while also ensuring that the glucocorticoid receptor is in the optimal conformation for high affinity binding of hormone (Bresnick *et al.* 1989, Nemoto *et al.* 1990). Upon steroid binding the complex undergoes a topological change that results in exposure of the DNA-binding domain of the glucocorticoid receptor (see below). Such topological change involves dissociation of the complex's subunits, eg hsp90, to leave activated glucocorticoid receptor with bound steroid. For detailed expositions of the postulated mechanisms of glucocorticoid–receptor complex activation the reader is refered to the articles of Wright *et al.* (1993), Muller & Renkawitz (1991) and Bodine & Litwick (1990). The activated steroid–glucocorticoid subunit is now able to dimerize and translocate to the nucleus. Nuclear translocation of the receptor is under hormonal control with hormone-dependent nuclear localization sequences closely associated with the steroid-binding domain (see below) of the protein (Picard & Yamamoto 1987).

Molecular biology has afforded isolation and cloning of the glucocorticoid receptor gene and, through site-directed mutagenesis techniques, permitted structure–activity investigations of the respective glucocorticoid receptor protein. The glucocorticoid receptor itself is a polypeptide of ~90 kDa mass (Bodine & Litwick 1990) (human protein comprises 777 amino acid residues) with the amino acid sequences highly conserved – particularly at the DNA-binding domain – between species (Muller and Renkawitz 1991). Activated glucocorticoid receptor dimers contact the DNA mainly in the major groove of the double helix (Chalepakis *et al.*

1988), with dimerization facilitating binding of the protein to the DNA in a co-operative manner. Dimer co-operativity in binding is achieved through glucocorticoid receptor protein–protein interactions rather than through a change in the structure of DNA induced by the binding of the first glucocorticoid receptor molecule (Dahlman-Wright *et al.* 1990). The glucocorticoid receptor protein itself has a modular structure (Figure 4.7) such that different functions are mediated by distinct domains. Thus the steroid-binding, DNA-binding and transactivating functions of the protein are physically separated from one another (Encio & Detera-Wadleigh 1991). The steroid binding domain of the glucocorticoid receptor (amino acids 501–777) is located at the carboxyl terminus and is rich in hydrophobic amino acids with the presence of at least two reduced cysteine residues – cysteine 656 and cysteine 754 (Simons *et al.* 1987, Carlstedt-Duke *et al.* 1988) – considered vital for steroid binding. If disulphide bond formation between critical cysteine residues occurs then steroid binding is blocked. However, if the respective sulphydryl groups are reduced, then glucocorticoid binding can occur even if vicinal sulphydryl groups are oxidised (Miller & Simons 1988). The steroid binding domain also contains regions involved in transactivation, nuclear localization and hsp90 interaction. The DNA binding domain of the glucocorticoid receptor is located in a central basic region of the protein (amino acids 390–500) rich in cysteine residues, and is found to be highly conserved within the entire steroid receptor family, as well as between species. It is the DNA-binding domain that is responsible for the targeting of the activated glucocorticoid receptor to specific glucocorticoid responsive genes. This domain contains two motifs, 'zinc-finger' motifs (Freedman *et al.* 1988) – so called because co-ordination of four cysteine residues with one zinc atom is able to fold the peptide chain into a finger-shaped conformation. Each of the two 'zinc-fingers' consists of a loop of ~15 amino acid residues. They are involved in distinction between glucocorticoid and other steroid (eg oestrogen, thyroid hormone) responsive elements. The carboxy-terminal finger appears important in stabilizing the DNA–receptor interaction (Danielsen *et al.* 1989, Mader *et al.* 1989, Umesono & Evans 1989). Structural NMR data examining glucocorticoid receptor–DNA binding indicates that one of the two α-helical regions (lying in a region between the two zinc-fingers)

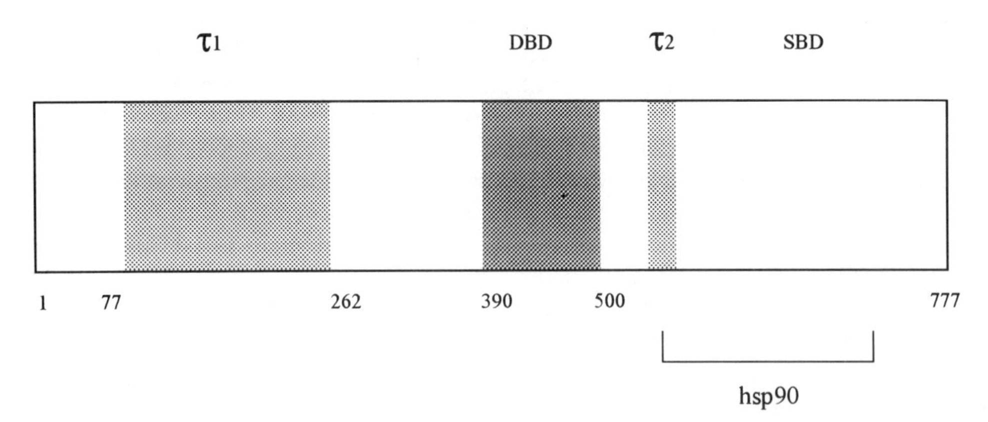

Figure 4.7 Schematic representation of domain structure of human glucocorticoid receptor protein. The numbers indicate the amino acid positions in the protein. DBD, DNA binding domain; SBD, steroid binding domain; τ_1 and τ_2, transactivating domains; hsp90-region contacting the heat shock protein of 90 kDa

of the DNA-binding domain serves as a point of contact of the receptor with the glucocorticoid response elements (Hard *et al.* 1990a). The DNA-binding domain alone, however, cannot regulate transcription, and likewise the remainder of the protein has no activity in the absence of the DNA-binding domain. The trans-activating domain of the glucocorticoid receptor protein comprises two regions, τ_1 and τ_2 (Figure 4.7), the major one of which (τ_1) accounts for >90% of the receptor's transactivation activity (Hollenberg & Evans 1988) and is located at the N-terminus of the protein (amino acids 77–262), while τ_2 is located on the N-terminal side of the steroid-binding domain.

The glucocorticoid response element comprises a sequence of 15 base pairs which has been shown to bind the glucocorticoid homodimer (Tsai *et al.* 1988, Hard *et al.* 1990b). The consensus glucocorticoid binding site sequence is 5′-GGTACAnnnTGTTCT-3′ (where n represents any nucleotide), a sequence which can also mediate the action of progesterone, androgen and mineralocorticoid receptors (Arizza *et al.* 1987, Cato *et al.* 1986, Strahle *et al.* 1987). Glucocorticoid response elements are located from 2600 base pairs upstream to 100 base pairs downstream of the transcription initiation sites for glucocorticoid-responsive genes (Yamamoto 1985, Beato 1989, Beato *et al.* 1989); glucocorticoid response elements are examples of enhancer elements.[1] Like other gene regulatory proteins the exact manner in which the DNA binding of activated glucocorticoid receptor protein gives rise to effects on transcription is still unclear. However, *in vitro* studies have shown that the glucocorticoid receptor facilitates the formation of 'promoter–transcription factor' complexes (Freedman *et al.* 1989, Tsai *et al.* 1990). Gene regulatory proteins, even when bound some distance (either upstream or downstream) from the transcription start-site of a responsive gene, can achieve gene regulation through DNA looping mechanisms, bringing bound regulatory protein in close proximity to promoter sites (Atchison 1988). Glucocorticoid actions may arise from inducing disruption of chromatin structure in the vicinity of the regulated gene with subsequent opening of the DNA helix enabling transcription factors, such as CAAT-box and TATA-box binding proteins, to bind to the respective promoter sites of glucocorticoid responsive genes (Santoro *et al.* 1988, Horikoshi *et al.* 1989). Studies (Zaret & Yamamoto 1984) demonstrating altered *in vivo* sensitivity of mouse mammary tumour virus DNA to DNase I following steroid administration is considered to support such a steroid-induced disruption of chromatin as an important mechanism of action. Reik *et al.* (1991) have shown the relationship between glucocorticoid-induced reversible disruption of nucleosomes over an enhancer element and induction of tyrosine aminotransferase (TAT) enzyme. Hager and co-workers (Cordingley *et al.* 1987, Richard-Foy & Hager 1987, Cordingly & Hager

[1] Both enhancer and promoter elements contain nucleotide sequences that bind gene regulatory proteins. An enhancer element can be defined as a regulatory DNA sequence that binds respective regulatory proteins and effects transcription through a linked promoter. The term 'enhancer' is a misnomer since many regulatory proteins bound at this site have actions to switch genes off. A promoter can be defined as a DNA sequence generally extending for about 100 base pairs upstream from the transcriptional start-site, and is essential for binding of RNA polymerase and for signalling the site for RNA polymerase transcription initiation. Promoter regions possess nucleotide binding sequences for many transcription factors including, for example, TFIID an important factor binding to the 'TATA-box' sequence and which is essential for stabilizing the formation of the transcription complex. Transcription regulatory proteins bound at enhancer and promoter elements co-operate to switch genes on or off.

1988) and Perlmann & Wrange (1988) have provided evidence for binding of activated steroid-receptors to nucleosomes, for the presence of nucleosomal glucocorticoid response elements, and for modulation of glucocorticoid receptor binding affinity to its respective response element by the presence of translational positioned nucleosomes (Li & Wrange 1993). Recently, a delayed glucocorticoid response element has been identified which binds glucocorticoid receptors via sequence motif distinct from the protypical element described above (Hess & Payvar 1992). The delayed element reportedly confers delayed or secondary hormone inducibility upon a linked promoter, effects which are preceded by a time lag of several hours and blocked by protein synthesis. The implications of this are that the 'information content' of a hormonal pulse is retained or 'memorized' more persistently by a receptor binding site of the delayed element than those of the protypical element.

Synergistic activities of the activated glucocorticoid receptor with other transcription factors (Oshima & Simons 1993), as well as with other steroid receptors, are increasingly being shown. For example, detailed analysis of some glucocorticoid-regulated genes has revealed the presence of several glucocorticoid response elements associated with single genes, with progressive deletion of the elements resulting in gradual loss of inducibility. Observations indicating that several steroid binding sites can co-operate to yield strong transcriptional induction include, eg chicken lyzozyme gene (Renkawitz *et al.* 1984), rabbit uteroglobin gene (Cato *et al.* 1984), rat tyrosine aminotransferase gene (Jantzen *et al.* 1987) and the rat tryptophan oxygenase gene (Danesch *et al.* 1987). A well-studied example of functional co-operation between the glucocorticoid receptor and other transcription factors is that of the positive regulation of a major acute phase protein, α1-acid glycoprotein, by glucocorticoids. This regulation requires interaction between glucocorticoid receptors and multiple transcriptional factors (DiLorenzo *et al.* 1991, Ratajczak *et al.* 1992, Alam *et al.* 1993,). Indeed, Ingrassia *et al.* (1994) have recently characterized a transcriptional complex required for glucocorticoid regulation of α1-acid glycoprotein gene composed of three nuclear proteins binding to a regulatory element 23 nucleotides upstream from the glucocorticoid response element. Other examples of functional co-operation include, for example, co-operation between glucocorticoid and oestrogen receptors (Ankenbauer *et al.* 1988) and between glucocorticoid receptor and other transcription factors including, for example, glucocorticoid enhancement of phosphoenolpyruvate carboxykinase (PEPCK) – a rate-limiting enzyme in gluconeogenesis (Imai *et al.* 1993), regulation of trytophan oxidase (Schule *et al.* 1988) and hepatic regulation of angiotensinogen by both glucocorticoids and cytokines generated as products of the acute phase response (Brasier *et al.* 1990).

Some physiologically relevant genes are negatively-regulated by glucocorticoids, however, transcriptional repression has not been as intensively studied. Glucocorticoid inhibition of human glycoprotein-α-subunit has been shown to be dependent upon the presence of a cAMP-responsive element closely adjacent to, or overlapping, the glucocorticoid response element (Akerblom 1988). Thereby negative regulation in this instance, due to the glucocorticoid receptor binding to its response element sterically hindering the binding of a positive transcription factor to the cAMP-responsive element. Indeed, the contradictive results showing dexamethasone negative regulation of insulin gene expression in isolated islet cells, but of glucocorticoid enhancement of insulin biosynthesis *in vivo*, is considered to be due to differences in intracellular cAMP levels between *in vitro* and *in vivo* cells, with higher *in vivo* levels of intracellular cAMP countering the negative effects of steroid (Philippe *et al.* 1992). Dexamethasone transcriptional regulation of IL-2 gene in human T-lymphocytes is mediated mainly through inhibiting the binding of AP-1 to the IL-2 gene promoter (Paliogianni *et al.* 1993). An interesting example of the inhibitory effects of glucocorticoids upon protein expression is that of Fardel *et al.* (1993) who reported that dexamethasone displays negative regulation of the multidrug resistance plasma membrane transporter, P-glycoprotein, through specific action at the level of the gene, mdr1, in cultured hepatocytes. Great interest has focused on pharmacological modulation of P-glycoprotein expression to overcome the problem of clinical resistance of human tumors to anti-cancer agents (Ford & Hait 1990). Fardel *et al.* reported doxorubicin P-glycoprotein mediated efflux to be decreased in dexamethasone-treated cells with glucocorticoid regulation occurring at the level of the mdr1 gene. Other examples of negative regulation include, for example, bovine prolactin gene, rat α1-fetoprotein (Guertin *et al.* 1988, Rabek *et al.* 1994) and rat propiomelanocortin gene (Drouin *et al.* 1989). Indeed the negative regulation of propiomelanocortin (a 31 kDa precursor protein of ACTH) by glucocorticoids is proposed to be mediated by activated steroid–receptor complex binding to a specific DNA sequence known as the 'negative glucocorticoid-responsive element' or 'nGRE' which resides within the promoter region of the pro-opiomelanocortin gene the negative glucocorticoid-responsive element overlapping the binding site for a putative member of the COUP (chicken ovalbumin upstream promoter) transcription factor family. Inhibition of translation may therefore be inhibited by preventing the binding of a COUP-like transcription factor. A different type of negative regulation has been proposed which does not involve DNA binding of the glucocorticoid receptor. One example of this is the inhibition of the collagenase by direct binding interaction between the transcription factor AP1 (positive transcription factor for collagenase) and the glucocorticoid receptor protein to form a complex which is unable to activate transcription (Jonet *et al.* 1990). Furthermore, glucocorticoid receptors lacking the DNA binding domain can also function as repressors of transcription, thereby demonstrating that interaction between DNA is not necessary. Other examples of indirect effects upon transcription (Klein *et al.* 1987) and indeed post-translational effects (Simons *et al.* 1987, Weiner *et al.* 1987, Silver *et al.* 1990) exist. For example, glucocorticoids stabilize phosphoenolpyruvate carboxykinase mRNA by inducing the synthesis of another 'stabilizing' protein that binds to the 3′-end of the message.

In summary the molecular pharmacology of glucocorticoids is complex involving protein–DNA and protein–protein interactions. Glucocorticoid effects upon a 'target' gene can be direct and/or indirect in nature, be the net outcome of the actions of a number of regulatory proteins, and be dependent upon the target gene itself, the tissue in question and the status of the system as a whole. Sometime after nuclear translocation, the glucocorticoid receptor dissociates from the DNA binding site, the steroid is metabolized and the receptor itself may be degraded or recycled back to the cytosol as an unoccupied steroid receptor. The cellular half-life of unoccupied/unactivated glucocorticoid receptor is ~20 h, whereas the half-life of the occupied/activated receptor is ~10 h (McIntyre & Samuels 1985, Dong *et al.* 1988). The cellular half-life of the glucocorticoid receptor mRNA is ~5 h.

4.4 Glucocorticoid Physiologic and Pharmacologic Actions

In the following sections we will provide a review of the effects of glucocorticoids on carbohydrate, protein and lipid metabolism before dealing more specifically with the pharmacology of glucocorticoids on selected body systems.

4.4.1 Effects Upon Intermediary Metabolism

The effects of glucocorticoids on intermediary metabolism can be teleologically rationalized as a protective mechanism whereby glucose-dependent cerebral functions are maintained in the presence of depleted carbohydrate stores. This is achieved through glucocorticoid stimulation of gluconeogenesis, leading to increased glycogen deposition and diminuation in the peripheral utilization of glucose. Glucocorticoids therefore result in a tendency for hyperglycaemia and decreased glucose tolerance. However, in the non-diabetic individual exposed to excess glucocorticoids, increased insulin secretion will counterbalance glucocorticoid actions such that fasting blood glucose levels, although elevated, will nevertheless reside within the normal range. Conversely in adrenalectomized animals there is increased sensitivity to insulin, although no marked abnormality in carbohydrate metabolism is exhibited while food is regularly ingested. However, a brief period of starvation or exercise depletes carbohydrate reserves with ensuing hypoglycaemia. The most important overall action of hydrocortisone is to facilitate the conversion of protein to glucose and glycogen.

Glucocorticoids promote the mobilization of protein for gluconeogenesis by accelerating protein degradation and inhibiting protein synthesis; muscle, the protein matrix of bone, connective tissue and skin are those tissues primarily affected (Baxter & Tyrell 1995). Glucocorticoid atrophic actions upon skeletal muscle should be highlighted by the fact that muscle, because of its high mass, is a major source of gluconeogenic and of glucose utilization. The maintenance of normal skeletal muscle function requires adequate concentrations of corticosteroids, eg in adrenocortical insufficiency diminished work capacity of skeletal muscle is a prominent sign. However, in glucocorticoid excess, quite extensive muscle atrophy can be apparent – 'glucocorticoid myopathy'. Muscle weakness in glucocorticoid myopathy results from accelerated protein catabolism with more pronounced influences on the type II glycolytic fibres (white fibres) (Clark & Vignos 1979); the review by Seene (1994) addresses the turnover of skeletal muscle contractile proteins in glucocorticoid myopathy. As the content of lysosomes in skeletal muscle, particularly in type II glycolytic fibres, is relatively low, the non-lysosomal pathways make a significant contribution to the catabolism of contractile proteins, with increased muscle concentrations of certain proteolytic enzymes apparent (Clark & Vignos 1981). A number of more recent papers (Almon & Dubois 1990, Hickson *et al.* 1990, Max 1990, Falduto *et al.* 1992a,b) have addressed issues such as the biochemistry involved in the antagonism of glucocorticoid-induced myopathy by exercise and by androgens; fibre-type discrimination in glucocorticoid-induced muscle atrophy, and the inter-relationship between glucocorticoid-induced expression in muscle of glutamine synthetase and the development/susceptibility of certain muscle fibres to glucocorticoid myopathy. Glucocorticoids can also enhance the levels of amino acids available for gluconeogenesis by blocking the stimulatory effects of insulin upon amino acid uptake and protein synthesis in peripheral tissues. Indeed, a well-recognized manifestation of glucocorticoid excess is the induction of an insulin-resistant state, with such anti-insulin action reportedly mediated at the post-insulin-receptor level (Rizza *et al.* 1982, Lenzen & Bailey 1984). However, effects upon insulin receptor expression or affinity have yet to be discounted. Of related interest is a report by Zamir *et al.* (1993) implicating glucocorticoids in promoting the inhibitory actions of TNFα upon amino acid uptake into skeletal muscle during states of sepsis and endotoxemia. Furthermore, glucocorticoids can increase glycerol release from adipocytes (by stimulating lipolysis) and of lactate release from muscle, thereby enhancing the availability of gluconeogenic substrates.

In excess of 90% of gluconeogenesis takes place in the liver, and it is within this organ that the anabolic effects of glucocorticoids upon protein and RNA synthesis predominate. The hepatic synthesis of a number (if not all) enzymes involved in gluconeogenesis is induced, including several of the transaminases, eg alanine aminotransferase, tyrosine aminotransferase, and enzymes more distal in the gluconeogenic pathway including, for example, pyruvate carboxylase, phosphoenolpyruvate carboxykinase, phosphoglyceraldehyde dehydrogenase and fructose 1,6-diphosphatase (Munck *et al.* 1984, Pilkis *et al.* 1988). However, induction requires some hours and cannot account for the earliest effects of the hormones on gluconeogenesis. For example, dexamethasone has been shown to stimulate gluconeogensis within 20 min in a rat hepatocyte culture, an effect which is not blocked by inhibitors of protein synthesis (Sistare & Haynes 1985). Studies of the molecular biology of glucocorticoid action and interaction with other regulatory factors (eg repressor activity of insulin, stimulatory effects of exercise, tissue specificity) upon the transcription of gluconeogenic enzymes are now appearing in the literature (Hadley *et al.* 1990, Eisenberger *et al.* 1992, Oshima & Simons 1992, Aggerbeck *et al.* 1993, Imai *et al.* 1993, Gadson & McCoy 1993, Nitsch *et al.* 1993, Friedman 1994, Onoagbe 1994). Glucocorticoids also display secondary actions on carbohydrate metabolism through increasing the sensitivity of the liver to the gluconeogenic actions of glucagon and to catecholamines (Goldberg *et al.* 1980). For example glucagon- or α-adrenergic-mediated stimulation of hepatic gluconeogenesis is impaired by adrenalectomy. The exact mechanism by which glucocorticoids inhibit utilization of glucose in peripheral tissues is still under investigation. However, decreases in glucose transport in polymorphonuclear leukocytes (Okuno *et al.* 1993), in lymphocytes (Munck *et al.* 1979), in fibroblasts (Murrey *et al.* 1990), in adipose tissue (Fain 1979, Carter-Su & Ukamoto 1985) and in hippocampal astrocytes and neurons (Horner *et al.* 1990, Virgin *et al.* 1991) have been documented. The latter two reports are of interest in the context of the present discussion in that they implicate glucocorticoid (either exogenous, or endogenous resulting from a stress episode) mediated decrease in intracellular glucose uptake with initiation of hippocampal neurotoxicity. Glucocorticoid inhibition of glucose uptake is mediated through type II receptors, with oestrogen, progesterone, and testosterone unable to elicit any inhibitory action. The inhibition appeared regio-specific in that glucose transport in hypothalamic, cerebellar or cortical cultures were unaffected. The inhibition in glucose transport was dose-dependent (with steroid concentrations in the nanomolar range inhibiting uptake by $\sim 30\%$) and time-dependent (with a minimum of 4 h exposure to the steroid required to achieve inhibition). The comparative time- and concentration-dependency in inhibition of glucose transport in peripheral and hippocampal tissue clearly requires further investigation. Prolonged treatment with glucocorticoids has been found to elevate plasma concentrations of glucagon (Marco *et al.* 1973, Wise *et al.* 1973) and, in as much as glucagon itself stimulates gluconeogenesis, such a rise in glucagon will further contribute to the signals promoting enhanced glucose synthesis. This thus serves a physiologic function to protect against long-term food deprivation glucocorticoids promote glycogen deposition with deposition occurring predominantly in the liver. Increased glycogen deposition is mediated through a glucocorticoid induction of glycogen synthetase, an action that is insulin-dependent. The effects of glucocorticoids upon carbohydrate metabolism have been reviewed by McMahon *et al.* (1988) and Baxter & Tyrrell (1995).

Two well-established effects of excess glucocorticoids on lipid metabolism are the dramatic redistribution of body fat, and the facilitation of the actions of lipolytic agents such as catecholamines. Prolonged administration of large doses of corticosteroids or hypersecretion of hydrocortisone as occurs in Cushing's syndrome leads to increased fat deposition to the back of the neck, face and supraclavicular area and loss of fat from the extremities. This redistribution that has been postulated to be due to selective lipogenic and lipolytic actions of insulin (elevated by glucocorticoid-induced hyperglycaemia) and glucocorticoids, respectively (Fain & Czech 1975). Thus if regional differences existed in the relative sensitivity of adipose tissue to glucocorticoid and insulin there could be a net fat deposition at sites where insulin sensitivity dominates and fat depletion at sites where glucocorticoid influence is dominant. Indeed, although such fat redistribution could be mediated by indirect actions of the hormones, the presence of glucocorticoid receptors in human adipose tissue has been confirmed (Rebuffe-Scrive *et al.* 1990). Furthermore, more recent evidence from the same group (Rebuffe-Scrive *et al.* 1992) indicates, that at least in the rat, there does indeed exist regional differences in the adipose tissue metabolism in response to direct subcutaneous placement of glucocorticoid releasing implants. Glucocorticoids increase lipolysis and elevate plasma levels of free fatty acids via their action to inhibit peripheral glucose uptake and glycerol production (Gaca & Bernend 1974, Fain 1979), and through permissive actions which facilitate lipolytic responses mediated via cAMP, eg stimulation by catecholamines (Birnbaum & Goodman 1973). Glucocorticoids also exert direct effects upon lipoprotein metabolism in man. A recent study by Berg & Nilsson-Ehle (1994) has contrasted the effects of exogenous administration for 4 days of glucocorticoid (dexamethasone) or ACTH upon plasma lipoprotein profiles in man. They reported that ACTH effects are primarily upon apo-B-containing lipoproteins with decreases in LDL cholesterol and plasma triglyceride and increases in HDL. Dexamethasone effects were primarily to decrease lipoprotein (A) concentrations, with increases observed in HDL and apolipoprotein A-I levels. Dexamethasone was found not to affect hepatic lipase expression. In contrast, Jansen *et al.* (1992), studying the effect of glucocorticoids upon hepatic lipase and lecithin-cholesterol acyltransferase (LCAT), reported in rats administered for 4 days the synthetic glucocorticoids dexamethasone and triamcinoline hepatic lipase expression to decrease and hepatic lecithin-cholesterol acyltransferase to increase; the resulting lowered lipase/LCAT ratio contributing to glucocorticoid increased plasma levels of HDL cholesterol. Although no clear evidence exists for glucocorticoid associated artheroslerosis, Aoki & Kawi (1993) have addressed the pathophysiologic implication of changes in lipoprotein serum profiles in patients receiving chronic synthetic glucocorticoid treatment, ie arthergenic actions of increased total serum cholesterol and serum apolipoprotein B levels, and artheroprotective actions of increased HDL cholesterol. Reports exist that may support a depleting effect of glucocorticoids upon the lipid content of biological membranes, with specific influences upon membrane cholesterol and sphingomyelin (Murrey *et al.* 1982, Nelson 1990). The consequence of these reports upon membrane function await further investigation. Although more pertinent to the role of glucocorticoids in cellular and tissue development, reports by Viscardi & Max (1993), Batenburg & Elfing (1992), Xu *et al.* (1990) and Gonzales *et al.* (1990) record some actions of glucocorticoids upon fatty acid and phophatidlycholine synthesis in fetal and undifferentiated tissue.

4.4.2 Effects Upon Metabolic Systems Directly Involved in Xenobiotic Disposition

Given the vast array of effects of glucocorticoids upon protein expression it is inevitable that those enzymes commonly involved in xenobiotic disposition will be affected. Indeed from a developmental perspective, circulating glucocorticoids may well be required to achieve full phenotypic expression. Reports over the last 5 years include glucocorticoid promotion of phenobarbital and polycyclic aromatic hydrocarbon inductive effects upon a range of cytochrome P450 isozymes, in addition to the inductive actions of the glucocorticoid itself (Sherratt *et al.* 1990, Waxman *et al.* 1990, Shaw *et al.* 1993); inductive and development aspects of glucocorticoid regulation of cytochrome P4501A1 and glutathione-S-transferase Ya isozyme expression (Devaux *et al.* 1992, Linder & Prough 1993); glucocorticoid induction of flavin mono-oxygenase isozymes in lung and kidney tissue (Lee *et al.* 1993); and the dependence and regulation of glutathione-S-transferase expression upon glucocorticoid (Fan *et al.* 1992, Lu *et al.* 1992).

4.4.3 Immune and Anti-inflammatory Effects

Glucocorticoids are among the most widely used drugs for the treatment of inflammatory diseases. They are prominent agents in the management of a variety of acute and chronic inflammatory conditions such as asthma, allergic rhinitis, arthritis, systemic lupus erythematosus (SLE) and other connective tissue disorders. The glucocorticoids inhibit not only the early phenomena of the inflammatory process (oedema, fibrin deposition, capillary dilatation, leucocyte migration into inflammed tissues, and phagocytic activity) but also the later manifestations (proliferation of capillaries and fibroblasts, deposition of collagen). The immunosuppressive and anti-inflammatory actions of the glucocorticoids are inextricably linked because both result in large part from inhibition of specific leucocyte functions, either directly, or indirectly through modulating the intercellular humoral cytokine signalling network (Almawi *et al.* 1991). The basic and clinical correlates of glucocorticoid therapy in inflammatory and other immune-mediated diseases have been addressed in the more recent reviews of Boumpas *et al.* (1991, 1993), de Waal (1994), Kimberly (1994), Pytsky (1993), Wick *et al.* (1993), Stein & Miller (1993), Pruett *et al.* (1993), Thompson (1993), Goulding & Guyre (1992, 1993), Cohen (1993), Goust *et al.* (1993), Schleimer (1993), Stam *et al.* (1993), Morand & Goulding (1993), Ray & Sehgal (1992), Hickman (1992) and Russo-Marie (1992). The anti-inflammatory actions of the glucocorticoids result from quantitative effects upon the levels of circulating leucocytes, and modulation of the synthesis and action of vasoactive, chemotactic and cell-activating agents leading to reduced leucocyte recruitment to inflammatory sites and inhibition of certain leucocyte functions. In contrast to the immunosuppressive and anti-inflammatory actions associated with relatively high circulating therapeutic levels of glucocorticoids, little is known about the role of basal levels of glucocorticoids upon the immune system, although a modulating action can be expected. For example, apoptosis is induced in immature thymocytes by physiologic levels of glucocorticoids (Cohen 1992, Migliorati *et al.* 1992, Iwata *et al.* 1994), and it has been suggested that basal glucocorticoid levels may have some role in thymic selection of T-cells.

Glucocorticoids are effective anti-inflammatory agents as a result of two specific actions described in the following paragraphs.

4.4.3.1 *Quantitative effects upon leucocyte populations*

Glucocorticoid administration to man causes an increase in the number of circulating neutrophils and a decrease in number of most other leucocytes, including lymphocytes, monocytes and eosinophils (Baxter 1979, Parillo & Fauci 1979, Altman *et al.* 1981, Fahey *et al.* 1981, Shoenfeld *et al.* 1981). After an acute single dose administration, such changes are maximal at 6 h and are generally dissipated by 24 h; however, the changes persist with continued steroid administration. The increase in neutrophils is due to their increased rate of entrance into the circulation from the bone marrow, diminished rate of removal from the circulation, and increased release from vascular walls. The cytokine IL-1 is a powerful hematopoietic promoter, and the studies of Dubois *et al.* (1993) have shown that glucocorticoids synergize with granulocyte- and granulocyte/monocyte-, colony-stimulating factors (G-CSF and GM-CSF, respectively) to upregulate IL-1 receptors on myeloid cells, most notably myelocytes, promyelocytes and metamyelocytes – precursor cells of the neutrophil lineage. They concluded that the ability of IL-1 to stimulate granulocyte hematopoiesis *in vivo* is partly due to the synergistic actions of colony-stimulating factors and hydrocortisone to upregulate IL-1 receptors. The other leucopenic manifestations are mostly due to redistribution of cells from the circulation to other sites such as bone marrow, spleen, lymph nodes and thoracic duct. Selective decreases are observed in subsets of lymphocytes, with cells of the T lineage decreased proportionally greater than cells of the B lineage (Parrillo & Fauci 1979, Schleimer 1990, Cupps & Fauci 1992). Although any acute effects of glucocorticoids upon circulating lymphocytes are due to sequestration from blood rather than lymphocytolysis, glucocorticoids can cause rapid lysis of lymphocytes in rats and mice, and evidence for glucocorticoid activation of programmed cell death in human mature T-cells exists (Nieto & Lopez-Rivas 1992). Indeed, it is well known that acute lymphoblastic leukemia cells, and cells of other lymphatic malignancies, can undergo programmed cell death in response to glucocorticoid (Caron-Leslie & Cidlowski 1991, Alnemri *et al.* 1992, Hickman 1992, Caron-Leslie *et al.* 1994).

4.4.3.2 *Modulation of the synthesis and action of vasoactive, chemotactic and cell-activating agents*

Inhibition of cell–cell activation, leading to a reduced leucocyte recruitment to inflammatory sites and inhibition of certain leucocyte functions, is a consequence of glucocorticoid interference with an array of intercellular humoral signalling molecules involved in immune responses.

Increased capillary blood flow and permeability play an important role in the acute inflammatory response, influencing the transport of leucocytes and plasma proteins to localized tissue sites of inflammation. Indeed, the earliest changes at a site of inflammation include local capillary dilatation, enhanced vascular permeability and tissue oedema. Such changes are mediated by a variety of molecules but a prominant role exists for arachidonic acid metabolites, kinins and histamine (Fahey *et al.* 1981, Larsen & Henson 1983, Munck *et al.* 1984, Naray-Fejes-Toth *et al.* 1984, Johnson & Baxter 1985, Beutler & Cerami 1986, Flower 1986, Goodwin *et al.* 1986). Arachidonic acid is derived directly from phosphatidylcholine by the actions of phospholipase A_2 and is a precursor of a number of molecules collectively called eicosanoids, ie prostaglandins, thromboxanes, hydroperoxyeicosatetraenoic acids (HPETEs), hydroxyeicosatetraenoic acids (HETs) and leukotrienes. Eicosanoids have widespread actions in both the innate and acquired immune responses, and glucocorticoids can profoundly inhibit induced phospholipase A_2 activity (Goppelt-Struebe & Rehfeldt 1992), with cells of the macrophage phagocytic system one the main targets (De Caterina *et al.* 1993). Recent considerations of mechanism(s) by which glucocorticoids inhibit phospholipase A_2 have invoked post-transcriptional events (Nakano *et al.* 1990, Bailey 1991) which may nevertheless involve the protein or family of proteins, lipocortin (DiRosa *et al.* 1985, Flower 1988, Peers *et al.* 1993) (see section 4.3). Furthermore, glucocorticoids also regulate cyclooxygenase activity and thereby can selectively disrupt processing of arachidonic acid to prostaglandin and thromboxane metabolites, eg in glomerular mesangial cells glucocorticoids abolish IL-1- and TNF-stimulated prostaglandin synthesis at the level of both phospholipase A_2 and cyclooxygenase (Coyne *et al.* 1992). Two cDNAs, a 2.8- and a 4.1-kilobase sequence, have been isolated and cloned which code for constitutive and inducible cyclooxygenase enzymes, respectively (O'Banion *et al.* 1992). It is the latter 4.1-kilobase inducible gene that is regulated by glucocorticoids. Under normal conditions glucocorticoids will maintain tonic inhibition of inducible cyclooxygenase expression (Masferrer *et al.* 1992). With depletion of glucocorticoid or exposure to inflammatory signals, the negative regulation of the enzyme is diminished or overcome and excess prostaglandin and thromboxane metabolites are generated, leading to inflammatory sequelae. However, not all modes of induction of arachidonic acid metabolism appear susceptible to inhibition by glucocorticoids, eg in alveolar macrophages glucocorticoids fail to inhibit arachidonic acid metabolism stimulated by hydrogen peroxide (Sporn *et al.* 1990).

Glucocorticoids are potent inhibitors of the release of histamine and leukotrienes (Schleimer *et al.* 1981, Schleimer 1985) from basophils. They also stabilize lysosomes and thereby reduce the local release of proteolytic enzymes and hyaluronidase that contribute to tissue oedema. Although human mast cell function is relatively resistant to the effects of glucocorticoids, mast cell numbers and their tissue migration can be affected (Schleimer *et al.* 1983, Schleimer *et al.* 1989, Turner & Spannhake 1990). Reports of the effect of glucocorticoids upon complement activation are equivocal: Zach *et al.* (1992, 1993) reported that in the alveolar cell line, A549, dexamethasone stimulates elevations in C3 mRNA and increases the expression of a functionally active C3 protein. While Dauchel *et al.* (1990) reported that consistent with its anti-inflammatory activity, dexamethasone decreases C3, and Factor B expression in human umbilical vein endothelial cells (HUVECs).

Recent studies have shown glucocorticoids to have effects upon neuropeptide driven inflammatory responses – 'neurogenic inflammation'. For example, glucocorticoids can induce the expression of the plasma membrane enzyme, neutral endodpeptidase 24.11, an enzyme which degrades, among other molecules, Substance P, a recognized neuropeptide responsible for increased vascular permeability and cellular infiltration (Piedimonte *et al.* 1990, Borson & Gruenert 1991). Furthermore, of great significance in recent years has been the discovery that nitric oxide serves as a short-lived mediator in some immune responses. Induction of glucocorticoid-regulated, calcium-dependent inducible nitric oxide synthase (iNOS) in vascular smooth muscle and endothelial cells contributes significantly to the vascular reactivity and hypotension elicited following exposure to bacterial endotoxin (Knowles *et al.* 1990, Radomski *et al.* 1990, Rees *et al.* 1990, Moritoki *et al.* 1992, Szabo *et al.* 1993, Berrazueta *et al.* 1994, Szabo *et al.* 1994). Transcriptional upregulation of the iNOS is subject to a variety of stimuli including lipopolysaccharide (LPS) and cytokines INFγ, TNF and IL-1 (Geller *et al.* 1993), whereas glucocorticoids down-regulate iNOS without effects upon arginase activity (Pittner & Spitzer 1993); glucocorticoids do not inhibit the expression of constitutive nitric oxide synthase (Rdomski *et al.* 1990). Glucocorticoid prevention of iNOS induction is an important component in the therapeutic management of endotoxin shock. Similarly, cardiovascular tolerance to endotoxin may be explained by attenuated induction of iNOS due to elevated endogenous glucocorticoid levels.

While the mechanism by which glucocorticoids inhibit infiltration to tissue sites has not been fully elucidated, the role of glucocorticoids in inhibiting the actions of leucocyte chemo-attractants and of endothelial-activating cytokines (Wahl *et al.* 1975, Staruch & Wood 1985, Bochner *et al.* 1987, Guyre *et al.* 1988, Wallace & Whittle 1988, Wu *et al.* 1991, Kerner *et al.* 1992) is of prime significance. Lipid mediators such as leukotrienes B_4, C_4, D_4, and cytokines such as IL-8, PDGF and macrophage inflammatory proteins (MIP), function as leucocyte chemo-attractants. Endothelial activation by cytokines such as IL-1, IL-4, IFNγ and TNF induces expression of endothelial cell surface adhesion molecules, such as P and E-selectin, ICAM-1 and VCAM-1 which can bind to respective ligands, eg sialylLewis-X–E-Selectin and CD11a/CD18–ICAM-1, expressed on the surface of activated leucocytes. Such endothelial–leucocyte interactions promote leucocyte adhesion and transendothelial migration of leucocytes at sites of inflammation. The end result of glucocorticoid inhibition of chemotaxis and endothelial activation is the disruption of the margination process whereby leucocytes are arrested on the vasculature wall prior to tissue infiltration. Glucocorticoids also modify the permeability and function of mucosal epithelium (Lundgren *et al.* 1988), derangement of which is evident in a number of inflammatory disorders.

A number of specific leucocyte functions are blocked by the glucocorticoids (Nowell 1961, Gillis *et al.* 1979, Lotta-Larsson 1980), functions which are pertinent to both the anti-inflammatory and immunosuppressive effects of glucocorticoids. Macrophage functions are relatively sensitive to glucocorticoid inhibitory actions (Russo-Marie 1992). The steroids suppress committed marrow-forming monocyte stem cells and block the differentiation of monocytes to macrophages (Bar-Shavit *et al.* 1984). Glucocorticoids also inhibit the action of macrophage migration inhibitory factor (MIF), thereby promoting the egress of macrophages from inflammatory sites. Glucocorticoids effectively repress major histocompatibility complex (MHC) class II gene expression (Celada *et al.* 1993), for MHC proteins both constitutively expressed in B-lymphocytes, and those regulated by IFNγ-induction as in macrophages. The synthesis and secretion of IFNγ by activated T_{helper}-lymphocytes is inhibited by glucocorticoids with consequent disruption of the facilitating effects of IFNγ upon macrophage expression of surface Fc receptors, and macrophage antigen processing and presentation in association with MHC proteins (Larsen & Henson 1983, Gerrard *et al.* 1984, Mokoena & Gordon 1985). The reduced expression of Fc receptors limits binding of antibody-bound micro-organisms or immune complexes, and cell macrophage activation with decreased phagocytic activity and exocytosis of proteases such as collagenase and elastase, and of plasminogen activator. However, the role of glucocorticoids upon Fc receptor expression highlights the complexity of glucocorticoid pharmacology in that glucocorticoids not only inhibit IFNγ synthesis, but can also possess opposing actions by acting synergistically with IFNγ to enhance gene transcription, eg synergistic induction by glucocorticoids and IFNγ of the mRNA for the high affinity IgG Fc receptor (FcγRI) on macrophages (Sivo *et al.* 1993). Glucocorticoids have also been shown to inhibit the LPS and IFNγ stimulation of the induction of iNOS in macrophages (DiRosa *et al.* 1990). Nitric oxide production in macrophages is implicated in cell activation and the 'respiratory burst' to generate reactive oxygen species. Not all glucocorticoid actions are to inhibit gene products, eg glucocorticoids increase the expression of mannose receptors on the macrophage plasma membrane thereby enhancing uptake and internalization of lysomal enzymes (Johnson & Baxter 1985), an effect which nevertheless 'dampens' the inflammatory response. Glucocorticoids inhibit macrophage synthesis and release of the eicosenoids, leukotriene B_4 and prostaglandin E_2, and the cytokines GM-CSF, IL-6, TNF and IL-1 (Dinarello & Mier 1987, Lew *et al.* 1988, Linden 1992, Kato & Schleimer 1994, Linden & Brattsand 1994). Cytokines incite many of the processes of inflammation, eg GM-CSF is an important haematopoietic growth factor inducing proliferation and activation of inflammatory cells, TNF and IL-1 activate increased expression of endothelial cell adhesion molecules and induce the hepatic synthesis of the acute phase proteins. In particular, IL-1 participates in the proliferation of B-lymphocytes and their ultimate differentiation into immunoglobulin-secreting plasma cells. However, despite marked effects upon B-lymphocyte function and immunoglobulin secretion *in vitro* (Garvy & Fraker 1991, Nusslein *et al.* 1994), the effects of glucocorticoids upon B-lymphocyte functions at normal therapeutic concentrations *in vivo* are generally minimal. Therapeutic doses do not significantly decrease either the concentration of antibodies in the circulation (Butler & Rossen 1973, Parrillo & Fauci 1979) (contrary reports exist; Wira & Rossell 1990) or ordinarily affect antibody responses to antigenic challenges (Butler 1975). Nevertheless, glucocorticoids can suppress humoral immune responses, for example, IgE antibody-mediated cutaneous anaphylaxis (Inagaki *et al.* 1992, Miura *et al.* 1992) through indirect actions upon cytokine mediators. More importantly macrophage-secreted IL-1 participates in activation of the T_{helper}-lymphocytes during the antigen presentation process. The activated T_{helper}-lymphocyte is central to immune responses elaborating a series of humoral factors facilitating differentiation of B-cells into plasma cells (mediators IL-2, IL-4, IL-5, IL-6), directing T-lymphocytes to undergo clonal expansion, and induction of cytotoxic T-lymphocyte populations (IL-2) (Gatti *et al.* 1986, Papa *et al.* 1986), and chemotaxis and stimulation of granuloctyes and monocytes (IL-3, GM-CSF); Stam *et al.* (1993) have characterized at least two subtypes of T_{helper}-lymphocytes with respect to the pattern of cytokine secretion and its differential modulation by glucocorticoids. In addition to inhibiting activation of T_{helper}-lymphocytes by IL-1, glucocorticoids also have direct actions upon the synthesis of at least IL-4 (Wu *et al.* 1991) and IL-2 (Horst & Flad 1987) by T_{helper}-cells, suppressing the amplification of the cell-mediated immune response by inhibiting the expression of the IL-2 gene in

T-lymphocytes (possibly mediated through suppression of leukotriene B_4 synthesis; Godwin *et al.* (1986) and by disrupting IL-2 binding with its membrane-bound receptor, and in some instances by modulating circulating levels of soluble IL-2 receptor (Sauer *et al.* 1993). Not surprisingly, glucocorticoids have been reported to enhance *in vitro* pathways promoting lymphocyte proliferation, ie glucocorticoid inhibition of cytokine (IFNγ, IL-2)-mediated activation of bone marrow derived natural suppressor cells, a function of these cells being the non-specific suppression of lymphocyte proliferative responses (Rodriguez *et al.* 1994). Kam *et al.* (1993) have addressed the mechanisms contributing to persistent T-cell activation and poor response to glucocorticoids in chronic inflammatory diseases such as steroid resistant asthma. They found that functional resistance to glucocorticoid suppression of T-cell proliferation was associated with reduced binding affinity of the glucocorticoid to the nuclear receptor, an effect that could be induced *in vitro* by exposure of T-cells to a combination of IL-2 and IL-4 and which could be blocked by IFNγ. These observations are consistent with glucocorticoids exerting their suppressive effects at an early stage of T-lymphocyte proliferation (Haczku *et al.* 1994). In recent years, the phenomenon of glucocorticoid resistance has been reviewed by Arai & Chrousos (1994), Kaspers *et al.* (1994), Chrousos *et al.* (1993) and Cypcar & Busse (1993).

In general, granulocyte functions are less susceptible to the actions of glucocorticoids, for example, glucocorticoids at concentrations normally achieved *in vivo* do not inhibit chemotaxis, phagocytosis or the production of arachidonic acid metabolites in neutrophils, or chemotaxis in eosinophils (Schleimer *et al.* 1989). However, glucocorticoids can inhibit the priming effect of certain cytokines in both these cell types, eg glucocorticoids inhibit the activating action of IFNγ to increase expression of Fc receptors on neutrophils (Petroni *et al.* 1988). Similarly, eosinophil survival can be potentiated by IL-3, IL-5 and GM-CSF, cytokines whose synthesis is inhibited by glucocorticoids (Lamas *et al.* 1989, 1991, Cox *et al.* 1991, Wallen *et al.* 1991). Glucocorticoids inhibit both specific matrix cell activities of fibroblasts, such a collagen and glycoaminoglycan formation, and the overall function of the cell (Aronow 1979, Furcht *et al.* 1979, Cuttroneo *et al.* 1981, Sterling *et al.* 1983, Smith 1984). Inhibitory actions upon fibroblasts lead to easy bruising and poor wound healing, and is one of the many contributing factors to thinning of the bone matrix and osteoporosis seen in glucocorticoid excess conditions. For recent work dealing with clinical and basic aspects of glucocorticoid induced osteoporosis the reader is refered to the papers of Kerstjens *et al.* (1994), Reid *et al.* (1994), Lian & Stein (1993), Morrison & Eisman (1993), Russell (1993), Nielsen *et al.* (1991), Mitchell & Lyles (1990), Hodgson (1990), Lukert & Raisz (1990) and Bockman & Weinerman (1990).

In summary, the maze of immunological responses to humoral mediators combined with the complex nature of glucocorticoid molecular interactions leads to seemingly paradoxical effects of glucocorticoids on the immune system. The predominant actions of glucocorticoids on the immune system are inhibitory in nature, consistent with the major therapeutic role of glucocorticoids as anti-inflammatory and immunosuppressive agents. Glucocorticoids do not cause permanent damage to the immune system. In humans the steroids affect cellular more than humoral immune responses and macrophages more than polymorphonuclear leucocytes, and are more effective if given prior to, or concurrent with, an antigenic challenge, rather than after.

4.4.4 Central Nervous System Effects

Glucocorticoids affect a variety of events in the central nervous system (CNS) (Duope & Patterson 1982, Baxter & Tyrrell 1995) including aspects of behavioural and psychological development (Carpenter & Gruen 1982, Oitzl & de Kloet 1992). Changes in psychological development observed in children and adolescents receiving chronic glucocorticoid treatment have been reviewed by Satel (1990). Although the earliest manifestations of high plasma glucocorticoid concentrations is euphoria, chronic high levels of glucocorticoids can be accompanied by depression, sleep disturbances and psychoses. The role of glucocorticoids in depression has been reviewed by Lesch & Lerer (1991) and Murphy (1991). A prominent CNS research area is that of glucocorticoid modulation of neuroendocrine physiology, and in particular the regulation of neuropeptides. For example, this list includes mechanistic aspects of glucocorticoid inhibition of growth hormone release (Giustina *et al.* 1990, Martinoli *et al.* 1991, Lima *et al.* 1993, Popovic *et al.* 1993, Giustina *et al.* 1994, Thakore & Dinan 1994); glucocorticoid upregulation of prepro-neuropeptide Y and Y1 receptors in the arcuate nucleus (Larsen *et al.* 1994); glucocorticoid induction of peptidyl-glycine alpha-amidating monooxygenase (post-translational processing enzyme catalysing formation of alpha-amidated peptides) expression (Grino *et al* 1990) and atrial natriuretic peptide (Huang *et al.* 1991); and effects of glucocorticoids on the expression of vasoactive intestinal peptide (VIP) (Watanobe 1990), an established prolactin-releasing factor, in the brain, and divergent effects of glucocorticoids on VIP expression in cerebral cortex and pituitary regions (Lam *et al.* 1992).

Glucocorticoids regulate a variety of CNS enzyme activities. Thus, glucocorticoids regulate increased hippocampal expression of glycerol phosphate dehydrogenase in response to stress events (Masters *et al.* 1994). This enzyme fulfils a role in CNS development and participates in oligodendrocyte response to stress in the adult brain. Indeed, glucocorticoids participate in oligodendrocyte differentiation (Barras *et al.* 1994). Consistent with their peripheral tissue effects, glucocorticoids inhibit the inducible expression of cyclooxygenase in the CNS (Yamagata *et al.* 1993). The inducible cyclooxygenase gene (COXII) is expressed throughout the forebrain and enriched in neurons of the cortex and hippocampal regions. Neuronal expression of cyclooxygenase is rapidly and transiently induced in seizures or NMDA-dependent synpatic activity. Cyclooxygenase activity in the CNS potentially fulfils a regulatory function in neuronal signaling. Regulation of catecholamine biosynthesis is crucial in the adaptation to various CNS conditions, eg stress and depression. MacMahon & Sabban (1992) have reported glucocorticoid induction of dopamine-β-hydroxylase, the enzyme catalysing formation of noradrenaline from dopamine, while Biron *et al.* (1992) have investigated the effects of adrenolectomy and glucocorticoid replacement therapy on the density of dopamine receptors in different regions of the brain. Although different regions of the brain were more sensitive to the regulatory effects of glucocorticoids, in general decreased glucocorticoids downregulate dopamine receptor density *and vice versa.* Glucocorticoids modulate synaptic membrane biochemistry and function. For example, synaptic plasma membranes in the brain possess specific binding sites for glucocorticoids, glucocorticoids regulate Ca^{2+} uptake in brain synaptosomes and promote calmodulin binding to the synaptic membrane (Sze & Iqbal 1994a,b) – a variety of biochemical processes associated with synaptic membranes are dependent upon calmodulin regulation. Glucocorticoids affect the electrical activity of the brain (Chen *et al.* 1991), eg high circulating levels of glucocorticoids lower the threshold for electrically-induced convulsions, while adrenal insufficiency causes marked slowing of the alpha rhythm of the EEG.

One of the most intense areas of steroid–CNS research in recent years is that of the inter-relationship between stress-induced release of high levels of glucocorticoid and promotion of age-related central neuron degeneration and intensification of some acute neuronal toxicities arising from metabolic insults – see review articles by Landfield (1994), McEwen *et al.* (1992), Hall (1990, 1993) and Stein-Behrens & Sapolsky (1992). Indeed, even the clinical manifestations of glucocorticoid effects upon neuronal pathophysiology are still equivocal, with some studies demonstrating that glucocorticoid treatment may attenuate post-traumatic and post-ischaemic central neuronal damage. While deleterious effects of glucocorticoids upon chronic and acute neuronal degeneration are thought to be mediated through the glucocorticoid receptor, eg disruption of ATP cellular metabolism in hippocampal neurons during metabolic insults – aglycaemia (Lawrence & Sapolsky 1994), protective effects appear to involve an intrinsic ability of certain glucocorticoids to inhibit oxygen free radical-induced lipid peroxidation independently of glucocorticoid receptor interaction (Hall 1990, 1993).

4.4.5 Effects Upon the Cardiovascular System

By far the most important action of glucocorticoids upon the cardiovascular system is their effect upon blood pressure. Glucocorticoid excess as in Cushing's syndrome is associated with hypertension in at least 70% of patients, independently of the subtype (pituitary or adrenal) (Mantero & Boscaro 1992), and hypertension occurs in 20% of patients treated with glucocorticoids. Indeed, the mortality of patients with Cushing's syndrome is approximately four times that of the general population when matched for age and sex, and much of this excess mortality is the result of cardiovascular disease. In Cushing's patients and patients receiving chronic glucocorticoid therapy, the circadian blood pressure variations are absent or reversed, consistent with a role of glucocorticoids in the control of circadian blood pressure rhythm (Panarelli *et al.* 1990). Both hydrocortisone and ACTH raise blood pressure in normal subjects, the effect of ACTH probably being mediated through the raised levels of hydrocortisone. High doses of hydrocortisone (200 mg/day) or maximal stimulation by ACTH rapidly (3–5 days) raise blood pressure in subjects on a normal sodium intake, and this is accompanied by a marked sodium retention and a negative potassium balance. Increased total sodium is responsible for suppressing renin and aldosterone levels while that of atrial natriuretic peptide (ANP) increases. Expansion of extracellular fluid and plasma volumes is evident, and both glomerular filtration rate (GFR) and renal vascular resistance rise (Kenyon & Fraser, 1992). Clearly high dose hydrocortisone raises blood pressure by actions arising from interaction with both type I (mineralocorticoid) and type II (glucocorticoid) receptors. However, that glucocorticoids can raise blood pressure independently of activating the type I receptor is evident by the hypertensive effects of synthetic glucocorticoids, such as dexamethasone, methyprednisolone and triamcinolone, which are devoid of mineralocorticoid actions upon electrolyte and water balance (Whitworth *et al.* 1989). Synthetic glucocorticoids such as these give rise to hypertension which is of rapid onset and independent of salt intake. While this iatrogenic hypertension cannot be controlled by spironolactone, it can be inhibited by glucocorticoid-specific receptor antagonists such as RU486. Although the mechanism of glucocorticoid-induced hypertension in man remains poorly defined, it appears not to reflect sodium retention or volume expansion. Current views are that increased pressor responsiveness is likely to be an important factor. Glucocorticoid hypertension and the evoking mechanism have been reviewed in the more recent articles of Whitworth (1994), Mantero & Boscaro (1992) and Grunfeld (1990). Glucocorticoid actions to increase glomerular filtration rate have been reviewed by Baylis *et al.* (1990). Examples of modulation of vascular smooth muscle biochemical responses through glucocorticoid activation of type II receptors are evident. Receptors for both mineralocorticoids (type I) and glucocorticoids (type II) are present in arterial smooth muscle cells, and recent work by Kornel *et al.* (1993a,b) has shown that glucocorticoids can markedly increase Na^+ and Ca^{2+} influx into arterial vascular smooth cells; mineralocorticoids only increase Na^+ influx. Glucocorticoid modulation of Na^+ and Ca^{2+} influx was shown to be mediated via type II receptors and inhibited by the type II receptor antagonist, RU486. Glucocorticoid increase in vascular smooth muscle uptake of Ca^{2+} is possibly mediated by an increase in the number of dihydropyridine-sensitive Ca^{2+} channels (Hayashi *et al.* 1991). Glucocorticoids regulate the contractility of vascular smooth muscle to noradrenaline, specifically to enhance vascular responsiveness to the actions of noradrenaline. Liu *et al.* (1992) have shown glucocorticoids to affect both alpha-1 adrenoreceptor number and coupling to G-proteins, with a consequent increase in the formation of second-messenger production, dexamethasone enhancing noradrenaline-induced inositol monophosphate, biphosphate and triphosphate formation up to two-fold. This indicates that the action of glucocorticoids upon vascular smooth muscle contractility is at least partly through modulation of alpha-1 adrenoreceptor-mediated second-messenger production. Additionally, in vascular smooth muscle cells glucocorticoids have been shown to increase the maximal binding efficiency of dopamine for its DA_1 receptors (Yasunari *et al.* 1994), and induce secretion of endothelin, a potent vasconstrictor (Kanse *et al.* 1991).

Among the other actions of glucocorticoids to have received attention in recent years is the action of glucocorticoids to directly modulate renal transporter systems through type II binding, including stimulation of Na^+-K^+-ATPase, the Na^+-H^+ exchanger and Na^+-dependent phosphate transporter (Kinsella 1990, Arruda *et al.* 1993, Baum *et al.* 1993, Celsi & Wang 1993, Wang *et al.* 1994).

4.4.6 *Quantitative Structure-Activity Relationships*

Modifications to the structure of hydrocortisone have led to increases in the ratio of anti-inflammatory to Na^+-retaining potency. While mineralocorticoid and glucocorticoid activity can be selected for by chemical modifications to the basic steroid nucleus, the anti-inflammatory actions of glucocorticoids and their effects upon carbohydrate and protein metabolism parallel each other closely. Marked selectivity between mineralocorticoid and glucocorticoid actions reflects the structure–activity requirements in molecule interaction with the respective steroid binding sites on the corticosteroid receptors. Relative differences in, for example, anti-inflammatory activity within a series of steroidal analogues potentially reflecting not only differences in 'pharmacological' receptor interaction but also differences in the absorption and disposition of the analogues, eg extent of oral bioavailability, extent of tissue distribution, and residence time in the body. Figure 4.8 shows the structural modifications to the hydrocortisone molecule that have been exploited in the development of some synthetic corticosteroid analogues, and the resulting impact upon glucocorticoid or mineralocorticoid activity. Important structural modifications leading to quantitative differences in activity include those described below.

4.4.6.1 Ring A

The 4,5 double bond and the 3-ketone are both necessary for typical adrenocorticosteroid activity. A double bond between C_1 and C_2 selectivity increases glucocorticoid activity. The modification also decreases the rate of metabolic degradation, eg prednisolone is metabolized more slowly than is hydrocortisone.

4.4.6.2 Ring B

Fluorination in the 9α position enhances all biological activities of the corticosteroids, although glucocorticoid activity to a greater extent. The effect of this structural alteration is considered to be mediated through the greater electron-withdrawing effect on the adjacent 11β-hydroxyl group. Substitution at the 6α position has unpredictable effects, eg 6α methylation of hydrocortisone increases both anti-inflammatory and Na^+-retaining actions, while the same modification made to prednisolone produces slightly greater anti-inflammatory actions and less electrolyte-regulating potency compared to prednisolone itself.

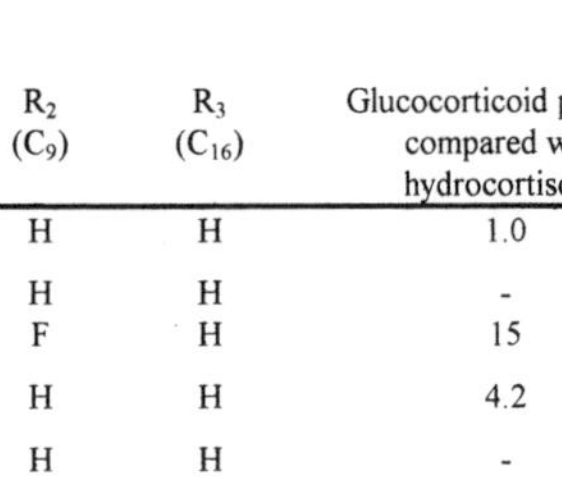

	R_1 (C_{11})	C_1-C_2 bond	R_2 (C_9)	R_3 (C_{16})	Glucocorticoid potency compared with hydrocortisone	Mineralocorticoid potency compared with hydrocortisone
Hydrocortisone	- OH	saturated	H	H	1.0	1.0
Cortisone	= O	saturated	H	H	-	-
Fludrocortisone	- OH	saturated	F	H	15	150
Prednisolone	- OH	unsaturated	H	H	4.2	0.8
Prednisone	= O	unsaturated	H	H	-	-
Triamcinolone	- OH	unsaturated	F	- OH	5	nil
Betamethasone	- OH	unsaturated	F	- CH_3	30	nil
Dexamethasone	- OH	unsaturated	F	- CH_3	30	nil

Figure 4.8 Structural determinants and relative glucocorticoid and mineralocorticoid potencies of some natural and synthetic corticosteroids

4.4.6.3 Ring C

A hydroxyl group at C_{11} is essential for glucocorticoid activity. For example 11-deoxycorticosterone has powerful mineralocorticoid activity but is virtually devoid of glucococorticoid activity, whereas corticosterone possesses both gluco- and mineralocorticoid actions. The presence of a keto group instead of a hydroxyl group at C_{11} removes corticosteroid activity. The effectiveness of, for example, prednisone is dependent upon enzymatic reduction of the keto group to a hydroxyl group.

4.4.6.4 Ring D

Substitution of a hydroxyl or methyl group at C_{16} selectively decreases mineralocorticoid actions without significant effects upon glucocorticoid actions.

All current steroidal anti-inflammatory agents are 17α-hydroxy compounds, although some carbohydrate-regulating and anti-inflammatory effects may occur in 17-deoxy compounds, eg corticosterone.

4.5 Conclusion

This review has identified the profound effect of glucocorticoids on vital control mechanisms of many cellular systems. Under normal physiological conditions these actions have a central role in maintaining homeostasis and under pathophysiological stress work towards restoring it. However, in situations where defects arise in the pituitary-adrenocortical axis, as for example by xenobiotic-induced adrenal toxicity, it should not be surprising that the adverse sequelae may be profound and widespread. Also, where the glucocorticoids are employed as pharmacological agents a correct understanding of their manifold influences on cellular processes by precise molecular mechanisms is essential if adverse reactions are to be avoided or minimized.

References

AGGERBECK, M., GARLATTI, M., FEILLEUX-DUCHE, S., VEYSSIER, C., DAHESHIA, M., HANOUNE, J. & BAROUKI, R. (1993) Regulation of the cytosolic aspatate aminotransferase housekeeping gene promoter by glucocorticoids. *Biochemistry*, **32**, 9065–9072.

AKERBLOM, I. E., SLATER, E. P., BEATO, M., BAXTER, J. D. & MELLON, P. L. (1988) Negative regulation by glucocorticoids through interference with the cAMP responsive enhancer. *Science*, **241**, 350–353.

ALAM, T., AN, M. R., MIFFLIN, R. C., HSIEH, C. C., GE, X. & PAPCONSTANTINOU, J. (1993) Transactivation of alpha-1-acid glycoprotein gene acute phase responsive element by multiple isoforms of c/EBP and glucocorticoid receptor. *Journal of Biological Chemistry*, **268**, 15681–15688.

ALBERTA, J. A., EPSTEIN, L. F., PON, L. A. & ORME-JOHNSON, N. R. (1989) Mitochondrial localization of a phosphoprotein that rapidly accumulates in adrenal cortex cells exposed to ACTH or to cAMP. *Journal of Biological Chemistry*, **264**, 2368–2372.

ALMAWI, W. Y., HADRO, E. T. & STROM, T. B. (1991) Evidence that glucocorticosteroid-mediated immunosuppressive effects do not involve altering second messenger function. *Transplantation*, **52**, 133–140.

ALMON, R. R. & DUBOIS, D. C. (1990) Fiber-type discrimination in disuse and glucocorticoid-induced atrophy. *Medicine and Science of Sport and Exercise*, **22**, 304–311.

ALNEMRI, E. S., FERNANDES, T. F., HALDER, S., CROCE, C. M. & LITWACK, G. (1992) Involvement of BCL-2 in glucocorticoid-induced apoptosis of human pre-B-leukemias. *Cancer Research*, **52**, 491–495.

ALTMAN, L. C., HILL, J. S., HAIRFIELD, W. M. & MULLARKEY, M. F. (1981) Effects of corticosteroids on eosinophil chemotaxis and adherence. *Journal of Clinical Investigation*, **67**, 28–34.

ANKENBAUER, W., STRAHLE, U. & SCHUTZ, G. (1988) Synergistic actions of glucocorticoid and estradiol responsive elements. *Proceedings of the National Academy of Sciences, USA*, **85**, 7526–7530.

AOKI, K. & KAWAI, S. (1993) Glucocorticoid therapy decreases serum lipoprotein(a) concentration in rheumatic diseases. *Internal Medicine*, **32**, 382–386.

ARAI, K. & CHROUSOS, G. P. (1994) Glucocorticoid resistance. *Clinical Endocrinology and Metabolism*, **8**, 317–331.

ARONOW, L. (1979) Effects of glucocorticoids on fibroblasts. In BAXTER, J. D. & ROUSSEAU, G. G. (Eds), *Glucocorticoid Hormone Action*, pp. 327–338, New York: Springer.

ARRIZA, J. L., WEINBERGER, C., CERELLI, G., GLASER, T. M., HANDELIN, B. L., HOUSMAN, D. E. & EVANS, R. E. (1987) Cloning of human mineralocorticoid receptor complementary DNA: structure and functional kinship with the glucocorticoid receptor. *Science*, **237**, 268–275.

ARRUDA, J. A., WANG, L. J., PAHLAVAN, P. & RUIZ, O. S. (1993) Glucocorticoids and the renal Na^+-H^+ antiporter: role in respiratory acidosis. *Regulatory Peptides*, **48**, 329–336.

ATCHISON, M. L. (1988) Enhancers: mechanisms of action and cell specificity. *Annual Review of Cell Biology*, **4**, 127–153.

BAILEY, J. M. (1991) New mechanisms for effects of anti-inflammatory glucocorticoids. *Biofactors*, **3**, 97–102.

BAR-SHAVIT, Z., KAHN, A. J., PEGG, L. E., STONE, K. R. & TEITEBAUM, S. L. (1984) Glucocorticoids modulate macrophage surface oligosaccharides and their bone binding activity. *Journal of Clinical Investigation*, **73**, 1277–1282.

BARRES, B. A., LAZAR, M. A. & RAFF, M. C. (1994) A novel role for thyroid hormones, glucocorticoids and retinoic acid in timing oligodenrocyte development. *Development*, **120**, 1097–1108.

BATENBURG, J. J. & ELFING, R. H. (1992) Pre-translational regulation by glucocorticoid of fatty acid and phosphatidylcholine synthesis in type II cells from fetal lung. *FEBS Letters*, **307**, 164–168.

BAUM, M., CANO, A. & ALPERN, R. J. (1993) Glucocorticoids stimulate Na^+/H^+ antiporter in OKP cells. *American Journal of Physiology*, **264**, F1027–F1031.

BAXTER, J. D. (1979) Glucocorticoid hormone action. In GILL, G. N. (Ed.), *Pharmacology of Adrenal Cortical Hormones*, pp. 67–83, Oxford: Pergamon.

BAXTER, J. D. & TYRRELL, J. B. (1995) The adrenal cortex. In FELIG, P., BAXTER, J. D., BROADUS, A. E. & FROHMAN, L. A. (Eds), *Endocrinology and Metabolism*, pp. 385–511, New York: McGraw-Hill.

BAYLIS, C., HANDRA, R. K. & SORKIN, M. (1990) Glucocorticoids and control of glomerular filtration rate. *Seminars in Nephrology*, **10**, 320–329.

BEATO, M. (1989) Gene regulation by steroid hormones. *Cell*, **56**, 335–344.

BEATO, M., CHALEPAKIS, G., SCHAUER, M. & SLATER, E. P. (1989) DNA regulatory elements for steroid hormones. *Journal of Steroid Biochemistry*, **32**, 737–748.

BERG, A. L. & NILSSON-EHLE, P. (1994) Direct effects of corticotrophin on plasma lipoprotein metabolism in man – studies *in vivo* and *in vitro*. *Metabolism, Clinical and Experimental*, **43**, 90–97.

BERRAZUETA, J. R., SALAS, E., AMADO, J. A., SANCHEZ DE VEGA, M. J. & POVEDA, J. J. (1994) Induction of nitric oxide synthase in human mammary arteries *in vitro*. *European Journal of Pharmacology*, **251**, 303–305.

BESMAN, M. J., YANAGIBASHI, K. LEE, T. D., KAWAMURA, M., HALL, P. F. & SHIVELY, J. E. (1989) Identification of des-(gly-Ile)-endozepine as an effector of corticotrophin dependent adrenal steroidogenesis: stimulation of cholesterol delivery is mediated peripheral benzodiazepine receptor. *Proceedings of the National Academy of Sciences, USA*, **80**, 3531–3535.

BEUTLER, B. & CERAMI, A. (1986) Cachetin and tumor necrosis factor as two sides of the same biological coin. *Nature*, **320**, 584–585.

BIRNBAUM, R. S. & GOODMAN, H. M. (1973) Effects of hypophysectomy on cyclic AMP accumulation and action in adipose tissue. *Fed. Proc.* **32**, 535–561.

BIRON, D., DAUPHIN, C. & DI PAOLO, T. (1992) Effects of adrenalectomy and glucocorticoids on rat brain dopamine receptors. *Neuroendocrinology*, **55**, 468–476.

BISGAIER, C. L., CHANDERBAHN, R., HINES, R. & VAHOUNY, G. V. (1985) Adrenal cholesterol esters as substrate source for steroidogenesis. *Journal of Steroid Biochemistry*, **23**, 967–975.

BOCHNER, B. S., RUTLEDGE, B. K. & SCHLEIMER, R. P. (1987) Interleukin-1 production by human lung tissue II. Inhibition by anti-inflammatory steroids. *Journal of Immunology*, **139**, 2303–2307.

BOCKMAN, R. S. & WEINERMAN, S. A. (1990) Steroid-induced osteoporosis. *Orthopedic Clinics of North America*, **21**, 97–107.

BODINE, P. V. & LITWICK, G. (1990) The glucocorticoid receptor and its endogenous regulators. *Receptor*, **1**, 83–119.

BORSON, D. B. & GRUENERT, D. C. (1991) Glucocorticoids induce neutral endopeptidase in transformed human tracheal epithelial cells. *American Journal of Physiology*, **260**, L83–L89.

BOUMPAS, D. T., CHROUSOS, G.P., WILDER, R. L., CUPPS, T. R. & BALOW, J. E. (1993) Glucocorticoid therapy for immune-mediated diseases: basic and clinical correlates. *Annals of Internal Medicine*, **119**, 1198–1208.

BOUMPAS, D. T., CHROUSOS, G.P., WILDER, R. L., CUPPS, T. R. & BALOW, J. E. (1993) Glucocorticoid therapy for immune-mediated diseases: basic and clinical correlates. *Annals of Internal Medicine*, **119**, 1198–1208.

BRASIER, A. R., RON, D., TATE, J. E. & HABENER, J. F. (1990) Synergistic enhancers located within an acute phase enhancer modulate glucocorticoid induction of angiotensinogen gene transcription. *Molecular Endocrinology*, **4**, 1921–1933.

BREM, A. S. & MORRIS, D. J. (1993) Interactions between glucocorticoids and mineralocorticoids in the regulation of renal electrolyte transport. *Molecular and Cellular Endocrinology*, **97**, C1–C5.

BRESNICK, E. H., DALMAN, F. C., SANCHEZ, E. R. & PRATT, W. B. (1989). Evidence that the 90 kDa heat shock protein is necessary for the steroid binding conformation of the L cell glucorticoid receptor. *Journal of Biological Chemistry*, **264**, 4992–4997.

BRESNICK, E. H., DALMAN, F. C. & PRATT, W. B. (1990) Direct stoichiometric evidence that the untransformed MR 3000000 9S, glucocorticoid receptor is a core unit derived from a larger heteromeric complex. *Biochemistry*, **29**, 520–527.

BURSTEIN, S., MIDDLEDITCH, B. S. & GUT, M. (1974) Enzymatic formation of (20R, 22R) dihydroxycholesterol from cholesterol and a mixture of $(16)O_2$ and $(16)O_2$: random incorporation of oxygen atoms. *Biochemical and Biophysical Research Communications*, **61**, 692–697.

BUTLER, W. T. (1975) Corticosteroids and immunoglobulin synthesis. *Transplantation Proceedings*, **7**, 49–53.

BUTLER, W. T. & ROSSEN R. D. (1973) Effects of corticosteroids on immunity in man. I. Decreased serum IgG concentration caused by 3 or 5 days of high doses of methyprednisolone. *Journal of Clinical Investigation*, **52**, 2629–2633.

CARLSTEDT-DUKE, J., STROMSTEDT, P. E., WRANGE, O., BERGMAN, T., GUSTAFSSON, J. A. & JORNVALL, H. (1988) Identification of hormone-interacting amino-acid residues within the steroid-binding domain of the glucocorticoid receptor in relation to other steroid hormone receptors. *Journal of Biological Chemistry*, **263**, 6842–6846.

Caron-Leslie, L. A. & Cidlowski, J. A. (1991) Similar actions of glucocorticoids and calcium on the regulation of apoptosis in S49 cells. *Molecular Pharmacology*, **5**, 1169–1179.

Caron-Leslie, L. A., Evans, R. B. & Cidlowski, J. A. (1994) Bcl-2 inhibits glucocorticoid-induced apoptosis but only partially blocks calcium ionophore or cycloheximide-regulated apoptosis in S49 cells. *FASEB Journal*, **8**, 639–645.

Carpenter, W. T. & Gruen, P. H. (1982) Cortisol's influence on human mental functioning. *Journal of Clinical Psychopharmacology*, **2**, 91–101.

Carter-Su, J. & Okamoto, K. (1985) Effect of glucocorticoids on hexose transport in rat adipocytes. *Journal of Biological Chemistry*, **260**, 11091–11095.

Cato, A. C. B., Geisse, S., Wenz, M., Westphal, H. M. & Beato, M. (1984) The nucleotide sequences recognized by the glucocorticoid receptor in the rabbit uterogloin gene region are located far upstream from the initiation of transcription. *EMBO Journal*, **3**, 2771–2778.

Cato, A. C. B., Miksicek, R., Schultz, G., Arnemann, J. & Beato, M. (1986) The hormone regulatory element of mouse mammary tumor virus mediates progesterone induction. *EMBO Journal*, **5**, 2237–2240.

Celada, A., McKercher, S. & Maki, R. A. (1993) Repression of major histocompatibility complex IA expression by glucocorticoids: the glucocorticoid receptor inhibits the DNA binding of the X box DNA binding protein. *Journal of Experimental Medicine*, **177**, 691–698.

Celsi, G. & Wang, Z. M. (1993) Regulation of Na^+-K^+-ATPase gene expression: a model to study terminal differentiation. *Pediatric Nephrology*, **7**, 630–634.

Chalepakis, G., Arnemann, J., Slater, E., Bruller, H. J., Gross, B. & Beato, M. (1988) Differential gene activation by glucocorticoids and progestins through hormone regulatory element of mouse mammary tumor virus. *Cell*, **53**, 371–382.

Chen, Y. Z., Hua, S. Y., Wang, C. A., Wu, L. G., Gu, Q. & Xing, B. R. (1991) An electrophysiological study on the membrane receptor-mediated action of glucocorticoids in mammalian neurons. *Neuroendocrinology*, **53**(Suppl.1), 25–30.

Chrousos, G. P., Detera-Wadleigh, S. D. & Karl, M. (1993) Syndromes of glucocorticoid resistance. *Annals of Internal Medicine*, **119**, 1113–1124

Chung, B., Mateson, K. J. & Miller, W. L. (1986) Structure of a bovine gene for $P450_{c21}$ (steroid 21-hydroxylase) defines a novel cytochrome P450 gene family. *Proceedings of the National Academy of Sciences, USA*, **83**, 4243–4247.

Clark, A. F. & Vignos, P. J. (1979) Experimental corticosteroid myopathy: effect on myofibrillar ATPase activity and protein degradation. *Muscle and Nerve*, **2**, 265–270.

Clark, A. F. & Vignos, P. J. (1981) The role of proteases in experimental glucocorticoid myopathy. *Muscle and Nerve*, **4**, 219–224.

Cohen, J. J. (1992) Glucocorticoid-induced apoptosis in the thymus. *Seminars in Immunology*, **4**, 363–369.

Cohen, J. J. (1993) Programmed cell death and apoptosis in lymphocyte development and function. *Chest*, **103**(Suppl. 2), S99–S101.

Conneely, O. M., Headon, D. R., Olson, C. D., Ungar, F. & Dempsey, M. E. (1984) Intramitochondrial movement of adrenal sterol carrier protein with cholesterol in response to corticotrophin. *Proceedings of the National Academy of Sciences, USA*, **81**, 2970–2974.

Cordingly, M. G. & Hager, G. L. (1988) Binding of multiple factors to the MMTV promoter in crude and fractionated nuclear extracts. *Nucleic Acids Research*, **16**, 609–628.

Cordingley, M. G., Riegel, A. T. & Hager, G. L. (1987) Steroid-dependent interaction of transcription factors with the inducible promoter of mouse mammary tumor virus *in vivo*. *Cell*, **48**, 261–270.

COX, G. OHTOSHI, T., VANCHERI, C., DENBERG, J. A., DOLOVICH, J., GAULDIE, J. & JORDANA, M. (1991) Promotion of eosinophil survival by human bronchial epithelial cells and its modulation by steroids. *American Journal of Respiratory Cell and Molecular Biology*, **4**, 525–531.

COYNE, D. W., NICKOLS, M., BERTRAND, W. & MORRISON, A. R. (1992) Regulation of mesangial cell cyclooxygenase synthesis by cytokines and glycocorticoids. *American Journal of Physiology*, **263**, F97–F102.

CUPPS, T. R. & FAUCI, A. S. (1982) Corticosteroid-mediated immunoregulation in man. *Immunological Reviews*, **65**, 133–154.

CURNOW, K. M., TUSIE-LUNA, M. T., PASCOE, L., NATARAJAN, R., GU, J. L., NADLER, J. L. & WHITE, P. C. (1991) Genetic analysis of the human type-1 angiotensin-II receptor. *Molecular Endocrinology*, **5**, 1513–1522.

CUTTRONEO, K. R., ROKOWSKI, R. & COUNTS, D. F. (1981) Glucocorticoids and collagen synthesis: comparison of *in vivo* and cell culture studies. *Coll. Relat. Res*, **1**, 557–563.

CYPCAR, D. & BUSSE, W. W. (1993) Steroid resistant asthma. *Journal of Allergy and Clinical Immunology*, **92**, 362–372.

DAHLMAN-WRIGHT, K., SILTALA-ROOS, H., CARLSTEDT-DUKE, J. & GUSTAFSSON, J.-A. (1990) Protein–protein interactions facilitate DNA binding by the glucocorticoid receptor DNA binding domain. *Journal of Biological Chemistry*, **265**, 14030–14035.

DANESCH, U., GLOSS, B., SCHMID, W., SCHUTZ, G., SCHULE, R. & REMKAWITZ, R. (1987) Glucocorticoid induction of the rat tryptophan oxygenase gene is mediated by two widely separated glucocorticoid responsive elements. *EMBO Journal*, **6**, 625–630.

DANIELSEN, M., HINCK, L. & RINGOLD, G. M. (1989) Two amino acids within the knuckle of the first zinc finger specify DNA response element activation by the glucocorticoid receptor. *Cell*, **67**, 1131–1138.

DAUCHEL, H., JULEN, N., LEMERCIER, C., DAVEAU, M., OZANNE, D., FONTAINE, M. & RIPOCHE, J. (1990) Expression of complement alternative pathway proteins by endothelial cells. Differential regulation by interleukin-1 and glucocorticoids. *European Journal of Immunology*, **20**, 1669–1675.

DE CATERINA, R., SICARI, R., GIANNESSI, D., PAGGIARO, P. L., PAOLETTI, P., LAZZERINI, G., BERNINI, W., SOLITO, E. & PARENTE, L. (1993) Macrophage-specific eicosanoid synthesis inhibition and lipocotin-1 induction by glucocorticoids. *Journal of Applied Physiology*, **75**, 2368–2375.

DENIS, M. & GUSTAFSSON, J.-A. (1989) Translation of glucocorticoid mRNA *in vitro* yields a nonactivated receptor. *Cancer Research*, **49**, 2275s–2281s.

DEVAUX, A., PESONEN, M., MONOD, G. & ANDERSSON, T. (1992) Glucocorticoid-mediated potentiation of P450 induction in primary culture of rainbow trout hepatocytes. *Biochemical Pharmacology*, **43**, 898–901.

DICKERMAN, D. R., FAIMAN, G. C. & WINTER, J. S. D. (1984) Intraadrenal steroid concentrations in man: zonal differences and developmental changes. *Journal of Clinical Endocrinology and Metabolism*, **59**, 1031–1038.

DILORENZO, D., WILLIAMS, P. & RINGOLD, G. (1991) Identification of two distinct nuclear factors within the glucocorticoid regulatory region of the rat alpha 1-acid glycoprotein promoter. *Biochemical and Biophysical Research Communications*, **173**, 1326–1332.

DINARELLO, C. A. & MIER, J. W. (1987) Lymphokines. *New England Journal of Medicine*, **317**, 940–945.

DIROSA, M., CALIGNANO, A., CARNUCCIO, R., IALENTI, A. & SAUTEBIN, L. (1985) Multiple control of inflammation by glucocorticoids. *Agents and Actions*, **17**, 284–289.

DIROSA, M., RADOMSKI, M., CARNUCCIO, R. & MONCADA, S. (1990) Glucocorticoids inhibit the induction of nitric oxide synthase in macrophages. *Biochemical and Biophysical Research Communications*, **172**, 1246–1252.

DONG, Y., POELLINGER, L., GUSTAFASSON, J. A. & OKRET, S. (1988) Regulation of glucocorticoid receptor expression: evidence for transcriptional and posttranslational mechanisms. *Molecular Endocrinology*, **2**, 1256–1264.

DROUIN, J., TRIFIRO, M. A., PLANTE, R., NERMER, M., ERIKSSON, P. & WRANGE, O. (1989) Glucocorticoid receptor binding to a specific DNA sequence is required for hormone-dependent repression of proopiomelanocortin gene transcription. *Molecular and Cellular Biology*, **9**, 5305–5314.

DUBOIS, C. M., NETA, R., KELLER, J. R., JACOBSON, S. E., OPPENHEIM, J. J. & RUSCETTI, F. (1993) Hematopoietic growth factors and glucocorticoids synergise to mimic the effects of IL-1 on granulocyte differentiation and IL-1 receptor induction on bone marrow cells. *Experimental Hematology*, **21**, 303–310.

DUOPE, A. J. & PATTERSON, P. H. (1982) Glucocorticoids and the developing nervous system. In GANTEN, D. & PFAFF, D. (Eds), *Current Topics in Neuroendocrinology*, pp. 23–41, New York: Springer.

EISENBERGER, C. L., NECHUSTAN, H., COHEN, H., SHANI, M. & RESHEF, L. (1992) Differential regulation of the rat phosphoenolpyruvate carboxykinase gene expression in several tissues of transgenic mice. *Molecular and Cellular Biology*, **12**, 1396–1403.

ENCIO, I. J. & DETERA-WADLEIGH, S. D. (1991) The genomic structure of the glucocorticoid receptor. *Journal of Biological Chemistry*, **266**, 7182–7188.

EVANS, R. M. (1988) The steroid and thyroid hormone receptor superfamily. *Science*, **247**, 889–895.

FAIN, N. J. (1979) Inhibition of glucose transport in fat cells and activation of lipolysis by glucocorticoids. In BAXTER, J. D. & ROUSSEAU, G. G. (Eds), *Glucocorticoid Hormone Action*, pp. 547–560, New York: Springer.

FAIN, N. J. & CZECH, M. P. (1975) Glucocorticoid effects on lipid mobilization and adipose tissue metabolism. In BLASKO, H. (Ed.), *Handbook of Physiology*, pp. 169–189, Washington, DC: American Physiological Society.

FAHEY, J. V., GUYRE, P. M. & MUNCK, A. (1981) Mechanisms of anti-inflammatory actions of glucocorticoids. In WEISSMAN, G. (Ed.), *Advances in Inflammatory Research*, pp. 21–42, New York: Raven Press.

FALDUTO, M. T., YOUNG, A. P. & HICKSON, R. C. (1992a) Exercise interupts ongoing glucocorticoid-induced muscle atrophy and glutamine synthetase induction. *American Journal of Physiology*, **263**, E1157–E1163.

FALDUTO, M. T., YOUNG, A. P. & HICKSON, R. C. (1992b) Exercise inhibits glucocorticoid-induced glutamine synthetase expression in red skeletal muscle. *American Journal of Physiology*, **262**, C214–C220.

FAN, W., TRIFILETTI, R., COOPER, T. & NORRIS, J. S. (1992) Cloning of a mu-class glutathione S-transferase gene and identification of the glucocorticoids regulatory domains in its 5′ flanking sequence. *Proceedings of the National Academy of Sciences, USA*, **89**, 6104–6108.

FARDEL, O., LECUREUR, V. & GUILLOUZO, A. (1993) Regulation of P-glycoprotein expression in cultured rat hepatocytes. *FEBS Letters*, **327**, 189–193.

FARESE, R. V. (1987) An update on the role of phospholipid metabolism in the action of steroidogenic agents. *Journal of Steroid Biochemistry*, **27**, 737–743.

FARKASH, Y., TIMBERG, R. & ORLY, J. (1986) Preparation of antiserum to rat cytochrome P450 cholesterol side-chain cleavage, and its use for ultrastructural localization of the immunoreactive enzyme by protein A gold technique. *Endocrinology*, **18**, 1353–1365.

FLOWER, R. J. (1986) The mediators of steroid action. *Nature*, **320**, 20–21.

FLOWER, R. J. (1988) Lipocortin and the mechanism of action of glucocorticoids. *British Journal of Pharmacology*, **94**, 987–1017.

FORD, J. M. & HAIT, W. N. (1990) Pharmacology of drugs that alter multidrug resistance in cancer. *Pharmacological Reviews*, **42**, 155–199.

FREEDMAN, L., LUISI, B., KORSZUN, Z., BASSAVAPPA, R., SIGLER, P. & YAMAMOTO, K. (1988) The function and structure of the metal coordination sites within the glucocorticoid receptor DNA binding domain. *Nature*, **334**, 543–546.

FREEDMAN, L., YOSHINAGA, S. K., VANDERBILT, J. N. & YAMAMOTO, K. R. (1989) *In vitro* transcription enhancement by purified derivatives of the glucocorticoid receptor. *Science*, **245**, 298–301.

FRIEDMAN, J. E. (1994) Role of glucocorticoids in activation of hepatic PEPCK gene transcription during exercise. *American Journal of Physiology*, **266**, E560–E566.

FURCHT, L. T., MOSHER, D. F., WENDELSCHAFTER-CRABB, G. & WOODBRIDGE, P. A. (1979) Dexamethasone induced accumulation of a fibronectin and collagen extracellular matrix in transformed human cells. *Nature*, **277**, 393–394.

GACA, G. & BERNEND, K. (1974) Plasma glucose, insulin and free fatty-acids during long-term coticosteroid treatment in children. *Acta Endocrinologica*, **77**, 699–674.

GADSON, P. & MCCOY, J. (1993) Differential expression of tyrosine aminotransferase by glucocorticoids and insulin. *Biochimica et Biophysica Acta*, **1173**, 22–31.

GARREN, L. D., GILL, G. N., MASUI, H. & WALTON, G. M. (1971) On the mechanism of action of ACTH. *Recent Progress in Hormone Research*, **27**, 433–478.

GARVY, B. A. & FRAKER, P. J. (1991) Suppression of the antigenic response of murine bone marrow B cells by physiological concentrations of glucocorticoids. *Immunology*, **74**, 519–523.

GATTI, G., CAVALLO, R., SARTORI, M. L., MARINONE, C. & ANGELI, A. (1986) Cortisol at physiological concentrations and prostaglandin E2 are active inhibitors of human natural killer activity. *Immunopharmacology*, **11**, 119–128.

GELLER, D. A., NUSSLER, A. K., DI SILVIO, M., LOWENSTEIN, C. J., SHAPIRO, R. A., WANG, S. C., SIMMONS, R. L. & BILLIAR, T. R. (1993) Cytokines, endotoxin and glucocorticoids regulate the expression of inducible nitric oxide synthase in hepatocytes. *Proceedings of the National Academy of Sciences, USA*, **90**, 522–526.

GERRARD, T. L., CUPPS, T. R., JURGENSEN, C. H. & FAUCI, A. S. (1984) Hydrocortisone-mediated inhibition of monocyte antigen presentation: dissociation of inhibitory effect and expression of DR antigens. *Cellular Immunology*, **85**, 330–339.

GILLIS, S., CRABTREE, G. R. & SMITH, K. A. (1979) Glucocorticoid-induced inhibition of T-cell growth factor prodution II. The effect on the *in vitro* generation of cytolytic T-cells. *Journal of Immunology*, **123**, 1632–1638.

GIUSTINA, A., GIRELLI, A., DOGA, M., BODINI, C., BOSSONI, S., ROMANELLI, G. & WEHRENBERG, W. B. (1990) Pyridostigmine blocks the inhibitory effect of glucocorticoids on growth hormone-releasing hormone stimulated growth hormone secretion in normal man. *Journal of Clinical Endocrinology and Metabolism*, **71**, 580–584.

GIUSTINA, A., BUSSI, A. R., DOGA, M. & WEHRENBERG, W. B. (1994) Effect of hydrocortisone on the growth hormone response to growth hormone releasing hormone in acromegaly. *Hormone Research*, **41**, 33–37.

GOLDBERG, A. L., TISCHLER, M., DEMORTINO, G. & GRIFFIN, G. (1980) Hormonal regulation of protein degradation and synthesis in skeletal muscle. *Federation Proceedings*, **39**, 31–43.

GONZALES, L. W., ERTSEY, R., BALLARD, P. L., FROH, D., GOERKE, J. & GONZALES, J. (1990) Glucocorticoid stimulation of fatty-acid synthesis in explants of human fetal lung. *Biochimica et Biophysica Acta*, **1042**, 1–12.

GOODWIN, J. S., ATLURU, D., SIERAKOWSKI, S. & LARSSON, E. A. (1986) Mechanism of action of glucocorticoids. Inhibition of T-cell proliferation and interleukin-2 production by hydrocortisone is reversed by leukotriene B_4. *Journal of Clinical Investigation*, **77**, 1244–1250.

GOPPELT-STRUEBE, M. & REHFELDT, W. (1992) Glucocorticoids inhibit TNF alpha-induced cytosolic phospholipase A2 activity. *Biochimica et Biophysica Acta*, **1127**, 163–167.

GOULDING, N. J. & GUYRE, P. M. (1992) Regulation of inflammation by lipocortin-1. *Immunology Today*, **13**, 295–297.

GOULDING, N. J. & GUYRE, P. M. (1993) Glucocorticoids, lipocortins and the immune response. *Current Opinion in Immunology*, **5**, 108–113.

GOUST, J. M., STEVENSON, H., GALBRAITH, R. & VIRELLA, G. (1993) Immunosuppression and immunomodulation. *Immunology Series*, **58**, 465–480.

GRINO, M., GUILLAUME, V., BOUDOURESQUE, F., CONTE-DEVOLX, B., MALTESE, J. Y. & OLIVER, C. (1990) Glucocorticoids regulate peptidyl–glycine alpha-amidating monooxygenase gene expression in the rat hypothalamic paraventricular nucleus. *Molecular Endocrinology*, **4**, 1613–1619.

GRUNFELD, J. P. (1990) Glucocorticoids in blood pressure regulation. *Hormone Research*, **34**, 111–113.

GUERTIN, M., LARUE, H., BERNIER, D., WRANGE, O., CHEVRETTE, M., GINGRAS, M. & BELANGER, L. (1988) Enhancer and promoter elements directing activation and glucocorticoid repression of α1-fetoprotein gene in hepatocytes. *Molecular and Cellular Biology*, **8**, 1398–1407.

GUSTAFSSON, J. A., CARLSTEDT-DUKE, J. & POELLINGER, L. (1987) Biochemistry, molecular biology and physiology of the glucocorticoid receptor. *Endocrine Reviews*, **8**, 185–234.

GUYRE, P. M., GIRAD, M. T., MORGANELLI, P. M. & MANGANIELLO, P. D. (1988) Glucocorticoid effects on the production and actions of immune cytokines. *Journal of Steroid Biochemistry*, **30**, 89–93.

HACZKU, A., ALEXANDER, A., BROWN, P., ASSOUFI, B., LI, B., KAY, A. B. & CORRIGAN, C. (1994) The effect of dexamethasone, cyclosporine and rapamycin on T-lymphocyte proliferation *in vitro*: comparison of cells from patients with glucocorticoid-sensitive and glucocorticoid-resistant chronic asthma. *Journal of Allergy and Clinical Immunology*, **93**, 510–519.

HADLEY, S. P., HOFFMAN, W. E., KUHLENSCHMIDT, M. S., SANECKI, R. K. & DORNER, J. L. (1990) Effects of glucocorticoids on alkaline phosphatase, alanine aminotransferase and gamma-glutamyltransferase in cultured dog hepatocytes. *Enzyme* **43**, 89–98.

HALL, E. D. (1990) Steroids and neuronal destruction or stabilization. *Ciba Foundation Symposium*, **153**, 206–219.

HALL, E. D. (1993) The effects of glucocorticoid and nonglucocorticoid steroids on acute neuronal degeneration. *Advances in Neurology*, **59**, 241–248.

HALL, P. F. (1984) Role of the cytoskeleton in hormone action. *Canadian Journal of Biochemistry and Cell Biology*, **62**, 653–665.

HARD, T., KELLENBACH, E., BOELENS, R., MALER, B., DAHLMAN, K., FREEDMAN, L. P., CARLSTEDT-DUKE, J., YAMAMOTO, K. R., GUSTAFSSON, J.-A. & KAPTEIN, R. (1990a) Solution structure of the glucocorticoid receptor DNA-binding domain. *Science*, **249**, 157–160.

HARD, T., DAHLMAN, K., CARLSTEDT-DUKE, GUSTAFSSON, J.-A. & RIGLER, R. (1990b) Cooperativity and specificity in the interaction between DNA and the glucocorticoid receptor DNA-binding domain. *Biochemistry*, **29**, 5358–5364.

HAYASHI, T., NAKAI, T. & MIYABO, S. (1991) Glucocorticoids increase Ca^{++} uptake and [3H]dihydropyridine binding in A7R5 vascular smooth muscle cells. *American Journal of Physiology*, **261**, C106–C114.

HERMUS, A. R. & SWEEP, C. G. (1990) Cytokines and the hypothalamic-pituitary-adrenal axis. *Journal of Steroid Biochemistry and Molecular Biology*, **37**, 867–871.

HESS, P. & PAYVAR, F. (1992) Hormone withdrawal triggers a premature and sustained gene activation from delayed secondary glucocorticoid response elements. *Journal of Biological Chemistry*, **267**, 3490–3497.

HICKMAN, J. A. (1992) Apoptosis induced by anticancer drugs. *Cancer and Metastasis Reviews*, **11**, 121–139.

HICKSON, R. C., CZERWINSKI, S. M., FALDUTO, M. T. & YOUNG, A. P. (1990) Glucocorticoid antagonism by exercise and androgenic-anabolic steroids. *Medicine and*

Science of Sports and Exercise, **22**, 331–340.

HODGSON, S. F. (1990) Corticosteroid-induced osteoporosis. *Endocrinology and Metabolism Clinics of North America*, **19**, 95–111.

HOLLENBERG, S. M. & EVANS, R. M. (1988) Multiple and cooperative transactivation domains of the human glucocorticoid receptor. *Cell*, **55**, 899–906.

HORIKOSHI, M., WANG, C. K., FUJI, H., CROMLISH, J. A., WEIL, P. A. & ROEDER, R. G. (1989) Cloning and structure of a yeast gene encoding a general transcription initiation factor TFIID that binds to the TATA box. *Nature*, **341**, 299–303.

HORNER, H. C., PACKAN, D. R. & SAPOLSKY, R. M. (1990) Glucocorticoids inhibit glucose transport in cultured hippocampal neurons and glia. *Neuroendocrinology*, **52**, 57–64.

HORST, H. J. & FLAD, H. D. (1987) Cortiosteroid-interleukin-2 interactions: inhibition of binding of interleukin 2 receptors. *Clinical and Experimental Immunology*, **68**, 156–161.

HUANG, W., CHOI, C. L., YANG, Z., COPOLOV, D. L. & LIM, A. T. (1991) Forskolin-induced immunoreactive atrial natriuretic peptide (ANP) secretion and pro-ANP mRNA expression of hypothalamic neurons in culture: modulation by glucocorticoids. *Endocrinology*, **128**, 2591–2600.

HUME, R. & BOYD, G. S. (1978) Cholesterol metabolism and steroid hormone production. *Biochemical Society Transactions*, **6**, 893–898.

IMAI, E., MINER, J. N., MITCHELL, J. A., YAMAMOTO, K. R. & GRANNER, D. K. (1993) Glucocorticoid receptor – cAMP response element-binding protein interaction and the response of the phosphoenolpyruvate carboxykinase gene to glucocorticoids. *Journal of Biological Chemistry*, **268**, 5353–5356.

INAGAKI, N., MIURA, T., NAKAJIMA, T., YOSHIDA, K. & KODA, A. (1992) Studies on the anti-allergic mechanism of glucocorticoids in mice. *Journal of Pharmacobio-Dynamics*, **15**, 581–587.

INGRASSIA, R., SAVOLDI, G. F., CARAFFINI, A., TIRONI, M., POIESI, C., WILLIAMS, P., ALBERTINI, A. & DILORENZO, D. (1994) Characterization of a novel transcription complex required for glucocorticoid regulation of the rat alpha-1-acid-glycoprotein gene. *DNA and Cell Biology*, **13**, 615–627.

IVARIE, R. D. & O'FARRELL, P. H. (1978) The glucocorticoid domain: steroid mediated changes in the rate of synthesis of rat hepatoma proteins. *Cell*, **13**, 41–55.

IWATA, M., ISEKI, R., SATO, K., TOZAWA, Y. & OHOKA, Y. (1994) Involvement of protein kinase C-epsilon in glucocorticoid-induced apoptosis in thymocytes. *International Immunology*, **6**, 431–438.

JANSEN, H., VAN TOL, A., AUWERX, J., SKRETTING, G. & STAELS, B. (1992) Opposite regulation of hepatic lipase and lecithin : cholesterol acyltransferase by glucocorticoids in rats. *Biochimica et Biophysica Acta*, **1128**, 181–185.

JANTZEN, H. M., STRAHLE, U., GLOSS, B., STEWART, F., SCHMID, W., BOSHART, M., MIKSICEK, R. & SCHUTZ, G. (1987) Cooperativity of glucocorticoid response elements located far upstream of the tyrosine aminotransferase gene. *Cell*, **49**, 29–38.

JOHN, M. E., JOHN, M. D., BOGGARAM, V., SIMPSON, E. R. & WATERMAN, M. R. (1986) Transcriptional regulation of steroid hydroxylase genes by corticotrophin. *Proceedings of the National Academy of Sciences, USA*, **83**, 4715–4719.

JOHNSON, L. K. & BAXTER, J. D. (1985) The mechanism of action of adrenocorticosteroids at the molecular level. In MCCARTY, D. (Ed.), *Landmark Advances in Rheumatology*, pp. 13–22, New York: Contact Associates.

JONET, C., RAHMSDORF, H. J., PARK, K. K., CATO, A. C. B., GEBEL, S., PONTA, H. & HERRLICH, P. (1990) Antitumor promotion and antiinflammation down modulation of AP-1 (Fos Jun) activity by glucocorticoid hormone. *Cell*, **62**, 1189–1204.

KAM, J. C., SZEFLER, S. J., SURS, W., SHER, W. & LEUNG, D. Y. (1993) Combination of IL-2 and IL-4 reduces glucocorticoid receptor binding affinity and T-cell response to glucocorticoids. *Journal of Immunology*, **151**, 3460–3466.

KANSE, S. M., TAKAHASHI, K., WARREN, J. B., GHATEI, M. & BLOOM, S. R. (1991) Glucocorticoids induce endothelin release from vascular smooth muscle cells but not endothelial cells. *European Journal of Pharmacology*, **199**, 99–101.

KASPERS, G. J., PIETERS, R., KLUMPER, E., DE WAAL, F. C. & VEERMAN, A. J. (1994) Glucocorticoid resistance in childhood leukemia. *Leukemia and Lymphoma*, **13**, 187–201.

KATO, M. & SCHLEIMER, R. P. (1994) Anti-inflammatory steroids inhibit granulocyte–macrophage colony stimulating factor production by human lung tissue. *Lung*, **172**, 113–124.

KENYON, C. J. & FRASER, R. (1992) Biochemistry of steroid hypertension. In JAMES, V. H. T. (Ed.), *The Adrenal Gland*, New York: Raven Press.

KERNER, B., TEICHMANN, B. & WELTE, K. (1992) Dexamethasone inhibits tumor necrosis factor-induced granulocyte colony stimulating factor production in human endothelial cells. *Experimental Hematology*, **20**, 334–348.

KERSTJENS, H. A., POSTMA, D. S., VAN DOORMAAL, J. J., VAN ZANTEN, A. K., BRAND, P. L., DEKHUIJZEN, P. N. & KOETER, G. H. (1994) Effects of short term and long term treatment with inhaled corticosteroids on bone metabolism in patients with airways obstruction. *Thorax*, **49**, 652–656.

KIMBERLY, R. P. (1994) Glucocorticoids. *Current Opinion in Rheumatology*, **6**, 273–280.

KIMURA, T. (1981) ACTH stimulation on cholesterol side-chain cleavage activity of adrenocortical mitochondria: Transfer of the stimulus from plasma membrane to mitochondria. *Molecular and Cellular Biochemistry*, **36**, 105–122.

KING, M. S. & BAERTSCHI, A. J. (1990) The role of intracellular messengers in adrenocorticotrophin secretion *in vitro*. *Experientia*, **15**, 26–40.

KINSELLA, J. L. (1990) Actions of glucocorticoids on proximal tubule transport systems. *Seminars in Nephrology*, **10**, 330–338.

KLEIN, E. S., REINKE, R., FEIGELSON, P. & RINGOLD, G. M. (1987) Glucocorticoid regulated expression from the 5′-flanking region of the rat alpha-1-acid glycoprotein gene: requirement for ongoing protein synthesis. *Journal of Biological Chemistry*, **262**, 520–523.

KNOWLES, R. G., SALTER, M., BROOKS, S. L. & MONCADA, S. (1990) Anti-inflammatory glucocorticoids inhibit the induction by endotoxin of nitric oxide synthase in the lung, liver and aorta of the rat. *Biochemical and Biophysical Research Communications*, **172**, 1042–1048.

KORNEL, L., NELSON, W. A., MANISUNDARAM, B., CHIGURUPATI, R. & HAYASHI, T. (1993a) Mechanism of the effect of glucocorticoids and mineralocorticoids on vascular smooth muscle contractility. *Steroids*, **58**, 580–587.

KORNEL, L., MANISUNDARAM, B. & NELSON, W. A. (1993b) Glucocorticoids regulate Na^+ transport in vascular smooth muscle through the glucocorticoid receptor-mediated mechanism. *American Journal of Hypertension*, **6**, 736–744.

LAM, K. S., SRIVASTAVA, K. S. & TAM, S. P. (1992) Divergent effects of glucocorticoid on the gene expression of vasoactive intestinal peptide in the rat cerebral cortex and pituitary. *Neuroendocrinology*, **56**, 32–37.

LAMAS, A. M., MARCOTTE, G. V. & SCHLEIMER, R. P. (1989) Human endothelial cells prolong eosinophil survival. Regulation by glucocorticoids. *Journal of Immunology*, **142**, 3978–3984.

LAMAS, A. M., LEON, O. G. & SCHLEIMER, R. P. (1991) Glucocorticoids inhibit eosinophil responses to granulocyte-macrophage colony stimulating factor. *Journal of Immunology*, **147**, 254–259.

LANDFIELD, P. W. (1994) The role of glucocorticoids in brain aging and Alzheimer's disease: an integrative physiological hypothesis. *Experimental Gerontology*, **29**, 3–11.

LAPLACE, J. R., HUSTED, R. F. & STOKES, J. B. (1992) Cellular responses to steroids in the enhancement of Na^+ transport by rat collecting duct cells in culture. Differences between glucocorticoid and mineralocorticoid hormones. *Journal of Clinical Investigation*, **90**, 1370–1378.

LARROQUE, C., ROUSSEAU, J. & VAN LIER, J. E. (1981) Enzyme bound sterols of bovine adrenocortical cytochrome $P450_{ccc}$. *Biochemistry*, **20**, 925–929.

LARSEN, G. L. & HENSON, P. M. (1983) Mediators of inflammation. *Annual Review of Immunology*, **1**, 335–347.

LARSEN, P. J., JESSOP, D. S., CHOWDREY, H. S., LIGHTMAN, S. L. & MIKKELSEN, J. D. (1994) Chronic administration of glucocorticoids directly upregulates prepro-neuropeptide Y and Y1 receptor mRNA levels in the arcuate nucleus of the rat. *Journal of Neuroendocrinology*, **6**, 153–159.

LAWRENCE, M. S. & SAPOLSKY, R. M. (1994) Glucocorticoids accelerate ATP loss following metabolic insults in cultured hippocampal neurons. *Brain Research*, **646**, 303–306.

LEE, M. Y., CLARKE, J. E. & WILLIAMS, D. E. (1993) Induction of flavin-containing monooxygenase (FMO B) in rabbit lung and kidney by sex steroids and glucocorticoids. *Archives of Biochemistry and Biophysics*, **302**, 332–336.

LENZEN, S. & BAILEY, C. J. (1984) Thyroid hormones, gonadal and adrenocortical steroids and the function of the Islets of Langerhans. *Endocrinology Reviews*, **5**, 411–421.

LESCH, K. P. & LERER, B. (1991) The 5-HT receptor-G-protein-effector system complex in depression. I. Effect of glucocorticoids. *Journal of Neural Transmission*, **84**, 3–18.

LEW, W., OPPENHEIM, J. J. & MATSUSHIMA, K. (1988) Analysis of the suppression of IL-1a and IL-1b production in human peripheral blood mononuclear adherant cells by a glucocorticoid hormone. *Journal of Immunology*, **140**, 1895–1902.

LI, Q. & WRANGE, O. (1993) Translational positioning of a nucleosomal glucocorticoid response element modulates glucocorticoid receptor affinity. *Genes and Development*, **7**, 2471–2481.

LIAN, J. B. & STEIN, G. S. (1993) The developmental stages of osteoblast growth and differentiation exhibit selective responses of genes to growth factors (TGFb1) and hormones (vitamin D and glucocorticoids). *Journal of Oral Implantology*, **19**, 95–105.

LIMA, L., ARCE, V., DIAZ, M. J., TRESGUERRES, J. A. & DEVESA, J. (1993) Glucocorticoids may inhibit growth hormone release by enhancing beta-adrenergic responsiveness in hypothalamic somatostatin neurons. *Journal of Clinical Endocrinology and Metabolism*, **76**, 439–444.

LINDEN, M. (1992) The effects of beta 2-adrenoceptor agonists and a corticosteroid, budesonide, on the secretion of inflammatory mediators from monocytes. *British Journal of Pharmacology*, **107**, 156–160.

LINDEN, M. & BRATTSAND, R. (1994) Effects of a corticosteroid, budesonide, on alveolar macrophage and blood monocyte secretion of cytokines: differential sensitivity of GM-CSF, IL-1β and IL-6. *Pulmonary Pharmacology*, **7**, 43–47.

LINDER, M. W. & PROUGH, R. A. (1993) Developmental aspects of glucocorticoid regulation of polycyclic aromatic hydrocarbon-inducible enzymes in the liver. *Archives of Biohemistry and Biophysics*, **302**, 92–102.

LIU, J., HAIGH, R. N. & JONES, C. T. (1992) Enhancement of noradrenaline induced in inositol polyphosphate formation by glucocorticoids in rat vascular smooth muscle cells. *Journal of Endocrinology*, **133**, 405–411.

LOMBES, M., KENOUCH, S., SOUQUE, A. FARMAN, N. & RAFESTIN-OBLIN, M. E. (1994) The mineralocorticoid receptor discriminates aldosterone from glucocorticoids independently of the 11 beta-hydroxysteroid dehydrogenase. *Endocrinology*, **135**, 834–840.

LOTTA-LARSSON, E. (1980) Cyclosporin A and dexamethasone suppress T-cell responses by selectively acting at distinct sites of the triggering process. *Journal of Immunology*, **124**, 2828–2833.

LU, S. C., GE, J. L., KUTHLENKAMP, J. & KAPLOWITZ, N. (1992) Insulin and glucocorticoid dependence of hepatic gamma-glutamylcysteine synthetase and glutathione synthesis in the rat. Studies in cultured hepatocytes and *in vivo*. *Journal of Clinical Investigation*, **90**, 524–532.

LUKERT, B. P. & RAISZ, L. G. (1990) Glucocorticoid induced osteoporosis: pathogenesis and management. *Annals of Internal Medicine*, **112**, 352–364.

LUNDGREN, J. D., HIRATA, F., MAROM, Z., LOGUN, C., STEEL, L., KALINER, M. & SHELHAMER J. (1988) Dexamethasone inhibits respiratory glycoconjugate secretion from feline airways *in vitro* by the induction of lipocortin synthesis. *American Review of Respiratory Disease*, **137**, 353–357.

MACMAHON, A. & SABBAN, E. L. (1992) Regulation of expression of dopamine-b-hydroxylase in PC12 cells by glucocorticoids and cyclic AMP analogues. *Journal of Neurochemistry*, **59**, 2040–2047.

MADER, S., KUMAR, V., DE VERNEUIL, H. & CHAMBON, P. (1989) Three aminoacids of the oestrogen receptor are essential to its ability to distinguish an oestrogen from a glucocorticoid-responsive element. *Nature*, **338**, 271–274.

MANTERO, F. & BOSCARO, M. (1992) Glucocorticoid dependent hypertension. *Journal of Steroid Biochemistry and Molecular Biology*, **43**, 409–413.

MARCO, J., CALLE, C., ROMAN, D., DIAZ-FERROS, M., VILLANUEVA, M. L. & VALVERDE, I. (1973) Hyperglucagonism induced by glucocorticoid treatment in man. *New England Journal of Medicine*, **309**, 21–24.

MARTINOLI, M. G., VEILLEUX, R. & PELLETIER, G. (1991) Effects of triiodothyronine, dexamethasone and estradiol-17 beta on GH mRNA in rat pituitary cells in culture as revealed by *in-situ* hybridization. *Acta Endocrinologica*, **124**, 83–90.

MASER, E. (1994) 11-Beta-hydroxysteroid dehydrogenase mediates reductive metabolism of xenobiotic carbonyl compounds. *Biochemical Pharmacology*, **47**, 1805–1812.

MASFERRER, J. L., SEIBERT, K., ZWEIFEL, B. & NEEDLEMAN, P. (1992) Endogenous glucocorticoids regulate an inducible cyclooxygenase enzyme. *Proceedings of the National Academy of Sciences, USA*, **89**, 3917–3921.

MASTERS, J. N., FINCH, C. E. & NICHOLS, N. R. (1994) Rapid increase in glycerol phosphate dehydrogenase mRNA in adult rat brain: a glucocorticoid-dependent stress response. *Neuroendocrinology*, **60**, 23–35.

MAX, S. R. (1990) Glucocorticoid-mediated induction of glutamine synthetase in skeletal muscle. *Medicine and Science of Sports and Exercise*, **22**, 325–330.

MCEWEN, B. S., GOULD, E. A. & SAKAI, R. R. (1992) The vulnerability of the hippocampus to protective and destructive effects of glucocorticoids in relation to stress. *British Journal of Psychiatry*, **15**(Suppl.), 18–23.

MCINTYRE, W. R. & SAMUELS, H. H. (1985) Triaminolone acetonide regulates glucocorticoid receptor levels by decreasing the half-life of the activated nuclear receptor form. *Journal of Biological Chemistry*, **260**, 418–427.

MCMAHON, M., GERICH, J. & RIZZA, R. (1988) Effects of glucocorticoids on carbohydrate metabolism. *Diabetes/Metabolism Reviews*, **4**, 17–30.

MEANEY, M. J., VIAU, V., BHATNAGER, S., BETITO, K., INY, L. J., O'DONNELL, D. & MITCHELL, J. B. (1991) Cellular mechanisms underlying the development and expression of individual differences in the hypothalamic-pituitary-adrenal stress response. *Journal of Steroid Biochemistry and Molecular Biology*, **39**, 265–274.

MENDEL, D. B. & ORTI, E. (1988) Isoform composition and stoichiometry of the ~90 kDA heat shock protein associated with glucocorticoid receptors. *Journal of Biological Chemistry*, **263**, 6695–6702.

MERTZ, L. M. & PEDERSON, R. C. (1989) The kinetics steroidogenesis activator polypeptide in the rat adrenal cortex. Effects of ACTH, cAMP, cycloheximide and circadian rhythm. *Journal of Biological Chemistry*, **264**, 15274–15279.

MIGLIORATI, G., PAGLIACCI, M. C., D'ADAMIO, F., CROCICCHIO, F., NICOLETTI, I. & RICCARDI, C. (1992) Glucocorticoid-induced DNA fragmentation: role of protein-kinase-C activity. *Pharmacological Research*, **26**(Suppl. 2), 5–9.

MILLER, N. R. & SIMONS, S. S. (1988) Steroid binding to hepatoma tissue culture cell glucocorticoid receptors involves at least two sulfhdryl groups. *Journal of Biological Chemistry*, **263**, 15217–15225.

MITCHELL, D. R. & LYLES, K. W. (1990) Glucocorticoid-induced osteoporosis: mechanisms for bone loss and evaluation of strategies for prevention. *Journal of Gerontology*, **45**, M153–M158.

MIURA, T., INAGAKI, N., YOSHIDA, K., NAKAJIMA, T., NAGAI, H. & KODA, A. (1992) Mechanisms for glucocorticoid inhibition of immediate hypersensitivity reactions in rats. *Japanese Journal of Pharmacology*, **59**, 77–87.

MOKOENA, T. & GORDON, S. (1985) Human macrophage activation. Modulation of mannosyl, fucosyl receptor activity *in vitro* by lymphokines, gamma and alpha interferons and dexamethasone. *Journal of Clinical Investigation*, **75**, 624–631.

MONDER, C., MIROFF, Y., MARANDICI, A. & HARDY, M. P. (1994) 11-Beta-hydroxysteroid dehydrogenase alleviates glucocorticoid-mediated inhibition of steroidogenesis in rat Leydig cells. *Endocrinology*, **134**, 1199–1204.

MORAND, E. F. & GOULDING, N. J. (1993) Glucocorticoids in rheumatoid arthritis – mediators and mechanisms. *British Journal of Rheumatology*, **32**, 816–819.

MORITOKI, H., TAKEUCHI, S., HISAYAMA, T. & KONDOH, W. (1992) Nitric oxide synthase responsible for L-arginine-induced relaxation of rat aortic rings *in vitro* may be an inducible type. *British Journal of Pharmacology*, **107**, 361–366.

MORNET, E., DUPONT, J., VITEK, A. & WHITE, P. C. (1989) Characterization of two genes encoding human steroid 11β-hydroxylase. *Journal of Biological Chemistry*, **264**, 20961–20967.

MORRISON, N. & EISMAN, J. (1993) Role of negative glucocorticoid regulatory element in glucocorticoid repression of the human osteocalcin promoter. *Journal of Bone and Mineral Metabolism*, **8**, 969–975.

MUKAI, K., IMAI, M., SHIMADA, H. & ISHIMURA, Y. (1993) Isolation and characterization of rat CYP11B genes involved in late steps of mineralo- and glucocorticoid syntheses. *Journal of Biological Chemistry*, **268**, 9130–9137.

MULLER, M. & RENKAWITZ, R. (1991) The glucocorticoid receptor. *Biochimica et Biophysica Acta*, **1088**, 171–182.

MUNCK, A., CRABTREE, G. R. & SMITH, K. A. (1979) Glucocorticoid receptors and actions in rat thymocytes and immunologically-stimulated human peripheral lymphocytes. In BAXTER, J. D. & ROUSSEAU, G. G. (Eds), *Glucocorticoid Hormone Action*, pp. 341–359, New York: Springer.

MUNCK, A., GUYRE, P. M. & HOLBROOK, N. J. (1984) Physiological functions of glucocorticoids in stress and their relation to pharmacological actions. *Endocrinology Reviews*, **5**, 25–44.

MURPHY, B. E. (1991) Steroids and depression. *Journal of Steroid Biochemistry and Molecular Biology*, **38**, 537–559.

MURREY, D. K., RUHMANN-WENNHOLD, A. & NELSON, D. H. (1982) Adrenolectomy decreases the sphigmyelin and cholesterol content of fat cell ghosts. *Endocrinology*, **111**, 452–459.

MURREY, D. K., HILL, M. E. & NELSON, D. H. (1990) Inhibitory action of sphingosine, sphinganine and dexamethasone on glucose uptake: studies with hydrogen peroxide and phorbol ester. *Life Sciences*, **46**, 1843–1849.

NAKANO, T., OHARA, O., TERAOKA, H. & ARITA, H. (1990) Glucocorticoids suppress group II phospholipase A2 production by blocking mRNA synthesis and post-transcriptional expression. *Journal of Biological Chemistry*, **265**, 12745–12748.

NARAY-FEJES-TOTH, A., FEJES-TOTH, G., FISCHER, C. & FROLICH, J. C. (1984) Effect of dexamethasone on *in vivo* protanoid production in the rabbit. *Journal of Clinical Investigation*, **74**, 120–125.

NEBERT, D. W. & GONZALEZ, F. J. (1987) P450 genes: Structure, evolution and regulation. *Annual Review of Biochemistry*, **56**, 945–993.

NELSON, D. H. (1980) Corticosteroid induced changes in phospholipid membranes as mediators of their action. *Endocrinology Reviews*, **1**, 180–191.

NEMOTO, T., OHARA, N. Y., DENIS, M. & GUSTAFSSON, J. A. (1990) The transformed glucocorticoid receptor has a lower steroid binding affinity than the nontransformed receptor. *Biochemistry*, **29**, 1880–1886.

NERI, G., MALENDOICZ, L. K., ANDREIS, P. & NUSSDORFER, G. G. (1993) Thyrotrophin-releasing hormone inhibits glucocorticoid secretion of rat adrenal cortex: *in vivo* and *in vitro* studies. *Endocrinology*, **133**, 511–514.

NIELSEN, H. K., LAURBERG, P., BRIXEN, K. & MOSEKILDE, L. (1991) Relationship between diurnal variations in serum osteocalcin, cortisol, parathyroid hormone and ionized calcium in normal individuals. *Acta Endocrinologica*, **124**, 391–398.

NIETO, M. A. & LOPEZ-RIVAS, A. (1992) Glucocorticoids activate a suicide program in mature T-lymphocytes: protective action of IL-2. *Annals of the New York Academy of Sciences*, **650**, 115–120.

NITSCH, D., BOSHART, M. & SCHUTZ, G. (1993) Activation of tyrosine aminotransferase gene is dependent upon synergy between liver-specific and hormone-responsive elements. *Proceedings of the National Academy of Sciences, USA*, **90**, 5479–5483.

NORMAN, A. W. & LITWACK, G. (1987) *Hormones*, New York: Academic.

NOWELL, P. C. (1961) Inhibition of human leucocyte mitosis by prednisolone *in vitro*. *Cancer Research*, **21**, 1518–1521.

NUSSLEIN, H. G., WEBER, G. & KALDREN, J. R. (1994) Synthetic glucocorticoids potentiate IgE synthesis. Influence of steroid and nonsteroid hormones on human *in vitro* IgE secretion. *Allergy*, **49**, 365–370.

O'BANION, M. K., WINN, V. D. & YOUNG, D. A. (1992) cDNA cloning and functional activity of a glucocorticoid-regulated inflammatory cyclooxygenase. *Proceedings of the National Academy of Sciences, USA*, **89**, 4888–4892.

OITZL, M. S. & DE KLOET, E. R. (1992) Selective corticosteroid antagonists modulate specific aspects of spatial orientation learning. *Behavioural Neuroscience*, **106**, 62–71.

OKUNO, Y., NISHIZAWA, Y., KAWAGISHI, T. & MORII, H. (1993) *In vivo* and *in vitro* effects of glucocorticoids on glucose transport in human polymorphonucelar leukocytes. *Hormone Metabolic Research*, **25**, 165–169.

ONOAGBE, I. O. (1994) Effects of glucocorticoids and insulin on tyrosine aminotransferase activity in isolated chick embryo hepatocytes and in intact embryos in ovo. *Archives of Biochemistry and Biophysics*, **309**, 58–61.

ORME-JOHNSON, N. R. (1990) Distinctive properties of adrenal cortex mitochondria. *Biochimica et Biophysica Acta*, **1020**, 213–231.

OSHIMA, H. & SIMONS, S. S. (1992) Modulation of glucocorticoid induction of tyrosine aminotransferase gene expression by variations in cell density. *Endocrinology*, **130**, 2106–2112.

OSHIMA, H. & SIMONS, S. S. (1993) Sequence-selective interactions of transcription factor elements with tandem glucocorticoid-responsive elements at physiological steroid concentrations. *Journal of Biological Chemistry*, **268**, 26858–26865.

PALIOGIANNI, F., RAPTIS, A., AHUJA, S. S., NAJJAR, S. M. & BOUMPAS, D. T. (1993) Negative transcriptional regulation of human interleukin-2 (IL-2) gene by glucocorticoids through interference with nuclear transcription factors AP-1 and NF-AT. *Journal of Clinical Investigation*, **91**, 1481–1489.

PANARELLI, M., TERZOLO, M., PIOVESAN, A., OSELLA, G., PACCOTTI, P., PINNA, G. & ANGELI, A. (1990) 24-Hour profiles of blood pressure and heart rate in Cushing's syndrome. Evidence for differential control of cardiovascular variables by glucocorticoids. *Annali Italiani di Medicina Interna*, **5**, 18–25.

PAPA, M. Z., VETTO, J. T., ETTINGHAUSEN, S. E., MULE, J. J. & ROSENBERG, S. A. (1986) Effect of corticosteroid on the anti-tumor activity of lymphokine-activated killer cells and interleukin-2 in mice. *Cancer Research*, **46**, 5618–5623.

PARRILLO, J. E. & FAUCI, A. S. (1979) Mechanisms of glucocorticoid action on immune processes. *Annual Review of Pharmacology and Toxicology*, **19**, 179–185.

PERLMANN, T. & WRANGE, O. (1988) Specific glucocorticoid receptor binding to DNA reconstituted in a nucleosome. *EMBO Journal*, **7**, 3073–3079.

PEERS, S. H., SMILLIE, F., ELDERFIELD, A. J. & FLOWER, R. J. (1993) Glucocorticoid and non-glucocorticoid induction of lipocortins (annexins) 1 and 2 in rat peritoneal leucocytes *in vivo*. *British Journal of Pharmacology*, **108**, 66–72.

PETRONI, K. C., SHEN, L. & GUYRE, P. M. (1988) Modulation of human polymorphonuclear leucocyte IgG Fc receptors and Fc receptor-mediated functions by IFN-γ and glucocorticoids. *Journal of Immunology*, **140**, 3467–3472.

PHILIPPE, J., GIORDANO, E., GJINOVCI, A. & MEDA, P. (1992) Cyclic adenosine monophosphate prevents the glucocorticoid mediated inhibition of insulin gene expression in rodent islet cells. *Journal of Clinical Investigation*, **90**, 2228–2233.

PICARD, D. & YAMAMOTO, K. R. (1987) Two signals mediate hormone-dependent nuclear localization of the glucocorticoid receptor. *EMBO Journal*, **6**, 3333–3340.

PICARD, D., KUMAR, V., CHAMBON, P. & YAMAMOTO, K. R. (1990) Signal transduction by steroid hormones: nuclear localization is differentially regulated in estrogen and glucocorticoid receptors. *Cell Regulation*, **1**, 291–299.

PIEDIMONTE, G. D., MCDONALD, M. & NADEL, J. A. (1990) Glucocorticoids inhibit neurogenic plasma extravasation and prevent virus potentiated extravasation in the rat trachea. *Journal of Clinical Investigation*, **86**, 1409–1415.

PILKIS, S. J., EL-MAGHRABI, M. R. & CLAUS, T. H. (1988) Hormone regulation of hepatic gluconeogensis and glycolysis. *Annual Review of Biochemistry*, **57**, 755–783.

PITTNER, R. A. & SPITZER, J. A. (1993) Steroid hormones inhibit induction of spontaneous nitric oxide production in cultured hepatocytes without changes in arginase activity or urea production. *Proceedings of the Society of Experimental Biology and Medicine*, **202**, 499–504.

POPOVIC, V., DAMJANOVIC, S., MICIC, D., MANOJLOVIC, D., MICIC, J. & CASANUEVA, F. F. (1993) Modulation by glucocorticoids of growth hormone secretion in patients with different pituitary tumors. *Neuroendocrinology*, **58**, 465–472.

PRATT, W. B. & WELSH, M. J. (1994) Chapterone functions of the heat shock proteins associated with steroid receptors. *Seminars in Cell Biology*, **5**, 83–93.

PRIVALLE, C. T., CRIVELLO, J. F. & JEFCOATE, C. L. (1983) Regulation of intramitochondrial cholesterol transfer to sidechain cleavage cytochrome P450 in rat adrenal gland. *Proceedings of the National Academy of Sciences, USA*, **80**, 702–706.

PRUETT, S. B., ENSLEY, D. K. & CRITTENDEN, P. L. (1993) The role of chemical-induced stress responses in immunosuppression: a review of quantitative associations and cause–effect relationships between chemical-induced stress responses and immunosuppression. *Journal of Toxicology and Environmental Health*, **39**, 163–192.

PYTSKY, V. I. (1993) New aspects in mechanisms of action of glucocorticoids. *Journal of Investigative Allergy and Clinical Immunology*, **1**, 359–367.

RABEK, J. P., ZHANG, D. E., TORRES-RAMOS, C. A. & PAPACOONSTANTINOU, J. (1994) Analysis of the mechanisms of glucocorticoid-mediated downregulation of mouse alpha-fetoprotein gene. *Biochimica et Biophysica Acta*, **1218**, 136–144.

RADOMSKI, M. W., PALMER, R. M. & MONCADA, S. (1990) Glucocorticoids inhibit the expression of an inducible, but not the constitutive, nitric oxide synthase in vascular endothelial cells. *Proceedings of the National Academy of Sciences, USA*, **87**, 10043–10047.

RAE, P. A., GUTMANN, N. S., TSAO, J. & SCHIMMER, B. P. (1979) Mutations in cyclic AMP dependent protein kinase and corticotrophin (ACTH) sensitive adenylate cyclase affect adrenal steroidogensis. *Proceedings of the National Academy of Sciences, USA*, **76**, 1896–1900.

RAINEY, W. E., BIRD, I. M. & MASON, J. I. (1994) The NCI-H295 cell line: a pluripotent model for human adrenocortical studies. *Molecular and Cellular Endocrinology*, **100**, 45–50.

RATAJCZAK, T., WILLIAMS, P. M., DILORENZO, D. & RINGOLD, G. M. (1992) Multiple elements within the glucocorticoidregulatory unit of rat a1-acid glycoprotein gene are recognition sites for C/EBP. *Journal of Biological Chemistry*, **267**, 1111–1119.

RAY, A. & SEHGAL, P. B. (1992) Cytokines and their receptors: molecular mechanism of IL-6 gene repression by glucocorticoids. *Journal of the American Society of Nephrology*, **2**, S214–S221.

REBUFFE-SCRIVE, M., BRONNEGARD, M., NILSSON, A., ELDH, J., GUSTAFSSON, J. A. & BJORNTORP, P. (1990) Steroid hormone receptors in human adipose tissues. *Journal of Clinical Endocrinology and Metabolism*, **71**, 1215–1219.

REBUFFE-SCRIVE, M., WALSH, U. A., MCEWEN, B. & RODIN, J. (1992) Effect of chronic stress and exogenous glucocorticoids on regional fat distribution and metabolism. *Physiology and Behaviour*, **52**, 583–590.

REES, D. D., CELLEK, S., PALMER, R. M. & MONCAD, S. (1990) Dexamethasone prevents the induction by endotoxin of a nitric oxide synthase and the associated effects upon vascular tone: an insight into endotoxin shock. *Biochemical and Biophysical Research Communications*, **173**, 541–547.

REID, I. R., VEALE, A. G. & FRANCE, J. T. (1994) Glucocorticoid osteoporosis. *Journal of Asthma*, **31**, 7–18.

REIK, A., SCHTZ, G. & STEWART, A. F. (1991) Glucocorticoids are required for establishment and maintenance of an alteration in chromatin structure: induction leads to a reversible disruption of nucleosomes over an enhancer. *EMBO Journal*, **10**, 2569–2576.

RENKAWITZ, R., SCHUTZ, G., VON DER AHE, D. & BEATO, M. (1984) Sequences in the promoter region of the chicken lysozyme gene required for steroid regulation and receptor binding. *Cell*, **37**, 503–510.

RENOIR, J. M., RADANYI, C., FABER, L. E. & BAULIEU, E. E. (1990) The non-DNA binding heterooligomeric form of mammalian steroid hormone receptors contains a HSP90 bound 59 kDa protein. *Journal of Biological Chemistry*, **265**, 10740–10745.

RICHARD-FOY, H. & HAGER, G. L. (1987) Sequence specific positioning of nucleosomes over the steroid inducible MMTV promoter. *EMBO Journal*, **6**, 2321–2328.

RIZZA, R. A., MANDARINO, L. J. & GERICH, J. E. (1982) Cortisol-induced insulin resistance in man: impaired suppression of glucose production and stimulation of glucose utilization due to post-receptor event. *Journal of Clinical Endocrinology and Metabolism*, **54**, 131–139.

RODRIGUEZ, R., ANGULO, I., VINUELA, J. E., MEDINA, M., GIL, J. & SUBIZA, J. L. (1994) Inhibition of bone marrow-derived natural suppressor activity by glucocorticoids and its reversal by IFNg or IL-2. *Transplantation*, **58**, 511–517.

RUSSELL, R. G. (1993) Cellular regulatory mechanisms that may underlie the effects of corticosteroids on bone. *British Journal of Rheumatology*, **32**(Suppl. 2), 6–10.

RUSSO-MARIE, F. (1992) Macrophages and glucocorticoids. *Journal of Neuroimmunology*, **40**, 281–286.

SABLONNIERE, B., ECONOMIDIS, I. V., LEFEBVRE, P., PLACE, M., RICHARD, C., FORMSTECHER, P., ROUSSEAU, G. G. & DAUTREVAUX, M. (1988) RNA binding to the untransformed glucocorticoid receptor. Sensitivity to substrate specific ribonucleases and characterization of a ribonucleic acid associated with the purified receptor. *European Journal of Biochemistry*, **177**, 371–382.

SANTORO, C., MERMOD, N., ANDREWS, P. C. & TIJAN, R. (1988) A family of human CCAAT-box-binding proteins active in transcription and DNA replication: Cloning and expression of multiple cDNAs. *Nature*, **334**, 218–224.

SATEL, S. L. (1990) Mental status changes in children receiving glucocorticoids. Review of the literature. *Clinical Pediatrics*, **29**, 383–388.

SAUER, J., RUPPRECHT, M., ARZT, E., STALLA, G. K. & RUPPRECHT, R. (1993) Glucocorticoids modulate soluble interleukin-2 receptor levels *in vivo* depending on the state of immune activation and the duration of glucocorticoid exposure. *Immunopharmacology*, **25**, 269–276.

SCHLEIMER, R. P. (1989) The effects of glucocorticoids on mast cells and basophils. In SCHLEIMER, R. P., CLAMAN, H. N. & ORONSKY, A. (Eds), *Anti-inflammatory steroid action. Basic and clinical aspects*, pp. 226–258, San Diego: Academic Press.

SCHLEIMER, R. P. (1990) Effects of glucocorticoids on inflammatory cells relevant to their therapeutic applications in asthma. *American Review of Respiratory Disease*, **141**, S59–S69.

SCHLEIMER, R. P. (1993) An overview of glucocorticoid anti-inflammatory actions. *European Journal of Clinical Pharmacology*, **45**(Suppl. 1), S3–S7; discussion S43–S44.

SCHLEIMER, R. P., LICHTENSTEIN, L. M. & GILLEPSIE, E. (1981) Inhibition of basophil histamine release by anti-inflammatory steroids. *Nature*, **292**, 454–455.

SCHLEIMER, R. P., SCHULMAN, E. S. & MACGLASHAN, D. W. (1983) Effects of dexamethasone on mediator release from human lung fragments and purified human lung mast cells. *Journal of Clinical Investigation*, **71**, 1830–1835.

SCHLEIMER, R. P., DAVIDSON, D. A., PETERS, S. P. & LICHTENSTEIN, L. M. (1985) Inhibition of basophil leukotriene releases by anti-inflammatory steroids. *International Archives of Allergy and Applied Immunology*, **77**, 241–243.

SCHLEIMER, R. P., FREELAND, H. S., PETERS, S. P., BROWN, K. E. & DERSE, C. P. (1989) An assessment of the effects of glucocorticoids on degranulation, chemotaxis, binding to vascular endothelial cells and formation of leukotriene B4 by purified hyamn neutrophils. *Journal of Pharmacology and Experimental Therapeutics*, **250**, 598–605.

SCHULE, R., MULLER, M., OTSUKA-MURAKAMI, H. & RENKAWITZ, R. (1988) Cooperativity of the glucocorticoid receptor and the CACC-box binding factor. *Nature*, **332**, 87–90.

SEENE, T. (1994) Turnover of skeletal muscle contractile proteins in glucocorticoid myopathy. *Journal of Steroid Biochemistry and Molecular Biology*, **50**, 1–4.

SHAW, P. M., ADESNIK, M., WEISS, M. C. & CORCOS, L. (1993) The phenobarbital-induced transcriptional activation of cytochrome P450 genes is blocked by the glucocorticoid progesterone antagonist RU486. *Molecular Pharmacology*, **44**, 775–783.

SHERRATT, A. J., BANET, D. E. & PROUGH, R. A. (1990) Glucocorticoid regulation of polycyclic aromatic hydrocarbon induction of cytochrome P450IA1, glutathione S-transferases and NAD(P)H : quinone oxidoreductase in cultured fetal rat hepatocytes. *Molecular Pharmacology*, **37**, 198–205.

SHOENFELD, Y., GUREWICH, Y., GALLANT, L. A. & PINKHAS, J. (1981) Prednisone-induced leukocytosis: influence of dosage, method and duration of administration on the degree of leukocytosis. *American Journal of Medicine*, **71**, 773–780.

SILVER, G., REID, L. M. & KRAUTER, K. S. (1990) Dexamethasone mediated regulation of 3-methylcholanthrene induced cytochrome P450D mRNA accumulation in primary rat hepatocyte cultures. *Journal of Biological Chemistry*, **265**, 3134–3138.

SIMONS, S. S., PUMPHREY, J. G., RUDIKOFF, S. & EISEN, H. J. (1987) Identification of cycteine-656 as the aminoacid of hepatoma tissue culture cell glucocorticoid receptors that is covalently labeled by dexamethasone 21-mesylate. *Journal of Biological Chemistry*, **262**, 9676–9680.

SIMPSON, E. R. & WATERMAN, M. R. (1983) Regulation by ACTH of steroid hormone biosynthesis in the adrenal cortex. *Canadian Journal of Biochemistry and Cell Biology*, **61**, 692–707.

SISTARE, F. D. & HAYNES, R. C. (1985) Acute stimulation of gluconeogenesis from lactate/pyruvate in isolated hepatocytes from adrenalectomized rats. *Journal of Biological Chemistry*, **260**, 12754–12760.

SIVO, J., POLITIS, A. D. & VOGEL, S. N. (1993) Differential effects of interferon-γ and glucocorticoids on FcγR gene expression in murine macrophages. *Journal of Leukocyte Biology*, **54**, 451–457.

SMITH, T. J. (1984) Dexamethasone regulation of glycoaminoglycan synthesis in cultured human skin fibroblasts. *Journal of Clinical Investigation*, **74**, 2157–2164.

SPINDLER, K. D., KANUMA, K. & GROSSMAN, D. (1991) Uptake of corticosterone into isolated rat liver cells: possible involvement of Na^+/K^+-ATPase. *Journal of Steroid Biochemistry and Molecular Biology*, **38**, 721–725.

SPORN, P. H., MURPHY, T. M. & PETERS-GOLDEN, M. (1990) Glucocorticoids fail to inhibit arachidonic acid metabolism stimulated by hydrogen peroxide in the laveolar macrophage. *Journal of Leukocyte Biology*, **48**, 81–88.

STAM, W. B., VAN OOSTERHOUT, A. J. & NIJKAMP, F. P. (1993) Pharmacologic modulation of Th-1 and Th-2 associated lymphokine production. *Life Sciences*, **53**, 1921–1934.

STARUCH, M. J. & WOOD, D. D. (1985) Reduction of serum interleukin-1-like activity after treatment with dexamethasone. *Journal of Leukocyte Biology*, **37**, 193-207.

STEIN, M. & MILLER, A. H. (1993) Stress, the hypothalamic-pituitary-adrenal axis and immune function. *Advances in Experimental Medicine and Biology*, **335**, 1–5.

STEIN-BEHRENS, B. A. & SAPOLSKY, R. M. (1992) Stress, glucocorticoids and ageing. *Ageing (Milano)*, **4**, 197–210.

STERLING, K. M., HARRIS, M. J., MITCHELL, J. J., DIPETRILLO, T. A., DELANEY, G. L. & CUTTRONEO, K. R. (1983) Dexamethasone decreases the amounts of type I procollagen mRNAs *in vivo* and in fibroblast cell cultures. *Journal of Biological Chemistry*, **258**, 7644–7649.

STRAHLE, U., KLOCK, G. & SCHUTZ, G. (1987) A DNA sequence of 15 base pairs is sufficient to mediate both glucocorticoid and progesterone induction of gene expression. *Proceedings of the National Academy of Sciences, USA*, **84**, 7871–7875.

SZABO, C., THIEMERMANN, C. & VANE, J. R. (1993) Inhibition of the production of nitric oxide and vasodilator prostaglandins attenuates the cardiovascular response to bacterial endotoxin in adrenolectomised rats. *Proceedings of the Royal Society of London (Series B: Biological Sciences)*, **253**, 233–238.

SZABO, C., THIEMERMANN, C., WU, C. C., PERETTI, M. & VANE, J. R. (1994) Attenuation of the induction of nitric oxide synthase by endogenous glucocorticoids accounts for endotoxin tolerance *in vivo*. *Proceedings of the National Academy of Sciences, USA*, **91**, 271–275.

SZE, P. Y. & IQBAL, Z. (1994a) Glucocorticoid actions on synaptic plasma membranes: modulation of calmodulin binding. *Journal of Steroid Biochemistry and Molecular Biology*, **48**, 179–186.

SZE, P. Y. & IQBAL, Z. (1994b) Glucocorticoid action on depolarisation-dependent calcium influx in brain synaptosomes. *Neuroendocrinology*, **59**, 457–465.

TAI, P. K. K., MAEDA, Y., NAKAO, K., WAKIM, N. G., DUHRING, J. L. & FABER, L. E. (1986) A 59 kDa protein associated with progestin, estrogen, androgen and glucocorticoid receptors. *Biochemistry*, **25**, 5269–5275.

THAKORE, J. H. & DINAN, T. G. (1994) Growth hormone secretion: the role of glucocorticoids. *Life Sciences*, **55**, 1083–1099.

THOMPSON, J. (1993) Role of glucocorticoids in the treatment of infectious diseases. *European Journal of Clinical Microbiology and Infectious Diseases*, **12**(Suppl. 1), S68–S72.

TSAI, S. Y., CARLSTEDT-DUKE, J., WEIGEL, N. L., DAHLMAN, K., GUSTAFSSON, J. A., TSAI, M. & O'MALLEY, B. W. (1988) Molecular interactions of steroid hormone receptor with its enhancer element: evidence for receptor dimer formation. *Cell*, **55**, 361–389.

TSAI, S. Y., SRINIVASAN, G., ALLAN, G. F., THOMPSON, E. B., O'MALLEY, B. W. & TSAI, M. J. (1990) Recombinant human glucocorticoid receptor induces transcription of hormone response genes *in vitro*. *Journal of Biological Chemistry*, **265**, 17055–17061.

TURNER, C. R. & SPANNHAKE, E. W. (1990) Acute topical steroid administration blocks mast increase and late asthmatic response of the canine peripheral airways. *American Review of Respiratory Disease*, **141**, 421–427.

UMESONO, K. & EVANS, R. M. (1989) Determinants of target gene specificity for steroid/thyroid hormone receptors. *Cell*, **57**, 1139–1146.

UNGER, A. L., UPPALURI, R., AHERN, S., COLBY, J. L. & TYMOCZKO, J. L. (1988) Isolation of RNA from unactivated rat liver glucocorticoid receptor. *Molecular Endocrinology*, **2**, 952–958.

VAMVAKOPOULOS, N. C. (1993) Tissue specific expression of heat shock proteins 70 and 90: potential implication for differential sensitivity of tissues to glucocorticoids. *Molecular and Cellular Endocrinology*, **98**, 49–54.

VAMVAKOPOULOS, N. C., MAYOL, V., MARGIORIS, A. N. & CHROUSOS, G. P. (1992) Lack of dexamethasone modulation of mRNAs involved in the glucocorticoid signal transduction pathway in two cell systems. *Steroids*, **57**, 282–287.

VEDECKIS, W. V. (1983) Subunit dissociation as a possible mechanism of glucocorticoid receptor activation. *Biochemistry*, **22**, 1983–1989.

VIARD, I., HALL, S. H., JAILLARD, C., BERTHELON, M. C. & SAEZ, J. M. (1992) Regulation of c-fos, c-jun and jun-B mRNA by angiotensin II and corticotrophin in ovine and bovine adrenocortical cells. *Endocrinology*, **130**, 1193–1200.

VIRGIN, C. E., HA, T. P., PACKAN, D. R., TOMBAUGH, G. C., YANG, S. H., HORNER, H. C. & SAPOLSKY, R. M. (1991) Glucocorticoids inhibit glucose transport and glutamate uptake in hippocampal astrocytes: implications for glucocorticoid neurotoxicity. *Journal of Neurochemistry*, **57**, 1422–1428.

VISCARDI, R. M. & MAX, S. R. (1993) Unsaturated fatty-acid modulation of glucocorticoid receptor binding in L2 cells. *Steroids*, **58**, 357–361.

DE WAAL, R. M. (1994) The anti-inflammatory activity of glucocorticoids. *Molecular Biology Reports*, **19**, 81–88.

WAHL, S. M., ALTMANN, L. C. & ROSENSTREICH, D. L. (1975) Inhibition of *in vitro* lymphokine synthesis by glucocorticoids. *Journal of Immunology*, **115**, 476–481.

WALKER, B. R., YAU, J. L., BRETT, L. P., SECKL, J. R., MONDER, C., WILLIAMS, B. C. & EDWARDS, C. R. (1991) 11β-Hydroxysteroid dehydrogenase in vascular smooth muscle and heart: implications for cardiovascular responses to glucocorticoids. *Endocrinology*, **129**, 3305–3312.

WALLACE, J. L. & WHITTLE, B. J. (1988) Gastro-intestinal damage induced by platelet-activating factor. Inhibition by the corticoid, dexamethasone. *Digestive Diseases and Sciences*, **33**, 225–232.

WALLEN, N., KITA, H., WEILER, D. & GLEICH, G. J. (1991) Glucocorticoids inhibit cytokine mediated eosinophil survival. *Journal of Immunology*, **147**, 3490–3495.

WANG, Z. M., YASUI, M. & CELSI, G. (1994) Glucocorticoids regulate the transcription of Na^+-K^+-ATPase genes in the infant kidney. *American Journal of Physiology*, **267**, C450–C455.

WATANOBE, H. (1990) The immunostaining for the hypothalamic vasoactive intestinal peptide, but not for beta-endorphin, dynorphin-A or methionine-enkephalin, is affected by the glucocorticoid milieu in the rat: correlation with prolactin secretion. *Regulatory Peptides*, **28**, 301–311.

WATERMAN, M. R. & SIMPSON, E. R. (1985) Regulation of the biosynthesis of cytochromes P450 involved in steroid hormone synthesis. *Molecular and Cellular Endocrinology*, **39**, 81–89.

WAXMAN, D. J., MORRISSEY, J. J., NAIK, S. & JAUREGUI, H. O. (1990) Phenobarbital induction of cytochromes P450. High level long term responsiveness of primary rat hepatocyte cultures to drug induction and glucocorticoid dependence of the phenobarbital response. *Biochemical Journal*, **271**, 113–119.

WEINER, F. R., CZAJA, M. J., JEFFERSON, D. M., GIAMBRONE, M. A., TUR-KASPA, R., REID, L. & ZERN, M. A. (1987) The effects of dexamethasone on *in vitro* collagen gene expression. *Journal of Biological Chemistry*, **262**, 6955–6958.

WHITWORTH, J. A. (1994) Studies on the mechanisms of glucocorticoid hypertension in humans. *Blood Pressure*, **3**, 24–32.

WHITWORTH, J. A., GORDON, D., ANDREWS, J. & SCOGGINS, B. A. (1989) The hypertensive effects of synthetic glucocorticoids in man: role of sodium and volume. *Clinical and Experimental Pharmacology and Physiology*, **13**, 353–358.

WICK, G., HU, Y., SCHWARTZ, S. & KROEMER, G. (1993) Immunoendocrine communication via the hypothalamo-pituitary-adrenal axis in autoimmune diseases. *Endocrine Reviews*, **14**, 539–563.

WIRA, C. R. & ROSSELL, R. M. (1991) Glucocorticoid regulation of the humoral immune system: dexamethasone stimulation of secretory component in serum, saliva and bile. *Endocrinology*, **128**, 835–842.

WISE, J. K., HENDLER, R. & FELIG, P. (1973) Influence of glucocorticoids on glucagon secretion and plasma aminoacids. *Journal of Clinical Investigation*, **52**, 2774–2782.

WRIGHT, A. P. H., ZILLIACUS, J., MCEWAN, I. J., DAHLMAN-WRIGHT, K., ALMLOF, T., CARLSTEDT-DUKE, J. & GUSTAFSSON, J.-A. (1993) Structure and function of the glucocorticoid receptor. *Journal of Steroid Biochemistry and Molecular Biology*, **47**, 11–19.

WU, C. Y., FARGEAS, C., NAKAJIMA, T. & DELESPESSE, G. (1991) Glucocorticoids suppress the production of interleukin-4 by human lymphocytes. *European Journal of Immunology*, **21**, 2645–2647.

XU, Z. X., SMART, D. A. & ROONEY, S. A. (1990) Glucocorticoid induction of fatty-acid synthase mediates the stimulatory effect of the hormone on choline-phosphate cytidylly-transferase activity in fetal rat lung. *Biochemica et Biophysica Acta*, **1044**, 70–76.

YAMAGATA, K., ANDREASSON, K. I., KAUFMAN, W. E., BARNES, C. A. & WORLEY, P. F. (1993) Expression of a mitogen-inducible cyclooxygenase in brain neurons: regulation by synaptic activity and glucocorticoids. *Neuron*, **11**, 371–386.

YAMAMOTO, K. R. (1985) Steroid receptor regulated transcription of specific genes and gene networks. *Annual Review of Genetics*, **19**, 209–252.

YASUNARI, K., KOHNO, M., YOKOKAWA, K., HORIO, T. & TAKEDA, T. (1994) Dopamine DA1 receptors on vascular smooth muscle cells are regulated by glucocorticoid and sodium chloride. *American Journal of Physiology*, **267**, R628–R634.

ZACH, T. L., HILL, L. D., HERRMAN, V. A., LEUSCHEN, M. P. & HOSTETTER, M. K. (1992) Effect of glucocorticoids on C3 gene expression by the A549 human pulmonary epithelial cell line. *Journal of Immunology*, **148**, 3964–3969.

ZACH, T. L., HERRMAN, V. A., HILL, L. D. & LEUSCHEN, M. P. (1993) Effect of steroids on the synthesis of complement C3 in a human alveolar epithelial cell line. *Experimental Lung Research*, **19**, 603–616.

ZARET, K. S. & YAMAMOTO, K. R. (1984) Reversible and persistant changes in chromatin structure accompany activation of a glucocorticoid dependent enhancer element. *Cell*, **38**, 29–38.

ZAMIR, O., HASSELGREEN, P. O., JAMES, H., HIGASHIGUCHI, T. & FISCHER, J. E. (1993) Effect of tumor necrosis factor or interleukin-1 on muscle amino acid uptake and the role of glucocorticoids. *Surgery, Gynecology and Obstetrics*, **177**, 27–32.

SECTION THREE

The Adrenal Gland as a Target Organ

5

The Adrenal Cortex as a Toxicological Target Organ

HOWARD D. COLBY

Albany College of Pharmacy, New York

5.1 Introduction

The adrenal cortex is a complex and multifunctional endocrine organ that secretes a variety of steroid hormones. The complexity of the gland may be a contributing factor in its vulnerability to toxicants. Among the endocrine organs, the adrenal cortex is the one most commonly affected by toxic substances (Ribelin 1984). A wide variety of chemicals have been found to cause structural and/or functional changes in the gland. Some of the reasons for this relatively high frequency of toxicant-induced adrenal lesions are discussed later in this chapter.

The literature dealing with chemical-induced toxicities of the adrenal cortex is extensive and diverse. Several review articles in recent years have addressed various aspects of this topic (Colby 1988, Szabo & Lippe 1989, Hallberg 1990, Colby & Longhurst 1992). Most investigations on adrenal toxicants have focused on either the morphologic or functional consequences of the substance studied, but rarely both. As a result, for relatively few substances are structure–function relationships well defined. Reviews by Ribelin (1984) and by Szabo & Lippe (1989) contain excellent descriptions of the morphologic lesions caused by a number of adrenal toxicants. In this chapter, emphasis is placed on the functional aspects of adrenocortical toxicology with respect to both the actions of toxicants and the resulting impact on the gland. Some of the general characteristics of adrenal toxicants are described first, with the actions of specific toxicants and the toxicological processes involved then provided as examples. Where possible, mechanisms of action are indicated, but for most adrenal toxicants the information available in the literature remains largely of a descriptive nature.

The consequences of toxicant interactions with any target organ, including the adrenal cortex, are inextricably linked to the functional role of the organ. Accordingly, an understanding of adrenocortical toxicology requires some knowledge of the actions of adrenal hormones and of the biochemical mechanisms involved in their biosynthesis. The steroid hormones that are secreted by the adrenal cortex have far-reaching effects in the body and participate in the regulation of many important physiological processes (Colby, 1987). Glucocorticoid actions include the

Table 5.1 Physiological effects of adrenal corticosteroids

Hormone type	Physiological effects
Glucocorticoid (cortisol)	Stimulates gluconeogenesis Increases hepatic glycogen Increases blood glucose Enhances lipolysis Promotes protein degradation
Mineralocorticoid (aldosterone)	Stimulates Na^+ reabsorption Stimulates K^+ excretion Stimulates H^+ excretion

modulation of carbohydrate, lipid and protein metabolism, and mineralocorticoids have significant roles in electrolyte and blood pressure homeostasis (Table 5.1). Adrenal androgens, although less potent than the testicular hormone, testosterone, have similar effects and are secreted in large amounts. As a result of the diversity of actions of adrenal hormones, disruption of normal adrenocortical function, whether caused by disease or toxicant-induced injury, can elicit a variety of functional changes.

5.2 Sites of Action of Adrenal Toxicants

The regulation of adrenocortical hormone secretion involves a number of extra-adrenal sites in addition to essential intra-adrenal processes. The hypothalamic-anterior pituitary axis has a central role in the control of glucocorticoid and androgen output (Figure 5.1) and the renin–angiotensin system is of primary importance in the regulation of mineralocorticoid secretion. As a result of the multiple levels of control of adrenocortical function, there are many potential sites of action

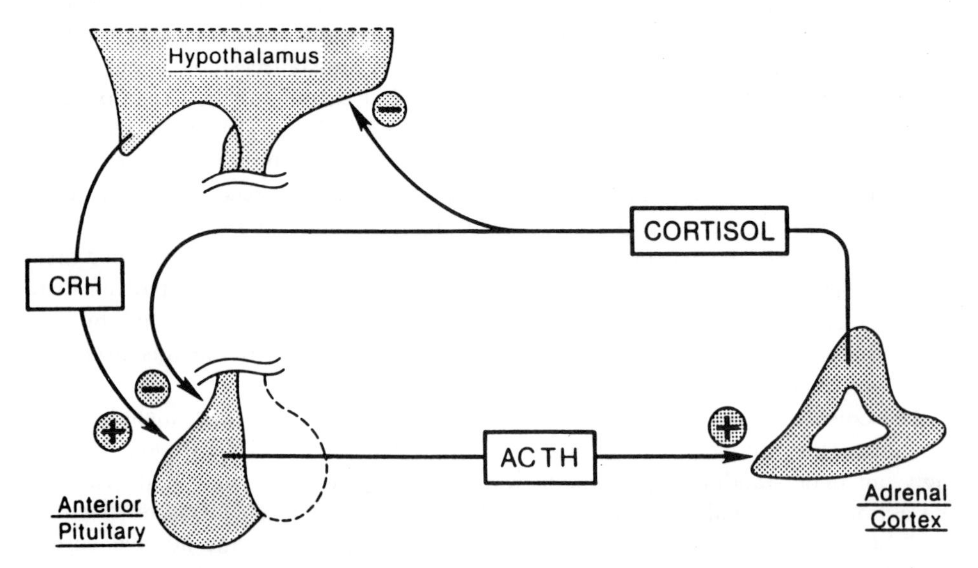

Figure 5.1 Regulation of cortisol secretion by the hypothalamo-pituitary axis. +, Stimulation; −, inhibition. From Colby (1987)

for toxicants that can effect disruption of adrenal homeostasis. For example, changes in ACTH secretion that result from chemical actions on the brain or pituitary gland will alter adrenal glucocorticoid and androgen secretion. Similarly, substances affecting any of the components of the renin–angiotensin system may perturb mineralocorticoid production. In addition, toxicants may influence the activity of adrenal negative feedback control loops by effects on hepatic metabolism of steroids or on the binding of steroid hormones to plasma proteins, thereby causing changes in adrenocortical secretion. Such alterations in adrenal function constitute secondary responses to toxicant actions at extra-adrenal sites; chemicals producing these effects may be considered indirect adrenal toxicants. There are also many chemicals that adversely affect the adrenal cortex by direct actions on the gland, and thus are primary or direct-acting adrenal toxicants. It is the latter group of substances that this chapter focuses on.

Within the adrenal cortex, there are a number of processes that are essential for the synthesis and secretion of steroid hormones (Figure 5.2). These include, but are not limited to, the following: interaction of adrenal regulatory factors (ACTH, angiotensin II) with target cell membrane receptors; synthesis of cyclic nucleotides in adrenocortical cells; activation or inhibition of protein kinases by second messenger systems; synthesis of steroidogenic enzymes; cellular uptake of plasma lipoproteins and intracellular transport of cholesterol for use in steroidogenesis (Colby 1987). Interference with any of these processes would be expected to adversely impact adrenocortical function. Although not all of these events have been definitively identified as sites of action of adrenal toxicants, it should be noted that the mechanisms of action of many adrenal toxicants have yet to be determined. Future studies may reveal mechanisms and target sites beyond those already described.

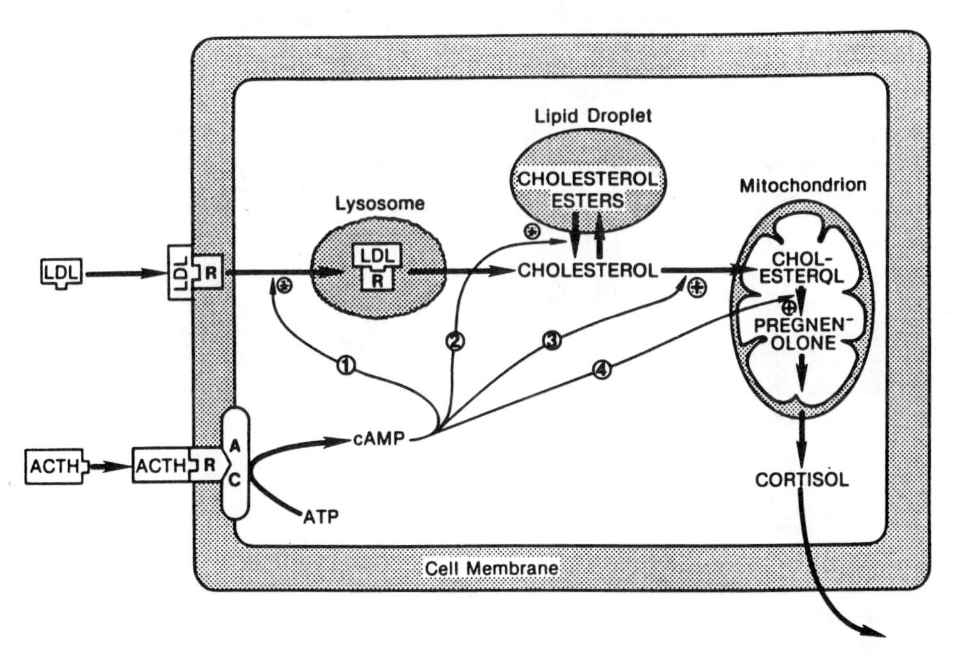

Figure 5.2 Mechanism of action of ACTH on adrenocortical steroidogenesis. The numbered arrows indicate the steps involved in cholesterol mobilization that are stimulated by ACTH. LDL, Low density lipoprotein; R, receptor; +, stimulation. From Colby (1987)

The most commonly described mechanism of action of adrenocortical toxicants involves interactions with one or more steroidogenic enzymes. The synthesis of adrenocortical hormones occurs by the sequential enzymatic modification of the precursor molecule, cholesterol (Figure 5.3). Because of its critical role in steroidogenesis, it should not be surprising that large amounts of cholesterol are stored by adrenocortical cells, principally as cholesterol esters. Many of the enzymatic reactions required for the conversion of cholesterol to the adrenal hormones are site-specific hydroxylations on the steroid nucleus that are catalysed by distinct cytochrome P450 isozymes (Hall 1986, Miller 1988). Some of these P450 isozymes are localized to mitochondrial membranes and others to the endoplasmic reticulum (ER) of adrenal cells, requiring the intracellular migration of precursor steroids in the course of hormone synthesis. As discussed later, the P450-catalysed steroid hydroxylases are of major importance in adrenal toxicology, not only as targets for some adrenal toxicants, but as the sources of oxygen radicals and reactive metabolites of xenobiotics that may mediate toxicity.

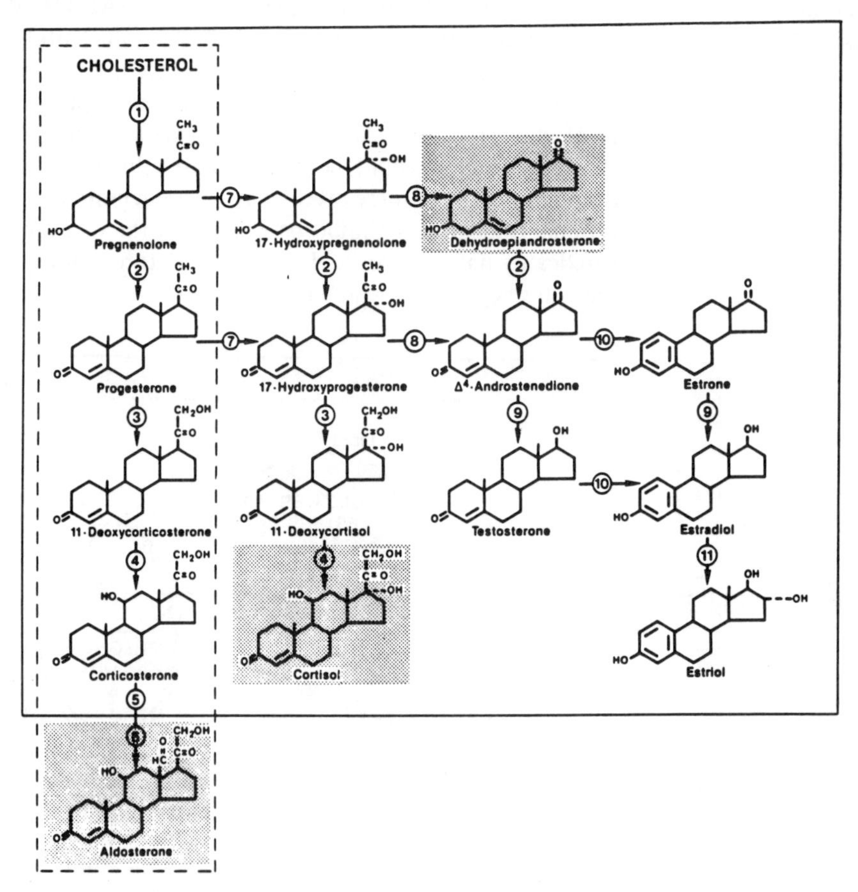

Figure 5.3 Steroidogenic pathways in the zona glomerulosa (within broken lines) and zona fasciculata-zona reticularis (within solid lines) of the human adrenal cortex. The major secretory products are shaded. Enzymes: 1, cholesterol side-chain cleavage; 2, 3β-hydroxysteroid dehydrogenase–isomerase; 3, 21-hydroxylase; 4, 11β-hydroxylase; 5, 18-hydroxylase; 6, 18-hydroxysteroid dehydrogenase; 7, 17α-hydroxylase; 8, C17–20 lyase; 9, 17-hydroxysteroid dehydrogenase; 10, aromatase; 11, 16α-hydroxylase. From Colby (1987)

The functional consequences of toxicant actions on the adrenal cortex, in addition to the anatomical regions of the gland that are affected by each toxicant, are determined in part by the metabolic capabilities of different adrenocortical cells. The profile of hormones that is synthesized by each steroid-secreting cell is dictated by the sites of expression of the different P450-catalysed steroid hydroxylases. For example, secretion of glucocorticoids and mineralocorticoids is limited to the adrenal cortex because several of the P450 isozymes required for the synthesis of these hormones are not expressed in other steroidogenic organs. Similarly, the differential expression of some P450 enzymes among the anatomical zones of the adrenal cortex results in zonal steroidogenic differences known as functional zonation. Thus, the mineralocorticoid, aldosterone, is secreted only by the cells of the zona glomerulosa, whereas the inner two zones of the gland, the zona fasciculata and zona reticularis, are the sites of glucocorticoid and androgen production. In addition to the uniqueness of the hormones produced by each zone, there are zone-specific regulatory mechanisms controlling steroid secretion (Colby 1987). It is important for the toxicologist to be aware of these zonal differences because, for example, the unique metabolic characteristics of certain adrenal cell types may at times be responsible for the formation of toxic substances, causing a cell-specific toxicity. In addition, the nature of any functional changes caused by toxicants depends upon the physiological role(s) of the adrenal zone(s) affected. As previously noted by Ribelin (1984), the zona fasciculata and zona reticularis are the target sites for adrenal toxicants far more often than is the zona glomerulosa. Some of the factors that may contribute to this differential susceptibility are discussed below.

5.3 Substances Causing Adrenal Toxicity

About 10 years ago, Ribelin (1984) reviewed the literature on chemical-induced lesions in endocrine organs, and found that the adrenal cortex was most frequently cited. He noted that many different types of substances caused morphological changes in one or more of the anatomical zones of the gland. Since then, additional reports in the literature indicate that many other chemicals may also be adrenal toxicants. This is particularly true if the definition of adrenal toxicant is taken broadly, as favoured by this author, to include all substances causing either morphologic or functional lesions in the gland. A number of adrenal toxicants have been discussed in earlier reviews (Ribelin 1984, Colby 1988, Szabo & Lippe 1989, Hallberg 1990, Colby & Longhurst 1992) and in-depth information on several of these compounds is available in those articles. The substances listed in Tables 5.2 and 5.3 include some whose actions on the adrenal have been relatively well investigated and others that are recent additions to the literature. There is obviously great diversity in the chemical characteristics of adrenal toxicants, making it difficult to classify them in accordance with any simple scheme. Table 5.2 arbitrarily includes only substances used for therapeutic purposes (drugs) and Table 5.3 contains non-therapeutic agents. It must be emphasized that the contents of Tables 5.2 and 5.3 are representative only and are not intended to be comprehensive. However, the adrenal toxicants included in Tables 5.2 and 5.3, in combination with those delineated in the earlier reviews cited above, provide some indication of the scope of compounds that can adversely affect the adrenal cortex.

Table 5.2 Therapeutic agents that cause functional and/or structural changes in the adrenal cortex

Drug	Reference
Adriamycin	Cuellar *et al.* (1984)
Chloramphenicol	Mazzocchi & Nussdorfer (1985)
Chlorphentermine	Hartmann & Jentzen (1979)
Cyclosporine-A	Mazzocchi *et al.* (1993)
Cyproterone acetate	Panesar *et al.* (1979)
DMNM	Yarrington *et al.* (1985)
Danazol	Barbieri *et al.* (1980)
Etomidate	Preziosi & Vacca (1988)
Insulin	Nestler *et al.* (1992)
Ketaconazole	Feldman (1986)
Melengestrol acetate	Robertson *et al.* (1989)
Mitotane (*o,p'*-DDD)	Hart *et al.* (1973)
Nifurtimox	DeCastro *et al.* (1990)
Phenytoin	Hirai & Ichikawa (1991)
RU-486	Albertson *et al.* (1994)
Spironolactone	Kossor *et al.* (1991)
Suramin	Ashby *et al.* (1989)
Tamoxifen	Lullman & Lullman-Rauch (1981)

All of the substances identified in Tables 5.2 and 5.3 are thought to have intra-adrenal sites of action, that is, are direct-acting adrenal toxicants. For many other toxicants, definitive conclusions about sites of action are precluded by the nature of the investigations that have been done. In some cases, only *in vivo* approaches were

Table 5.3 Non-therapeutic compounds that cause functional and/or structural changes in the adrenal cortex

Compound	Reference
Acrylonitrile	Szabo *et al.* (1980)
1-Aminobenzotriazole	Xu *et al.* (1994)
Cadmium	Nishiyama & Nakamura (1984)
Carbon tetrachloride	Brogan *et al.* (1984)
β-Carboline	Kühn-Velten (1993)
Chloroform	Hoerr (1931)
Copper	Veltman & Maines (1986)
3-Methylsulfonyl-DDE	Jönsson & Lund (1994)
7,12-Dimethylbenz(a)anthracene (DMBA)	Huggins & Morii (1961)
Gossypol	Gu & Lin (1991)
Kepone	Eroschenko & Wilson (1975)
Nicotine	Barbieri *et al.* (1987)
Nigericin	Cheng *et al.* (1993)
Polychlorinated biphenyls	Inao (1970)
PD132301-2	Vernetti *et al.* (1993)
TCDD	DiBartolomeis *et al.* (1987)
Δ9-Tetrahydrocannabinol	Warner *et al.* (1977)
Toxaphene	Mohammed *et al.* (1985)
Triaryl phosphates	Latendresse *et al.* (1993)

employed in assessing the actions of toxicants. Although *in vivo* investigations are essential for establishing the toxicity of chemicals on the adrenal cortex, it is difficult to determine the specific sites and mechanisms of action in such studies. Accordingly, *in vitro* adrenal preparations are often employed for more in-depth analyses of toxicant effects on the gland. The use of such *in vitro* techniques has contributed significantly to a fuller understanding of the mechanisms of action of adrenal toxicants and has provided relatively inexpensive and simple screening systems for substances that act directly on adrenocortical cells. However, studies that are limited to *in vitro* experimentation indicate only the potential for toxicity, which may or may not be manifested *in vivo*. There are many protective and/or compensatory factors applicable *in vivo* that cannot be predicted by *in vitro* studies alone. Thus, the optimal approach to investigating the actions of adrenal toxicants would include a combination of *in vitro* and *in vivo* approaches to establish both toxicological relevance and mechanism of action.

Where the actions of adrenal toxicants have been pursued in some depth, the results have revealed the same general mechanisms that are applicable to toxic substances in other organs. These include covalent interactions with critical cellular macromolecules, generation of highly reactive oxygen species, and interference with protein synthesis or other important metabolic processes. In fact, many adrenal toxicants have adverse effects in multiple organs, usually with the same mechanism of action at each site. It is the specialized function of the adrenal cortex (ie steroidogenesis) or unique metabolic characteristics of the gland that occasionally result in adrenal-specific lesions. For example, interactions of chemicals with certain steroidogenic enzymes might be expected to result in toxicities that are limited to the adrenal cortex. The absence or deficiency of some protective mechanisms in the adrenal could also be a contributing factor. In some cases, the reasons for adrenal-specific lesions have not yet been resolved. However, the rapidly increasing information available on the cellular and molecular events associated with the regulation and function of the adrenal cortex offers the prospect for a more thorough understanding of the actions of adrenal toxicants in the near future.

5.4 Adrenal Characteristics Contributing to Chemical-induced Toxicity

There appear to be many reasons for the relatively high frequency of chemical-induced toxicities of the adrenal cortex. Certainly, the diversity of adrenocortical functions and the associated extra-adrenal regulatory systems provide numerous target sites for toxicants, as already noted. In addition, there are a number of intra-adrenal characteristics that increase the vulnerability of the cortex to direct-acting toxicants (Table 5.4). Some of these characteristics may account for the large number of chemicals that adversely affect the adrenal cortex.

5.4.1 *Adrenal Accumulation of Xenobiotics*

Many xenobiotics accumulate at higher concentrations in the adrenal cortex than in most other organs (Pyykkö *et al.* 1977, Castracane *et al.* 1982, Darnerud & Brandt 1985, Mohammed *et al.* 1985, Brandt 1987, Lund *et al.* 1988, Town *et al.* 1993). There are several factors that probably contribute to this phenomenon. The high lipid content of the adrenal cortex promotes the deposition and accumulation of

Table 5.4 Adrenal characteristics contributing to toxicity

Accumulation of xenobiotics
high lipid content
blood supply
Bioactivation reactions
P450-derived oxygen radicals
Lipid peroxidation

lipophilic substances. Most xenobiotic chemicals that enter the body in significant quantities, including potential toxicants, must be somewhat lipophilic in order to cross cell membranes and be absorbed. Thus, the concentrations of such substances often reach higher levels in the adrenal cortex and are retained for longer periods of time than in other organs. In addition, many lipophilic substances circulate in the blood in association with lipoproteins. The extensive uptake of plasma lipoproteins by adrenocortical cells as a source of cholesterol for steroidogenesis may, therefore, promote the transport of xenobiotics into the gland.

The blood supply to the adrenal cortex may also contribute to the accumulation of toxicants in the gland. Blood flow to the adrenal is abundant, facilitating delivery of all substances, including toxicants. The adrenal receives blood from branches of the aorta, the phrenic arteries and the renal arteries. When calculated per unit weight of tissue, the adrenal has one of the highest rates of blood flow in the body. In addition, the unusual nature of the intra-adrenal vascular system may influence both the frequency and localization of toxicant-induced lesions within the cortex. Adrenocortical cells are supplied by a capillary system that drains from the outer to the inner regions of the gland before uniting in a plexus at the border with the medulla (Vinson & Kenyon, 1978, Hornsby, 1985). Thus, the blood supply to the inner zone(s) first perfuses the outer regions of the cortex. This vascular arrangement could result in steroid and/or xenobiotic metabolites that are produced by adrenal cells reaching very high concentrations in the inner regions of the cortex, thereby increasing the potential for toxicity. It is also possible that removal of toxic substances from the blood by outer zone cells would diminish the amounts that are delivered to the inner zone. In the latter case, the outer regions of the cortex would most likely be more vulnerable to toxicity. The opposite would pertain if there were extensive uptake of protective substances, such as antioxidants, in the outer regions of the cortex thereby compromising the amounts available to inner zone cells and increasing the vulnerability of the inner zones to toxicants. Thus, the nature of intra-adrenal blood flow may have a variety of effects on toxicity, depending at least partly upon the mechanism of action of the toxicant and the detoxication processes involved. Any of these effects might contribute to the zone-specific actions that are characteristic of many adrenal toxicants (Ribelin 1984).

5.4.2 *Adrenal Bioactivation Reactions*

Metabolism of exogenous substances that enter the body usually entails the enzymatic conversion of biologically active compounds, including toxicants, to products having little or no activity (Figure 5.4). The liver is the major site of such xenobiotic

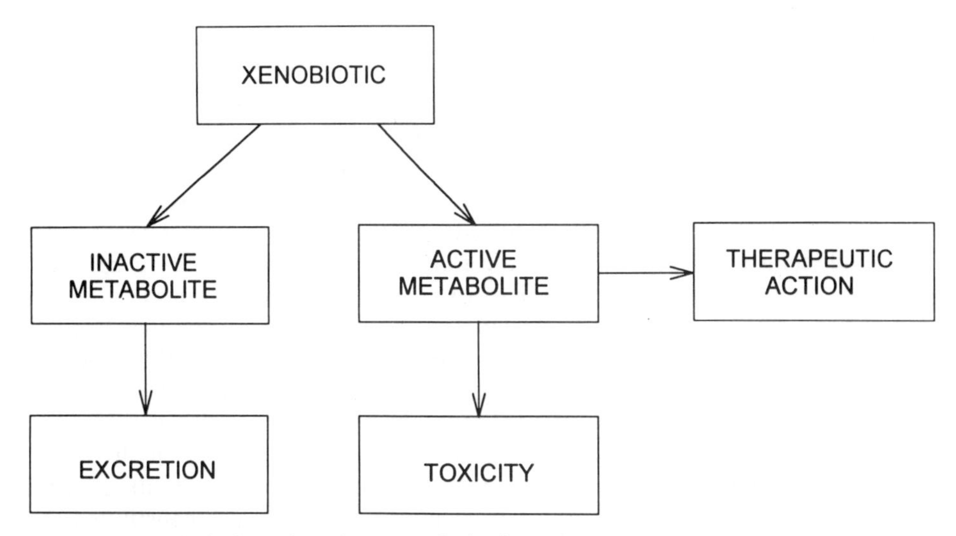

Figure 5.4 Metabolic fate of xenobiotics in the body

metabolism. However, it is now well established that many biologically inactive chemicals can be converted to active metabolites, some of which cause toxicity (Hinson *et al.* 1994) (Figure 5.4). These types of reactions take place in extra-hepatic tissues as well as in the liver. The process of active metabolite formation from inactive or less active precursors is known as bioactivation. Reactive metabolites that cause toxicity are usually produced directly in their target organs because their short half-lives and tendency to rapidly interact with cellular macromolecules preclude transport from organ to organ. Thus, the toxicity of some compounds is limited to sites where the activating enzyme(s) are expressed, accounting for some cases of organ-selective toxicities.

The enzymes most often implicated in bioactivation reactions that cause toxicity are cytochrome P450 isozymes (Hinson *et al.* 1994). Most of the P450-catalysed reactions that occur in the adrenal cortex are associated with steroid hormone synthesis (Hall 1986, Miller 1988, Simpson & Waterman 1988). However, adrenal metabolism and/or activation of xenobiotics has been demonstrated in many species and, in fact, is particularly prominent in the human foetal adrenal gland (Juchau & Pedersen 1973, Colby & Rumbaugh 1980, Colby 1988, Hallberg 1990, Colby & Longhurst 1992). The capacity for bioactivation of xenobiotics by the adrenal cortex probably contributes to the high frequency of chemical-induced toxicities of the gland. The adrenal toxicants listed in Tables 5.2 and 5.3 include some that require metabolic activation. Among those that must be activated for the expression of their toxicity are 1-aminobenzotriazole, dimethylbenz(a)anthracene (DMBA), *o,p*-DDD [1,1-dichloro-2,2-bis(4-chlorophenyl)-ethane; mitotane], carbon tetrachloride, chloroform and spironolactone. It is possible that metabolites mediate the actions of other adrenal toxicants, but for most the involvement of bioactivation has not been explored.

5.4.2.1 *Steroidogenic P450 isozymes*

Some adrenal bioactivation reactions are catalysed by steroidogenic P450 isozymes, but other isozymes have also been implicated. Among the steroidogenic isozymes, there is good evidence that activation by the P450c11 (11β-hydroxylase) accounts for the adrenotoxic effects of the DDT derivative, *o,p'*-DDD (Hart *et al.* 1973, Martz & Straw 1980). The latter would explain the selectivity of *o,p'*-DDD toxicity for the adrenal cortex, the sole site of P450c11 expression.

The most extensively investigated of the steroidogenic P450 isozymes, with respect to a role in xenobiotic metabolism, is the P450c17 (17α-hydroxylase/C17,20-lyase). This isozyme has been found to catalyse the metabolism of several xenobiotics and seems to be responsible for the conversion of the diuretic drug, spironolactone, to an adrenotoxic metabolite (Kossor *et al.* 1991). Suhara *et al.* (1984), using a purified and reconstituted enzyme preparation from pig testis, found that P450c17 catalysed N-demethylation, denitration, hydroxylation and O-dealkylation reactions with a variety of xenobiotic substrates. Similar studies were done by Mochizuki *et al.* (1988), but with an enzyme preparation that was obtained from guinea pig adrenal glands. These investigators found that their reconstituted 17α-hydroxylase/C17,20-lyase preparation catalysed aminopyrine N-demethylation, 2-nitropropane denitrification and benzo(a)pyrene hydroxylation. A similarly reconstituted 21-hydroxylase system had very little xenobiotic-activity (Mohizuki *et al.* 1988). The same investigators found that pre-incubation of guinea pig adrenal microsomes with anti-P450c17 antibody to inhibit the P450c17, decreased benzo(a)pyrene hydroxylase activity (Figure 5.5). These results suggest that at least some of the xenobiotic-metabolizing activity in guinea pig adrenal microsomes is catalysed by the P450c17. The guinea pig P450c17 has recently been cloned and sequenced by Tremblay *et al.* (1994) and by us (Figure 5.6), making it now possible through expression experiments to unequivocally determine the catalytic capabilities of this P450 isozyme.

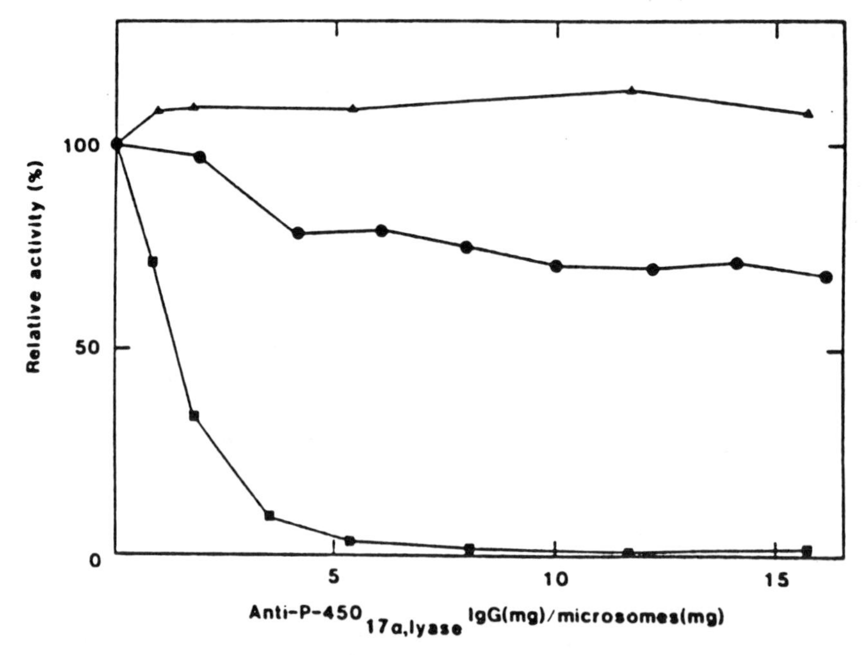

Figure 5.5 Effects of anti-P450c17 (17α, lyase) IgG on steroid and benzo(a)pyrene hydroxylase activities of guinea pig adrenal microsomes. ▲, 21-Hydroxylase; ●, benzo(a)pyrene hydroxylase; ■, 17α-hydroxylase. From Mochizuki *et al.* (1988)

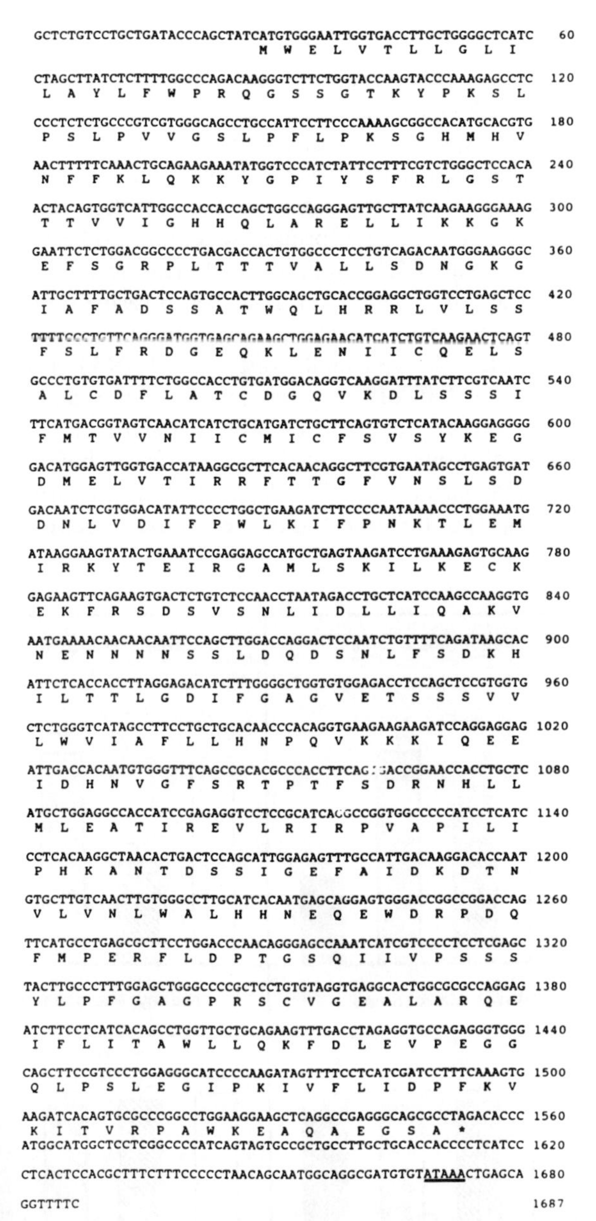

Figure 5.6 Nucleotide and deduced amino acid sequences of guinea pig P450c17 cDNA. The coding region stops at the position indicated by the asterisk. The one-letter amino acid code is used in the second position of each codon. The polyadenylation signal is underlined

One of the adrenal toxicants that is known to be activated by the P450c17 is the diuretic agent, spironolactone (Saunders & Alberti 1978). A number of investigations revealed that spironolactone administration to experimental animals inhibited adrenal steroidogenesis, apparently by decreasing cytochrome P450 concentrations and steroid hydroxylase activities in the gland (Menard *et al.* 1975, 1979a,b, Greiner *et al.* 1976, 1978). Clinical studies similarly indicated abnormal adrenocortical function in some patients treated with spironolactone (Sundsfjord *et al.* 1974, Abshagen *et al.* 1977, Tuck *et al.* 1981). The changes in adrenal steroid secretion in patients receiving spironolactone were consistent with inhibition of some P450-catalysed steroid hydroxylases (Tuck *et al.* 1981).

The results of *in vitro* experiments indicated that spironolactone was activated by adrenal microsomes and that a reactive metabolite was responsible for the P450 degradation and inhibition of steroidogenesis that resulted (Greiner *et al.* 1978, Menard *et al.* 1979a,b). The activation process was found to be NADPH-dependent and blocked by cytochrome P450 inhibitors, implicating a P450 isozyme in the pathway leading to the toxic metabolite. Further investigations demonstrated a close relationship between 17α-hydroxylase activity and the capacity for spironolactone activation. For example, spironolactone caused P450 degradation only in organs that expressed the P450c17 such as adrenal cortex and testis, but not in non-expressing organs like liver and kidney (Figure 5.7). In addition, degradation of adrenal P450 by spironolactone was limited to those species having 17α-hydroxylase activity in the cortex (Menard *et al.* 1975, 1979a,b). Thus, in cortisol producers like guinea pigs and dogs, spironolactone inhibited adrenal steroidogenesis, but in

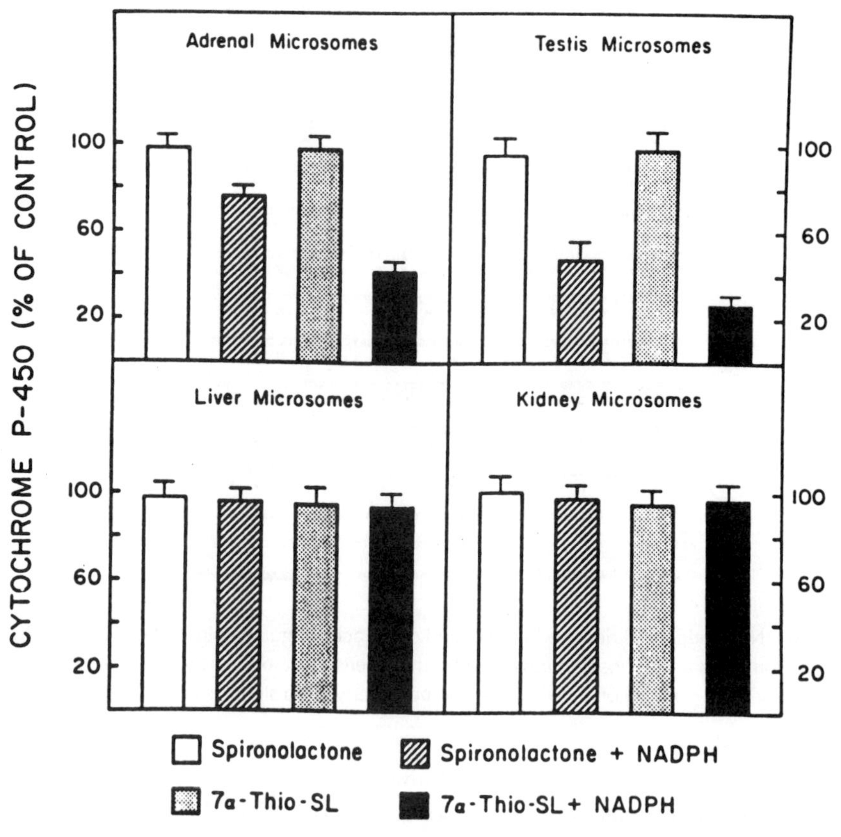

Figure 5.7 Effects of pre-incubating adrenal, testicular, hepatic or renal microsomes with spironolactone or 7α-thio-SL in the presence or absence of NADPH on cytochrome P450 concentrations. Values are expressed as a percentage of the corresponding control values (microsomes pre-incubated without drug or NADPH). Each value is the mean ± SE of 4–6 experiments. From Sherry *et al.* (1986)

corticosterone-producing species like rats and mice that do not express the P450c17 in adrenal glands, steroidogenesis was not affected by spironolactone. These and other observations indirectly linked the P450c17 to the activation and toxicity of spironolactone.

Studies by Sherry and co-workers (1986) demonstrated that the degradation of adrenal P450 by spironolactone required at least a two-step activation process. These investigations found that spironolactone was converted to its deacetylated metabolite, 7α-thiospirolactone (7α-thio-SL), by non-specific esterases prior to the P450-mediated activation step (Figure 5.8). In the presence of NADPH, the 7α-thio-SL was then further metabolized to cause the degradation of P450. Inhibition of 7α-thio-SL production by esterase antagonists such as diethyl-*p*-nitrophenyl phosphate (DPNP) completely prevented the NADPH-dependent degradation of P450 by spironolactone (Sherry *et al.* 1986). In contrast, DPNP had no effect on the loss of P450 caused by 7α-thio-SL. Thus, 7α-thio-SL was established as an essential intermediate in the activation pathway for spironolactone.

Definitive evidence for the involvement of the P450c17 in spironolactone-mediated P450 degradation was provided by the observations of Kossor and co-workers (1991). They found that pre-incubation of guinea pig adrenal microsomes with anti-P450c17 antibody to inhibit the 17α-hydroxylase prevented the subsequent degradation of microsomal P450 by spironolactone and by 7α-thio-SL (Figure 5.9). In contrast, inhibition of the 21-hydroxylase (P450c21) in the same microsomal preparations had no effect on P450 degradation by either compound. In addition, when a purified and reconstituted 17α-hydroxylase system was incubated with 7α-thio-SL plus NADPH there was a rapid loss of enzyme activity and of spectrally detectable P450 (Kossor *et al.* 1991). Under the same conditions, spironolactone had no effect on the P450c17, presumably because the purified enzyme preparation had no esterase activity, precluding the formation of 7α-thio-SL. Thus, these investigations clearly established that 7α-thio-SL was an obligatory intermediate in the actions of spironolactone on adrenal steroidogenesis and that the P450c17 catalyses the activation of 7α-thio-SL. The identity of the toxic metabolite of 7α-thio-SL remains unresolved, although the involvement of a highly reactive sulfenic acid derivative has been proposed (Kossor *et al.* 1991).

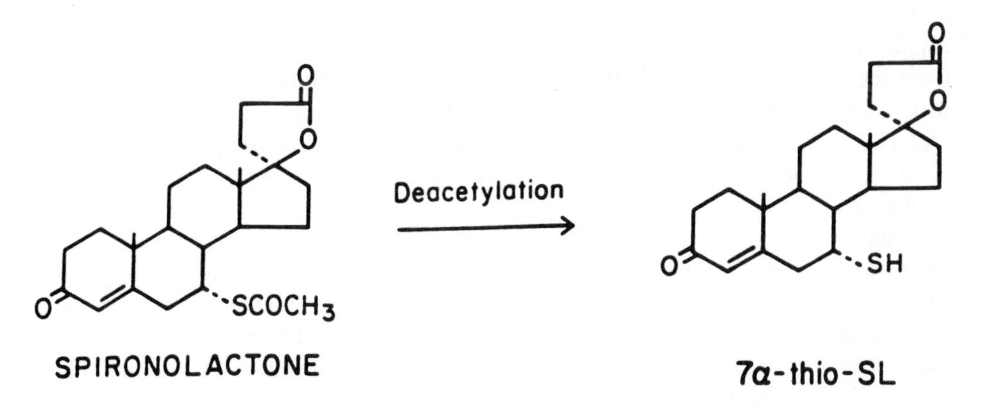

Figure 5.8 Chemical structures of spironolactone and its deacetylated metabolite, 7α-thio-SL

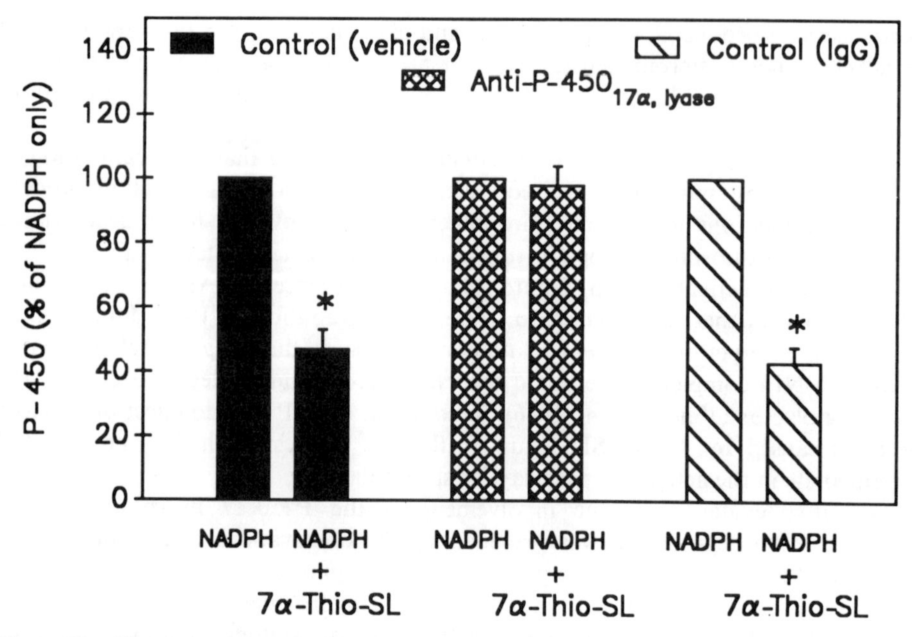

Figure 5.9 Effects of anti-P450c17 (17α, lyase) and control IgG on the 7α-thio-SL-mediated degradation of P450 in adrenal microsomes. Microsomal suspensions were pre-incubated with or without anti-P450c17 IgG or control IgG and then incubated with an NADPH-generating system in the presence or absence of 7α-thio-SL. * $p < 0.05$ versus corresponding NADPH-only value. From Kossor *et al.* (1991)

5.4.2.2 *Non-steroidogenic adrenal P450 isozymes*

Most adrenal xenobiotic metabolism seems to be catalysed by P450 isozymes that are distinct from those involved in steroidogenesis. Few of these enzymes have yet to be well characterized, but high activities have been demonstrated for some (Colby & Rumbaugh 1980). For example, it is widely recognized from the work of Juchau and co-workers (Juchau & Pedersen 1973) that the human foetal adrenal gland can metabolize a variety of foreign substances. Adrenal metabolism during foetal development may be of particular significance because of the low level of xenobiotic metabolism by the foetal liver. Many xenobiotic substrates are metabolized more rapidly by microsomal preparations from human foetal adrenal glands than those from foetal livers (Juchau & Pedersen 1973, Colby & Rumbaugh 1980). However, little is known about the P450 isozyme(s) involved in xenobiotic metabolism by the foetal adrenal gland. It is also not known if metabolism of foreign compounds to reactive metabolites by the foetal adrenal is a factor in the sensitivity of the foetus to toxicants.

It has been known for more than 30 years that administration of 7,12-dimethylbenz(a)anthracene (DMBA) to rats causes necrosis of the adrenal cortex (Huggins & Morii 1961). A number of experimental observations led some investigators to suspect that the adrenocorticolytic effect of DMBA was the result of adrenal production of a reactive metabolite. For example, certain inhibitors of P450-mediated xenobiotic metabolism prevented or diminished the severity of DMBA-induced adrenocortical necrosis. Some of the other evidence supporting this hypothesis has been discussed in depth by Hallberg (1990). The results of numerous experiments suggested that adrenal DMBA metabolism was catalysed by a unique P450 isozyme having a number of characteristics unlike those of known adrenal enzymes. For example, DMBA-induced adrenal necrosis occurs only in the rat and not in other common laboratory animals like mice, hamsters and guinea pigs, suggesting that expression of the activating enzyme is species-specific. An apparent ACTH requirement for the adrenal toxicity of DMBA further indicated that the activation process might be hormonally modulated (Hallberg 1990). Guenther and co-workers (1979) found that the adrenal toxicity of DMBA was highly correlated with the concentration of an apparent ACTH-inducible but unknown P450 isozyme in rat adrenal microsomes, and with the rate of adrenal microsomal metabolism of polycyclic aromatic hydrocarbons.

Recent findings by Jefcoate and co-workers may provide new insight into the mechanism of DMBA-induced adrenal toxicity (Otto *et al.* 1991). These investigators have partially purified a novel P450 (P450RAP) isozyme from rat adrenal microsomes that catalyses DMBA metabolism with a regioselectivity different than that for the major hepatic P450 isoforms. This newly identified P450 is immunochemically distinct from the adrenal steroid hydroxylases, is ACTH-dependent, and could not be detected in mouse or guinea pig adrenal microsomes. Taken together these observations strongly suggest a role for P450RAP in the conversion of DMBA to a reactive adrenal metabolite. It may now be possible to establish the identity of the toxic species produced and to more clearly define the mechanism(s) responsible for the resulting adrenal necrosis.

The guinea pig adrenal cortex also contains an apparently unique microsomal P450 isozyme that is associated with xenobiotic metabolism and bioactivation reactions. In fact, the rates of metabolism of many foreign compounds by guinea pig adrenal microsomes far exceed those by the corresponding hepatic preparations (Kupfer & Orrenius 1970, Colby & Rumbaugh, 1980). Comparison of the patterns of xenobiotic metabolism by adrenal and hepatic microsomal preparations suggested catalysis by different P450 isozymes in the two organs. For example, guinea pig adrenal and liver microsomes produced markedly dissimilar profiles of benzo(a)pyrene metabolites (Table 5.5), indicating the involvement of different enzymes (Colby *et al.* 1982a). Differences between liver and adrenal in the metabolism of chlorinated biphenyls have also been reported (Eacho *et al.* 1984). In addition, investigations on the physiological factors affecting adrenal xenobiotic metabolism clearly indicated that regulation was independent of both hepatic xenobiotic metabolism and adrenal steroid metabolism. For example, adrenal xenobiotic-metabolizing activities increase with ageing and are greater in male than female guinea pigs (Figure 5.10), but microsomal 17α-hydroxylase and 21-hydroxylase activities do not change with age and are not sex-dependent (Pitrolo *et al.* 1979, Colby *et al.* 1980, Martin & Black 1983). Similarly, neither age nor gender has any effect on the rates of hepatic xenobiotic metabolism in guinea pigs (Figure 5.10). In addition, ACTH treatment decreases the capacity for adrenal metabolism of foreign compounds, but increases 17α-hydroxylase activity in guinea pig adrenal microsomes (Figure 5.11) (Colby *et al.* 1982b, Black *et al.* 1989b, Colby *et al.* 1992). Hepatic metabolism of xenobiotics, in contrast, is unaffected by ACTH. Observations such as these led investigators to conclude that adrenal xenobiotic metabolism was catalysed by a different P450 isozyme than those involved in adrenal steroidogenesis or in hepatic metabolism of xenobiotics.

Table 5.5 Metabolism of benzo(a)pyrene (BP) by guinea pig adrenal and hepatic microsomes[a]

	Adrenal metabolism		Hepatic metabolism	
Metabolite	Rate (pmoles/min × mg protein)	Percentage of total	Rate (pmoles/min × mg protein)	Percentage of total
3-Hydroxy-BP	478 ± 58	64 ± 7	125 ± 19[b]	34 ± 5[b]
9-Hydroxy-BP	42 ± 5	6 ± 1	31 ± 4	9 ± 2
Quinones	29 ± 4	4 ± 1	17 ± 3[b]	5 ± 1
BP-7,8-diol	61 ± 7	8 ± 2	13 ± 3[b]	4 ± 2
BP-4,5-diol	14 ± 3	2 ± 1	139 ± 18[b]	39 ± 5[b]
BP-9,10-diol	83 ± 7	10 ± 2	14 ± 3[b]	4 ± 2[b]

[a] Values are means ± SE; 6–8 incubations per value.
[b] $p < 0.05$ (versus corresponding adrenal value).
From Colby *et al.* (1982a).

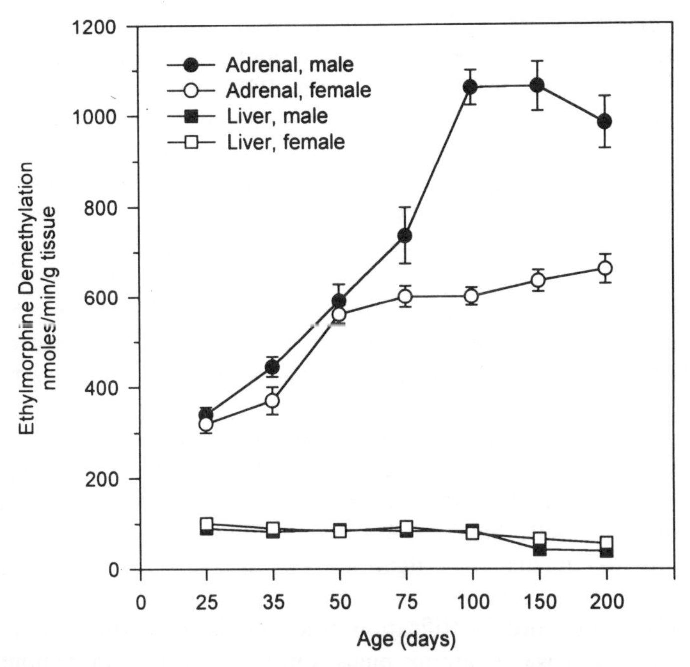

Figure 5.10 Maturational changes in adrenal and hepatic ethylmorphine demethylase activities in male and female guinea pigs. Based upon Pitrolo *et al.* (1979)

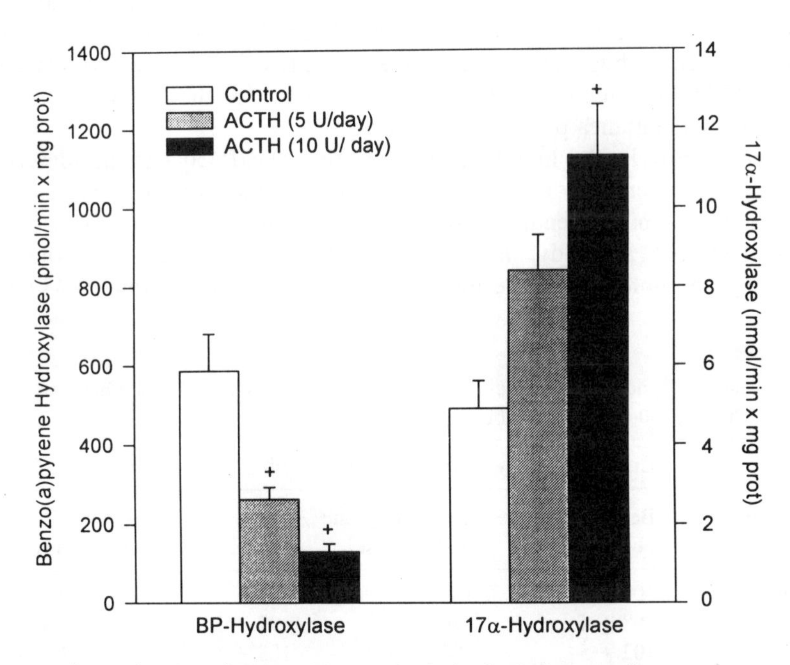

Figure 5.11 Effects of ACTH administration to guinea pigs for 7 days on microsomal benzo(a)pyrene (BP) hydroxylase and 17α-hydroxylase activities in the inner zone of the adrenal cortex. * $p < 0.05$ (versus corresponding control value). Based upon Colby *et al.* (1992)

Recent studies with the P450 suicide substrate, 1-aminobenzotriazole (ABT), further demonstrated a divergence of steroid and xenobiotic metabolism in the guinea pig adrenal cortex with respect to the P450 isozymes involved (Xu *et al.* 1994). Prior investigations with ABT in other organs indicated that this compound was a non-selective inhibitor of cytochromes P450. ABT undergoes a P450-catalysed oxidation, resulting in the formation of benzyne which alkylates the prosthetic heme group of cytochromes P450, causing irreversible loss of enzyme activity (see Murray & Reidy 1990). When guinea pig adrenal microsomes were incubated with ABT plus an NADPH-generating system (for activation of the inhibitor), there was almost complete loss of xenobiotic-metabolizing activities (Table 5.6). In contrast, ABT had no effect on the activities of the microsomal 17α-hydroxylase and 21-hydroxylase (Table 5.6). Similar experiments with adrenal mitochondrial preparations also demonstrated that ABT had no *in vitro* effects on mitochondrial P450-mediated steroid metabolism (Xu *et al.* 1994). These observations provide strong evidence that one or more non-steroidogenic P450 isozyme(s) is (are) responsible for most of the xenobiotic-metabolizing activity in the guinea pig adrenal gland. The apparent selectivity of ABT for the isozyme(s) involved should make this compound a useful probe for further investigations on the catalytic capabilities of the enzyme(s).

A major advance toward identification of a P450 isozyme that catalyses adrenal xenobiotic metabolism was made by Black and co-workers who demonstrated the presence of a 52 kDa protein in guinea pig adrenal microsomes that is immunochemically related to P4501A1/1A2 (Black *et al.* 1989a,b, Black 1990). This isozyme is highly localized to the zona reticularis of the cortex, the region of the gland in which xenobiotic-metabolizing activities are greatest. In addition, the apparent concentration of this P450 isozyme and the rates of adrenal xenobiotic metabolism are similarly affected by a number of physiological variables such as aging, gender and ACTH treatment. Thus Black and co-workers provided strong indirect evidence that the 52 kDa isozyme is at least partly responsible for the metabolism of foreign compounds by the guinea pig adrenal cortex.

A recent report in the literature suggests that the P450 isozyme described by Black and co-workers may now have been cloned and sequenced. Jiang *et al.* (1995) isolated a guinea pig adrenal microsomal protein that was immunoreactive with anti-P4501A1/1A2 antibody, and N-terminal microsequencing of the protein revealed high homology with the P4502D subfamily. Using a human P4502D6 full-

Table 5.6 Effects of incubating guinea pig adrenal microsomes with 1-aminobenzotriazole (ABT) and/or NADPH on P450-catalysed xenobiotic and steroid metabolism

	Enzyme activity (percentage of control)			
Incubation conditions	Benzphetamine N-demethylase	Benzo(a)pyrene hydroxylase	17α-Hydroxylase	21-Hydroxylase
Control	100	100	100	100
ABT	98 ± 4	96 ± 5	97 ± 5	96 ± 5
NADPH	102 ± 5	97 ± 4	103 ± 5	97 ± 4
ABT + NADPH	6 ± 1[a]	8 ± 2	98 ± 4	98 ± 5

[a] $p < 0.05$ (versus corresponding control value). Based upon Xu *et al.* (1994).

length cDNA as a probe, these investigators then isolated a number of positive clones from a guinea pig adrenal cDNA library. Sequencing of an apparent full-length clone (Figure 5.12) indicated that it was a member of the P4502D subfamily and was designated P4502D16 (CYP2D16) in accordance with the recommended nomenclature (Nelson *et al.* 1993). In order to confirm that the isozyme was expressed in guinea pig adrenal glands, Northern blotting was done with RNA obtained from adrenal inner (zona reticularis) and outer (zona fasciculata plus zona

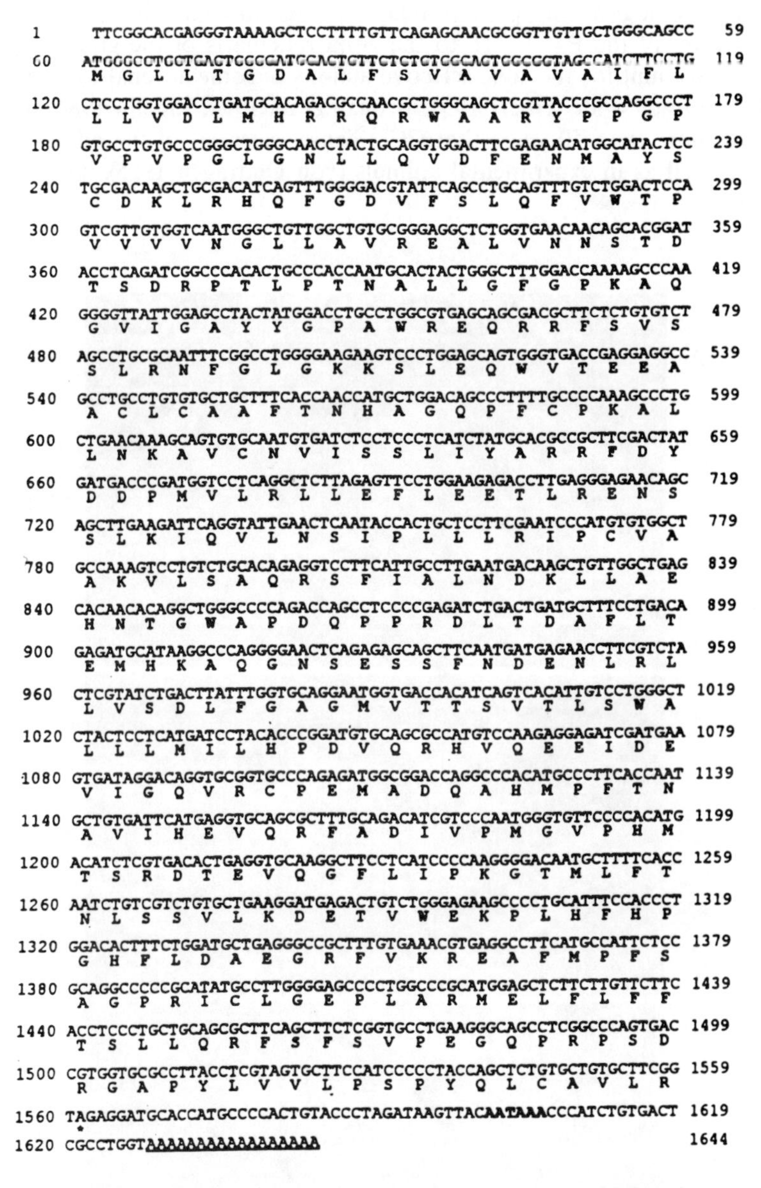

Figure 5.12 Nucleotide and deduced amino acid sequences of guinea pig adrenal P4502D16. The coding region stops at the position indicated by the asterisk. The one-letter amino acid code is used in the second position of each codon. The polyadenylation signal is bolded and the poly-A tail is underlined. From Jiang *et al.* (1995)

glomerulosa) zone tissue using the P4502D16 clone as a probe. A very strong band was detected with the inner zone RNA, but a relatively weak band with the outer zone RNA (Figure 5.13). Thus, the intra-adrenal site of expression of this P450 coincides with the location of xenobiotic-metabolizing activities and of the 52 kDa protein described by Black and co-workers (Black *et al.* 1989a,b, Black 1990). It should now be possible to determine the overall significance of P4502D16 in adrenal xenobiotic metabolism by protein expression and related experimental approaches. It would also be of interest to determine if this novel P450 isozyme has an as yet unknown physiological role, perhaps related to steroid metabolism.

Localization of the P4502D16 to the zona reticularis of the guinea pig adrenal cortex may account for the zone-selective actions of some adrenal toxicants, including carbon tetrachloride (Hoerr 1931, Brogan *et al.* 1984, Ribelin 1984). It has been known for more than 50 years that CCl_4 causes necrosis of the adrenal cortex in humans as well as in experimental animals (von Oettingen 1955). The adrenotoxic

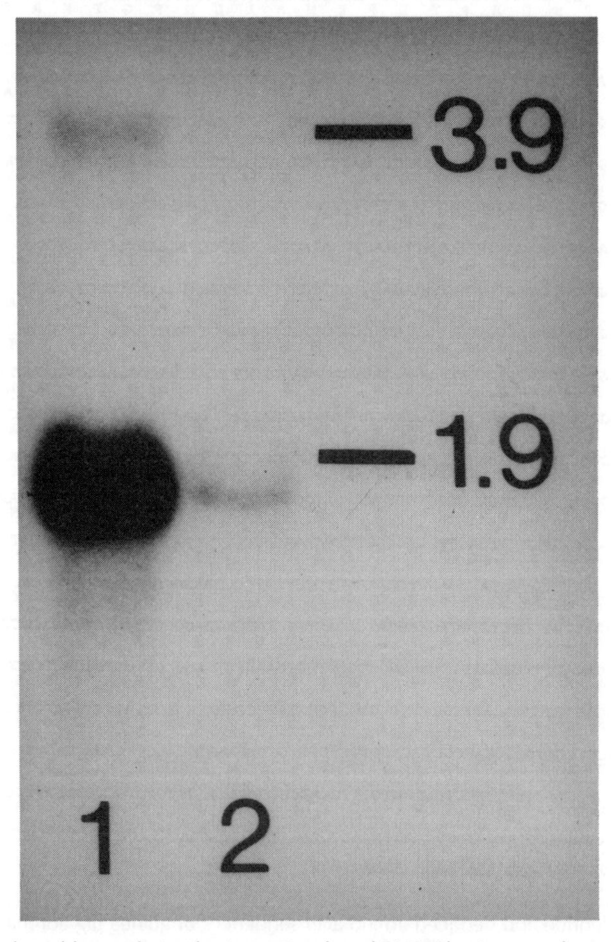

Figure 5.13 Northern blot analysis of guinea pig adrenal RNA. Blotting was done using a full-length P4502D16 cDNA as the probe. Lanes are labelled as follows: 1, adrenal inner zone RNA; 2, adrenal outer zone RNA. From Jiang *et al.* (1995)

effects of CCl_4 are localized to the zona reticularis, with little or no effect on the outer zones of the adrenal cortex (Brogan *et al.* 1984). Since the toxicity of CCl_4 requires its P450-mediated metabolism to a reactive product, probably the trichloromethyl radical (Recknagel *et al.* 1989), it was proposed by Colby and co-workers (1981) that activation occurred only in the zona reticularis of the gland. Subsequent investigations by Brogan *et al.* (1984) demonstrated that the conversion of CCl_4 to a metabolite that covalently bound to protein and initiated lipid peroxidation (Figure 5.14) was catalysed by microsomes from the zona reticularis but not from the outer zones of the guinea pig adrenal cortex. In addition, enzyme inhibition studies excluded a role for either the P450c17 or P450c21 in the adrenal activation of CCl_4 (Colby *et al.* 1994), implicating a non-steroidogenic microsomal P450 isozyme that is localized to the zonal reticularis. Collectively, these observations suggest that P4502D16 may be responsible for the zone-specific activation and toxicity of CCl_4 in the guinea pig adrenal cortex. Further investigations on this and other as yet uncharacterized adrenal P450 isozymes are now needed to determine the overall significance of xenobiotic metabolism in chemical-induced adrenocortical toxicity.

5.4.3 P450-Derived Oxygen Radicals and Lipid Peroxidation

The abundance of cytochromes P450 in the adrenal cortex may adversely influence cellular responses to toxicants or even cause toxicity. The catalytic mechanism for all P450-mediated reactions involves redox processes that generate reactive oxygen species such as superoxide radical and hydrogen peroxide. Some superoxide may be continually released into adrenal cells as a result of the normal ongoing catalytic cycles of steroidogenic P450 isozymes (Hornsby & Crivello 1983a,b). In addition, it has been proposed that the amounts of superoxide produced may be increased substantially by the interactions of certain steroid metabolites known as 'pseudo-substrates' with adrenal cytochromes P450 (Hornsby 1986, 1987, 1989).

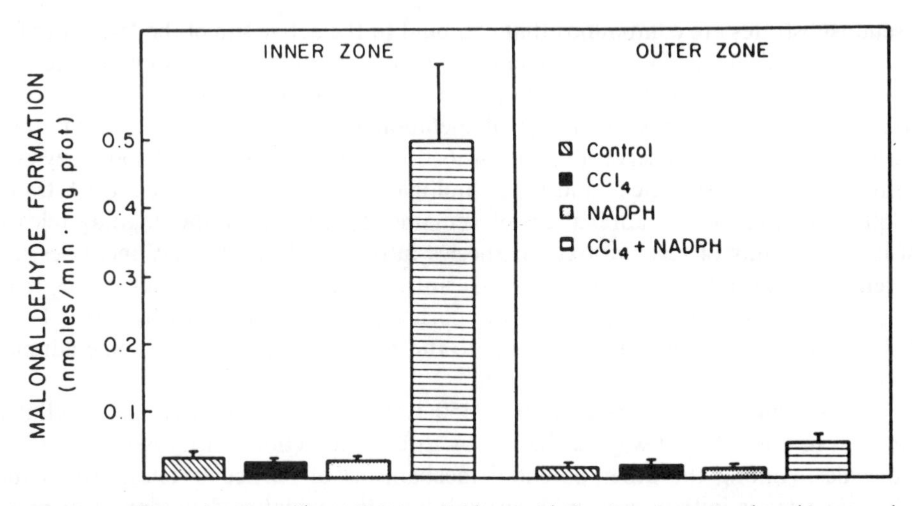

Figure 5.14 Effects of CCl_4 and/or NADPH on lipid peroxidation in guinea pig adrenal inner and outer zone microsomes. Microsomes were incubated for 30 min and lipid peroxidation was measured as the production of malonaldehyde. The data represent the means ± SE of six experiments. From Brogan *et al.* (1984)

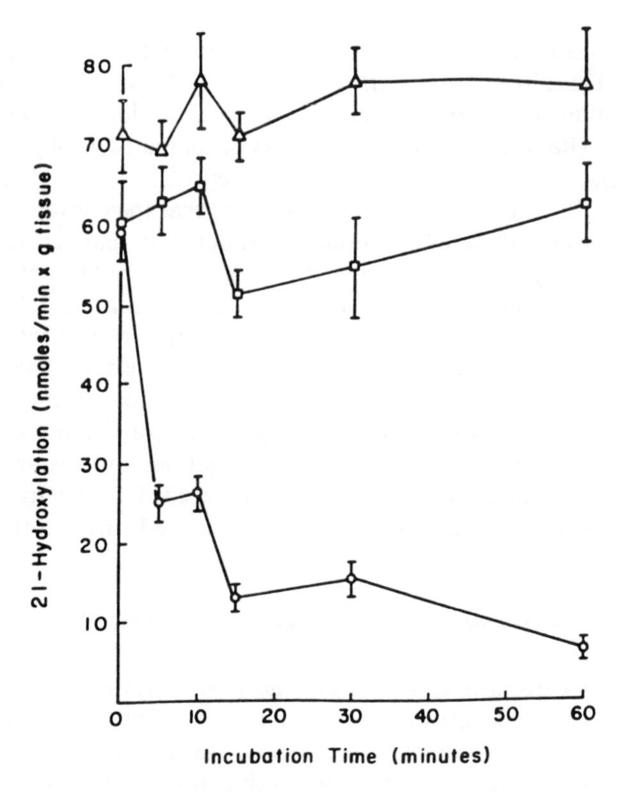

Figure 5.15 Time-course for the decrease in 21-hydroxylase activity in adrenal microsomes resulting from the initiation of lipid peroxidation by incubation with ferrous ion. Values are the means ± SE of 4–6 determinations. ○, Fe^{2+} present during incubation; □, Fe^{2+} added after incubation; △, control. From Brogan *et al.* (1983)

Pseudosubstrates are compounds that can bind to the active site of the P450, but for steric or chemical reasons are not metabolized and stimulate superoxide production. The adrenal cortex may be especially prone to such interactions because of the high concentrations of hydroxylated steroid metabolites that may accumulate in some parts of the gland and serve as pseudosubstrates for steroidogenic P450 isozymes. Although the physiological and/or pathological significance of pseudosubstrate-P450 interactions in the adrenal cortex remains to be resolved, the ongoing release of large amounts of reactive oxygen species into the cell most likely increases the potential for an adverse response to toxicants. Moreover, if pseudosubstrates tend to accumulate in certain regions of the cortex as proposed (Hornsby 1986, 1987, 1989), this phenomenon may contribute to the zone-specific actions of some adrenal toxicants.

The oxygen radicals produced by P450 isozymes can adversely affect cellular function in a number of ways including by direct interactions with essential macromolecules. Among the other potential mechanisms of particular relevance to the adrenal cortex is initiation of lipid peroxidation, the oxidative degradation of polyunsaturated fatty acids in cell membranes. This process, once initiated by free radicals, can become self-propagating, ultimately compromising membrane integrity and a host of membrane functions (Hornsby & Crivello, 1983a,b, Sevanian & Hochstein, 1985). One of the well-known consequences of lipid peroxidation is inactivation of membrane-bound enzymes, including various P450 isozymes. (Logani & Davies 1980, Cross *et al.* 1987, Tribble *et al.* 1987). Investigations in a number of laboratories have demonstrated that initiation of lipid peroxidation *in vitro* causes P450 degradation in adrenal mitochondria and microsomes and corresponding decreases in the activities of some steroidogenic enzymes (Figure 5.15) (Kitabchi 1967, Wang & Kimura 1976, Brogan *et al.* 1983, Imataka *et al.* 1985). Thus substances that initiate lipid peroxidation may directly impact adrenocortical function by inhibition of steroid hormone synthesis.

The adrenal cortex may be particularly susceptible to the adverse effects of lipid peroxidation because of its high content of unsaturated fatty acids, substrates for lipid peroxidation. It has been reported, for example, that the amount of lipid peroxidation resulting from a non-specific stimulus is far greater in adrenocortical than in hepatic mitochondrial or microsomal preparations (Staats *et al.* 1988a,b, 1989). In addition, a series of studies by Staats and co-workers (1988a,b, 1989) demonstrated that there are zonal differences in adrenal lipid peroxidation. These investigators found that lipid peroxidation *in vitro* was far greater with membrane preparations from the zona reticularis of the guinea pig adrenal cortex than with corresponding preparations from the outer zones of the gland (Figure 5.16). *In vivo*, adrenal concentrations of lipofuscin, thought to be indicative of peroxidative damage, are also greatest in the zona reticularis (Hornsby 1989). The apparent vulnerability of zona reticularis cells to peroxidative injury may be partly responsible for the frequency of chemical-induced injury in this region of the cortex (Hoerr 1931, Ribelin 1984).

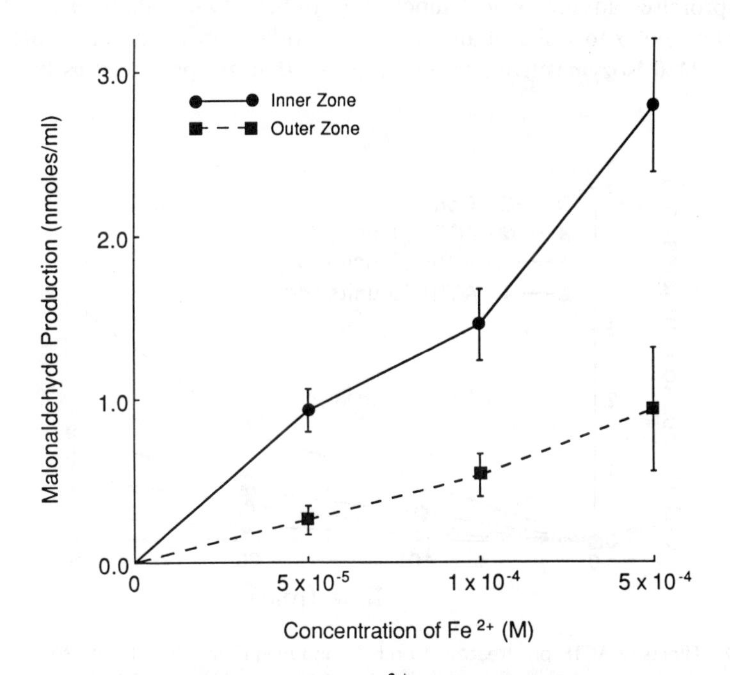

Figure 5.16 Concentration-dependent effects of Fe^{2+} on lipid peroxidation in adrenal inner and outer zone mitochondria. Lipid peroxidation was initiated with the concentration of Fe^{2+} (ferrous sulfate) indicated and incubations were done at 37°C for 30 min. Values are the means ± SE of 6–8 experiments. From Staats *et al.* (1988a)

Recent observations also suggest that lipid peroxidation in the adrenal cortex is hormonally modulated. It has been demonstrated that ACTH administration to guinea pigs causes a dose-dependent decline in ferrous ion-induced peroxidation *in vitro* in adrenal mitochondrial and microsomal preparations (Figure 5.17). This effect might be construed as another tropic action of ACTH, if by inhibition of lipid peroxidation, adrenal P450 isozymes and steroidogenic function are preserved. The physiological and toxicological significance of ACTH actions on adrenal lipid peroxidation, in addition to the mechanism of action, now need to be investigated.

Perhaps because of its vulnerability to lipid peroxidation and/or other oxidative damage, the adrenal cortex is extremely rich in antioxidant defense systems. As discussed by Hornsby & Crivello (1983a,b) and by Hallberg (1990), adrenocortical cells are relatively well equipped with a number of protective enzymes, including superoxide dismutase (SOD) and catalase. However, most notable are the very high concentrations of α-tocopherol and ascorbic acid found in the adrenals of many species. Adrenal concentrations of both antioxidants are greater than those in any other organ (Gallo-Torres 1980, Hornsby & Crivello 1983a,b). Nonetheless, very little information is available on the mechanisms responsible for this localization or on the physiological function of either antioxidant in the adrenal cortex. Both vitamins may, however, represent important components of the antioxidant armamentarium that protects the adrenal cortex from oxidant-induced injury.

α-Tocopherol, a fat-soluble vitamin, is localized principally in membranes and is a potent inhibitor of LP (Machlin & Bendich 1987, Niki 1987). It has been demonstrated by a number of investigators that α-tocopherol deficiency causes an increase in lipid peroxidation in adrenocortical mitochondria and microsomes (Figure 5.18) and compromises adrenocortical function (Kitabchi 1980, Staats *et al.* 1988b). An important role of α-tocopherol and other antioxidants in the adrenal cortex may be to protect P450 isozymes from oxygen radicals that are produced as byproducts of

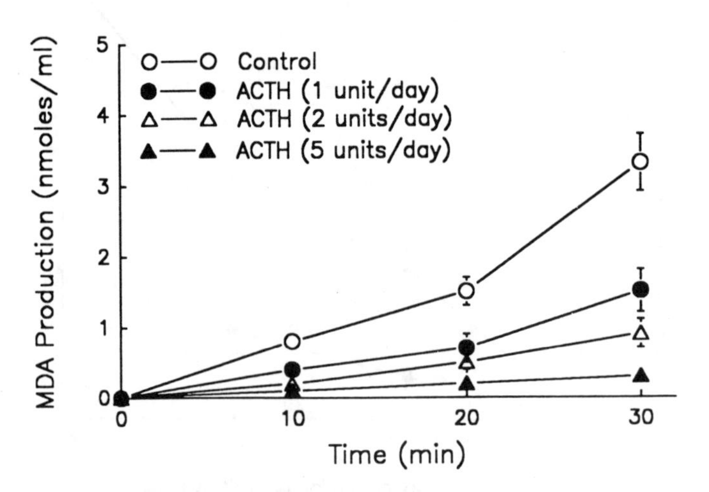

Figure 5.17 Effects of ACTH pre-treatment on Fe^{2+}-induced LP in adrenal microsomes. Male guinea pigs were pretreated with 1,2 or 5 units of ACTH (cortrophin-zinc) daily for 3 days. Adrenal microsomes were prepared and incubated with Fe^{2+} as described previously (Staats *et al.* 1988b). LP was assessed by the formation of malonaldehyde (MDA) equivalents. Values are means ± SE for four or five animals

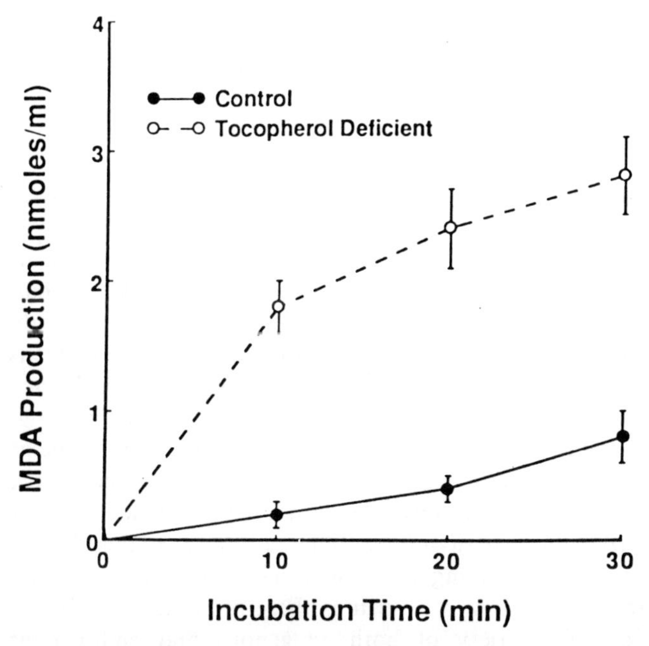

Figure 5.18 Time-courses for Fe^{2+}-induced lipid peroxidation in adrenal mitochondria from control and tocopherol-deficient guinea pigs. Mitochondrial preparations were incubated with 1.0 mM ferrous sulfate at 37°C for the times indicated and lipid peroxidation determined as malonaldehyde (MDA) equivalents. From Staats *et al.* (1988a)

steroidogenesis or in response to toxicants (Hornsby & Crivello 1983a,b, Hornsby 1989). In the guinea pig adrenal cortex, Staats and co-workers (1988a,b, 1989) found an inverse relationship between α-tocopherol concentrations and lipid peroxidation activity in the different zones of the gland, indicating a modulatory role for α-tocopherol. The results also suggest that regional differences in the distribution of α-tocopherol or other antioxidants within the adrenal cortex may contribute to the zone-selective effects of some toxicants.

Despite the abundance of ascorbic acid within adrenocortical cells, its significance, like that of α-tocopherol, remains uncertain. Over the years, numerous conflicting reports on the function of adrenal ascorbic acid have appeared in the literature. At the concentrations found in the adrenal cortex (Hornig 1975, Hornsby & Crivello 1983a,b), ascorbic acid would be expected to have potent antioxidant activity. In addition, although the water-soluble ascorbic acid is found in the cytosol of cells, it appears to be functionally linked with α-tocopherol. Tappel first proposed that ascorbic acid and α-tocopherol synergistically interact as part of an electron transport system that terminates free radical chains (Tappel 1968). According to his hypothesis, the tocopheroxyl radical formed in the course of free-radical scavenging is reduced back to α-tocopherol by ascorbic acid, thereby enhancing the antioxidant effectiveness of each α-tocopherol molecule. The high concentrations of both α-tocopherol and ascorbic acid in the adrenal cortex would, therefore, be expected to serve as a potent antioxidant system for the gland. However, the apparent amount of protection from free radicals must be considered in relation to the level of oxidative stress within the adrenal cortex which, as discussed above, may be substantial.

Thus, it is impossible to know if even the large amount of antioxidant protection available to the adrenal cortex is sufficient to meet its needs.

5.5 Factors Affecting Adrenocortical Toxicity

In addition to those characteristics that make the adrenal cortex uniquely vulnerable to toxicants, there are a number of other factors that may participate in the modulation of toxicity. For example, the overall impact of toxic substances in any organ, including the adrenal cortex, depends in part upon the capacity for detoxication reactions and other protective mechanisms. When metabolic activation occurs in a target organ, the balance between activating and inactivating processes, by controlling the steady-state level of the toxic metabolite, is often the critical determinant of toxicity. As already noted, cytochromes P450 may effect activation or detoxication, depending upon the substrates and P450 isozymes involved. Among the other major enzymatic detoxifying reactions are those catalysed by the phase II or conjugating enzymes, including the glucuronyl transferases, sulfotransferases and glutathione transferases. Enzymes such as these promote both the inactivation and excretion of a wide variety of both exogenous and endogenous compounds. Although some of these reactions have been demonstrated in the adrenal cortex, specific activities are generally far lower than in the liver, the major site of xenobiotic metabolism (Hornsby & Crivello 1983a,b, Hallberg 1990). It is possible that a relative deficiency of inactivating enzymes in the adrenal is a factor that contributes to the frequency of chemical-induced injuries in the gland. However, it should be noted that there has been a paucity of research done on adrenal detoxication processes and any definitive conclusions must await further investigation.

Many other factors, both physiological and pharmacological, also influence the actions of adrenal toxicants. For example, the effects of some toxicants are known to be species-specific, making it difficult to extrapolate the results of animal studies to humans, particularly in the absence of clearly defined mechanisms of action. Since many toxicants have been studied in only a single species, it is presently impossible to assess the overall significance of species differences as a factor affecting adrenocortical toxicity. Physiological variables such as age might also be expected to affect adrenal toxicity. In the human foetus, for example, adrenal xenobiotic-metabolizing activities are very high (Juchau & Pedersen 1973), increasing the potential for activation of protoxicants. Enzyme activities are apparently much lower in adrenals from adults (Hallberg 1990). In guinea pigs, in contrast, the rates of adrenal xenobiotic metabolism are relatively low in young animals but increase progressively with ageing (Pitrolo *et al.* 1979). Many other variables related to the actions of adrenal toxicants are similarly age-dependent. However, in neither humans nor any other species has the influence of age on the adrenal response to toxicants been systemically investigated.

Hormonal factors such as ACTH also affect a number of processes that influence the severity of chemical-induced adrenal toxicity. These include bioactivation and detoxication reactions, oxygen radical formation, lipid peroxidation, and antioxidant concentrations in the gland. Unfortunately, reports in the literature do not provide a sufficiently consistent pattern of effects to permit any general conclusions about the impact of ACTH on toxicity. This is probably not surprising because of the diversity of factors that may interact in determining the toxicity of any single compound. The simultaneous effects of multiple factors certainly complicate any attempt to resolve the influence of specific variables on adrenocortical responses to toxicants. However, such information is important for a better understanding of predisposing influences on adrenal toxicity and should constitute a productive area for future investigation.

5.6 Concluding Comments

Although the adrenal cortex has long been recognized as a toxicological target organ for numerous chemicals, research on the mechanisms involved has been slow to emerge. Most of the published research in this area has been of a descriptive nature, but signs of change are apparent. Recent investigations have begun to provide insight into the mechanisms responsible for the high incidence of chemical-induced lesions of the adrenal cortex. In addition, the rapid progress being made in our understanding of the molecular biology of adrenocortical cells offers the opportunity for new approaches to toxicological investigations. The utilization of molecular techniques in studies on the actions of adrenal toxicants is now beginning to occur, and should become the standard for further experimentation. Accordingly, it seems likely that important new advances will be forthcoming in the near future which should contribute to a more complete and in-depth understanding of adreno-cortical toxicology.

5.7 Acknowledgements

Investigations from the author's laboratory were supported by National Institutes of Health research grants CA22152, GM30261 and AG11987.

References

ABSHAGEN, V., SPORL, S., SCHONESHOFER, M. & OELKERS, W. (1977) Increased plasma 11-deoxycorticosterone during spironolactone medication. *Journal of Clinical Endocrinology and Metabolism*, **44**, 1190–1197.

ALBERTSON, B. D., HILL, R. B., SPRAGUE, K. A., WOOD, K. E., NIEMAN, L. K. & LORIAUX, D. L. (1994) Effect of the antiglucocorticoid RU486 on adrenal steroidogenic enzyme activity and steroidogenesis. *European Journal of Endocrinology*, **130**, 195–200.

ASHBY, H., DIMATTINA, M., LINEHAN, W. M., ROBERTSON, C. N., QUEENAN, J. T. & ALBERTSON, B. D. (1989) The inhibition of human adrenal steroidogenic enzyme activities by suramin. *Journal of Clinical Endocrinology of Metabolism*, **68**, 505–508.

BARBIERI, R. L., OSATHANONDH, R., CANICK, J. A., STILLMAN, R. J. & RYAN, K. J. (1980) Danazol inhibits human adrenal 21- and 11β-hydroxylation *in vitro*. *Steroids*, **35**, 251–263.

BARBIERI, R. L., YORK, C. M., CHERRY, M. L. & RYAN, K. J. (1987) The effects of nicotine, cotinine and anabasine on rat adrenal 11β-hydroxylase and 21-hydroxylase. *Journal of Steroid Biochemistry*, **28**, 25–28.

BLACK, V. H. (1990) Immunodetectable cytochromes P450I, II and III in guinea pig adrenal-hormone responsiveness and relationship to capacity for xenobiotic metabolism. *Endocrinology*, **127**, 1153–1159.

Black, V. H., Barilla, J. R., Russo, J. J. & Martin, K. O. (1989a) A cytochrome P450 immunochemically related to P450c,d (P450I) localized to smooth microsomes and inner zones of the guinea pig adrenal. *Endocrinology*, **124**, 2480–2493.

Black, V. H., Barilla, J. R. & Martin, K. O. (1989b) Effects of age, adrenocorticotropin and dexamethasone on a male-specific cytochrome P450 localized in the inner zone of the guinea pig adrenal. *Endocrinology*, **124**, 2494–2498.

Brandt, I. (1987) PCB methyl suphones and related compounds: identification of target cells and tissues in different species. *Chemosphere*, **16**, 1671–1676.

Brogan, W. C., Miles, P. R. & Colby, H. D. (1983) Effects of lipid peroxidation on adrenal microsomal monooxygenases. *Biochimica et Biophysica Acta*, **758**, 114–120.

Brogan, W. C., Eacho, P. I., Hinton, D. E. & Colby, H. D. (1984) Effects of carbon tetrachloride on adrenocortical structure and function in guinea pigs. *Toxicology and Applied Pharmacology*, **75**, 118–127.

Castracane, V. D., Allen-Rowlands, C. F., Hamilton, M. G. & Seifter, J. (1982) The effect of polybrominated biphenyl (PBB) on testes, adrenal and pituitary function in the rat. *Proceedings of the Society for Experimental Biology and Medicine*, **169**, 343–347.

Cheng, B., Horst, I. A. & Kowal, J. (1993) Nigericin inhibits ACTH- and dibutyryl-cAMP-stimulated steroidogenesis of cultured mouse adrenocortical tumor (Y1) cells. *Hormone and Metabolic Research*, **25**, 391–392.

Colby, H. D. (1987) The adrenal cortex. In Hedge, G. A., Colby, H. D. & Goodman, R. L., (Eds), *Clinical Endocrine Physiology*, pp. 127–159, Philadelphia: W. B. Saunders.

Colby, H. D. (1988) Adrenal gland toxicity: chemically induced dysfunction. *Journal of the American College of Toxicology*, **7**, 45–69.

Colby, H. D. & Longhurst, P. A. (1992) Toxicology of the adrenal gland. In Atterwill, C. K. & Flack, J. D. (Eds), *Endocrine Toxicology*, pp. 243–284, Cambridge: Cambridge University Press.

Colby, H. D. & Rumbaugh, R. C. (1980) Adrenal drug metabolism. In Gram, T. E. (Ed.), *Extrahepatic Metabolism of Drugs and Other Foreign Compounds*, pp. 239–266, Jamaica, NY: Spectrum Publications.

Colby, H. D., Rumbaugh, R. C. & Stitzel, R. E. (1980) Changes in adrenal microsomal cytochrome(s) P-450 with aging in the guinea pig. *Endocrinology*, **107**, 1359–1363.

Colby, H. D., Brogan, W. C. & Miles, P. R. (1981) Carbon tetrachloride-induced changes in adrenal microsomal mixed function oxidases and lipid peroxidation. *Toxicology and Applied Pharmacology*, **60**, 492–499.

Colby, H. D., Johnson, P. B., Pope, M. R. & Zulkoski, J. S. (1982a) Metabolism of benzo(a)pyrene by guinea pig adrenal and hepatic microsomes. *Biochemical Pharmacology*, **5**, 639–646.

Colby, H. D., Johnson, P. B., Zulkoski, J. S. & Pope, M. R. (1982b). Differential effects of adrenocorticotropic hormone on adrenal microsomal xenobiotic and steroid metabolism in guinea pigs. *Drug Metabolism and Disposition*, **10**, 326–329.

Colby, H. D., Levitt, M., Pope, M. R. & Johnson, P. B. (1992) Differential effects of adrenocorticotropic hormone on steroid hydroxylase activities in the inner and outer zones of the guinea pig adrenal cortex. *Journal of Steroid Biochemistry and Molecular Biology*, **42**, 329–335.

Colby, H. D., Purcell, H., Kominami, S., Takemori, S. & Kossor, D. C. (1994) Adrenal activation of carbon tetrachloride: role of microsomal P450 isozymes. *Toxicology*, **94**, 31–40.

Cross, C. E., Halliwell, B., Borish, E. T., Pryor, W. A., Ames, B. N., Saul, R. L., McCord, J. M. & Harman, D. (1987) Oxygen radicals and human disease. *Annals of Internal Medicine*, **107**, 526–545.

Cuellar, A., Escamilla, E., Ramirez, J. & Chavez, E. (1984) Adriamycin as an inhibitor of 11β-hydroxylase activity in adrenal cortex mitochondria. *Archives of Biochemistry and Biophysics*, **235**, 538–543.

DARNERUD, P. O. & BRANDT, I. (1985) Pitfalls of the interpretation of whole-body autoradiograms. *Acta Pharmacologica et Toxicologica*, **56**, 55–62.

DECASTRO, C. R., DETORANZO, E. G. D., CARBONE, M. & CASTRO, J. A. (1990) Ultrastructural effects of nifurtimox on rat adrenal cortex related to reductive biotransformation. *Experimental and Molecular Pathology*, **52**, 98–108.

DIBARTOLOMEIS, M. J., MOORE, R. W., PETERSON, R. E., CHRISTIAN, B. J. & JEFCOATE, C. R. (1987) Altered regulation of adrenal steroidogenesis in 2,3,7,8-tetrachlorodibenzo-*p*-dioxin-treated rats. *Biochemial Pharmacology*, **36**, 59–67.

EACHO, P. I., O'DONNELL, J. P. & COLBY, H. D. (1984) Metabolism of 4-chlorobiphenyl by guinea pig adrenocortical and hepatic microsomes. *Biochemical Pharmacology*, **33**, 3627–3632.

EROSCHENKO, V. P. & WILSON, W. O. (1975) Cellular changes in the gonads, livers and adrenal glands of Japanese quail as affected by the insecticide Kepone. *Toxicology and Applied Pharmacology*, **31**, 491–504.

FELDMAN, D. (1986) Ketoconazole and other imidazole derivatives as inhibitors of steroidogenesis. *Endocrine Reviews*, **7**, 409–420.

GALLO-TORRES, H. E. (1980) Vitamin E transport and metabolism. In MACHLIN, L. J. (Ed.), *Vitamin E: A Comprehensive Treatise*, pp. 193–267, New York: Marcel Dekker.

GREINER, J. W., KRAMER, R. E., JARRELL, J. & COLBY, H. D. (1976) Mechanism of action of spironolactone on adrenocortical function in guinea pigs. *Journal of Pharmacology and Experimental Therapeutics*, **198**, 709–715.

GREINER, J. W., RUMBAUGH, R. C., KRAMER, R. E. & COLBY, H. D. (1978) Relation of canrenone to the actions of spironolactone on adrenal cytochrome P-450-dependent enzymes. *Endocrinology*, **103**, 1313–1320.

GU, Y. & LIN, Y. C. (1991) Suppression of adrenocorticotropic hormone (ACTH)-induced corticosterone secretion in cultured rat adrenocortical cells by gossypol and gossypolone. *Research Communications in Chemical Pathology and Pharmacology*, **72**, 27–38.

GUENTHER, T., NEBERT, D. & MENARD, R. (1979) Microsomal aryl hydrocarbon hydroxylase in rat adrenal: regulation by ACTH but not by polycyclic hydrocarbons. *Molecular Pharmacology*, **15**, 719–728.

HALL, P. F. (1986) Cytochromes P-450 and the regulation of steroid synthesis. *Steroids*, **48**, 131–196.

HALL, P. F. (1991) Cytochrome P-450 C_{21scc}: one enzyme with two actions: hydroxylase and lyase. *Journal of Steroid Biochemistry and Molecular Biology*, **40**, 527–532.

HALLBERG, E. (1990) Metabolism and toxicity of xenobiotics in the adrenal cortex, with particular reference to 7,12-dimethylbenz(a)anthracene. *Journal of Biochemical Toxicology*, **5**, 71–90.

HART, M. M., REAGAN, R. L. & ADAMSON, R. H. (1973) The effect of isomers of DDD on the ACTH-induced steroid output, histology and ultrastructure of the dog adrenal cortex. *Toxicology and Applied Pharmacology*, **24**, 101–113.

HARTMANN, F. & JENTZEN, F. (1979) Effect of chlorphentermine on hormone content and function of the adrenal cortex in rats. *Hormone and Metabolic Research*, **11**, 158–160.

HINSON, J. A, PUMFORD, N. R. & NELSON, S. D. (1994) The role of metabolic activation in drug toxicity. *Drug Metabolism Reviews*, **26**, 395–412.

HIRAI, M. & ICHIKAWA, M. (1991) Changes in serum glucocorticoid levels and thymic atrophy induced by phenytoin administration in mice. *Toxicology Letters*, **56**, 1–6.

HOERR, N. (1931) The cells of the suprarenal cortex in the guinea pig. Their reaction to injury and their replacement. *American Journal of Anatomy*, **48**, 139–197.

HORNIG, D. (1975) Distribution of ascorbic acid, metabolites and analogues in man and animals. *Annals of the New York Academy of Sciences*, **258**, 103–118.

HORNSBY, P. J. (1985) The regulation of adrenocortical function by control of growth and structure. In ANDERSON, D. C. & WINTER, J. S. D. (Eds), *Adrenal Cortex*, pp. 1–31, London: Butterworth.

HORNSBY, P. J. (1986) Cytochrome P-450/pseudosubstrate interactions and the role of antioxidants in the adrenal cortex. *Endocrine Research*, **12**, 469–494.

HORNSBY, P. J. (1987) Physiological and pathological effects of steroids on the function of the adrenal cortex. *Journal of Steroid Biochemistry*, **27**, 1161–1171.

HORNSBY, P. J. (1989) Steroid and xenobiotic effects on the adrenal cortex: mediation by oxidative and other mechanisms. *Free Radical Biology and Medicine*, **6**, 103–115.

HORNSBY, P. J. & CRIVELLO, J. F. (1983a) The role of lipid peroxidation and biological antioxidants in the function of the adrenal cortex. Part 1: A background review. *Molecular and Cellular Endocrinology*, **30**, 1–20.

HORNSBY, P. J. & CRIVELLO, J. F. (1983b) The role of lipid peroxidation and biological antioxidants in the function of the adrenal cortex. Part 2. *Molecular and Cellular Endocrinology*, **30**, 123–147.

HUGGINS, C. & MORII, S. (1961) Selective adrenal necrosis and apoplexy induced by 7,12-dimethylbenz(a)anthracene. *Journal of Experimental Medicine*, **114**, 741–760.

IMATAKA, H., SUZUKI, K. & TAMAOKI, B. (1985) Effect of Fe^{2+}-induced lipid peroxidation upon microsomal steroidogenic enzyme activities of porcine adrenal cortex. *Biochemical and Biophysical Research Communications*, **128**, 657–663.

INAO, S. (1970) Adrenocortical insufficiency induced in rats by prolonged feeding of Kanechlor (chlorobiphenyl). *Kumamoto Medical Journal*, **23**, 27–31.

JIANG, Q., VOIGT, J. M. & COLBY, H. D. (1995) Molecular cloning and sequencing of a guinea pig cytochrome P4502D (CYP2D16): high level expression in adrenal microsomes. *Biochemical and Biophysical Research Communications*, **209**, 1149–1156.

JÖNSSON, C.-J. & LUND, B.-O. (1994) *In vitro* bioactivation of the environmental pollutant 3-methylsulphonyl-2,2-bis(4-chlorophenyl)-1,1-dichloroethene in the human adrenal gland. *Toxicology Letters*, **71**, 169–175.

JUCHAU, M. R. & PEDERSEN, M. G. (1973) Drug biotransformation reactions in the human fetal adrenal gland. *Life Sciences*, **12**, 193–204.

KITABCHI, A. E. (1967) Inhibition of steroid C-21 hydroxylase by ascorbate: alterations of microsomal lipids in beef adrenal cortex. *Steroids*, **10**, 567–576.

KITABCHI, A. E. (1980) Hormonal status in vitamin E deficiency . In MACHLIN, L. J. (Ed.), *Vitamin E: A Comprehensive Treatise*, pp. 348–371, New York: Marcel Dekker.

KOSSOR, D. C., KOMINAMI, S., TAKEMORI, S. & COLBY, H. D. (1991) Role of the steroid 17α-hydroxylase in spironolactone-mediated destruction of adrenal cytochrome P450. *Molecular Pharmacology*, **40**, 321–325.

KÜHN-VELTEN, W. N. (1993) Norharman (β-carboline) as a potent inhibitory ligand for steroidogenic cytochromes P450 (CYP11 and CYP17). *European Journal of Pharmacology*, **250**, R1–R3.

KUPFER, D. & ORRENIUS, S. (1970) Characteristics of guinea pig liver and adrenal monooxygenase systems. *Molecular Pharmacology*, **6**, 221–230.

LATENDRESSE, J. R., AZHAR, S., BROOKS, C. L. & CAPEN, C. C. (1993) Pathogenesis of cholesteryl lipidosis of adrenocortical and ovarian interstitial cells in F344 rats caused by tricresyl phosphate and butylated triphenyl phosphate. *Toxicology and Applied Pharmacology*, **122**, 281–289.

LOGANI, M. K. & DAVIES, R. E. (1980) Lipid oxidation: biological effects and antioxidants – a review. *Lipids*, **15**, 485–495.

LULLMAN, H. & LULLMAN-RAUCH, R. (1981) Tamoxifen-induced generalized lipidosis in rats subchronically treated with high doses. *Toxicology and Applied Pharmacology*, **61**, 138–146.

LUND, B., BERGMAN, A. & BRANDT, I. (1988) Metabolic activation and toxicity of a DDT-metabolite, 3-methyl-sulphonyl-DDE, in the adrenal zona fasciculata in mice. *Chemico-Biological Interactions*, **65**, 25–40.

MACHLIN, L. J. & BENDICH, A. (1987) Free radical tissue damage: protective role of antioxidant nutrients. *FASEB Journal*, **1**, 441–445.

MARTIN, K. O. & BLACK, V. H. (1983) Effects of age and adrenocorticotropin on microsomal enzymes in guinea pig adrenal inner and outer cortices. *Endocrinology*, **112**, 573–579.

MARTZ, F. & STRAW, J. A. (1980) Metabolism and covalent binding of 1-(O-chlorophenyl)-1(*p*-cholorphenyl)-2,2-dichloroethane (*o,p'*-DDD). *Drug Metabolism and Disposition*, **8**, 127–130.

MAZZOCCHI, G. & NUSSDORFER, G. G. (1985) Effects of chloramphenicol on the long term trophic action of ACTH on rat adrenocortical cells: a combined stereological and enzymological study. *Journal of Anatomy*, **140**, 607–612.

MAZZOCCHI, G., MARKOWOSKA, A., ANDREIS, P. G., TORTORELLA, C., NERI, G., GOTTARDO, G., MALENDOWICZ, L. K. & NUSSDORFER, G. G. (1993) Effects of cyclosporine-A on steroid secretion of dispersed rat adrenocortical cells. *Experimental Toxicology and Pathology*, **45**, 481–488.

MENARD, R. H., MARTIN, H. F., STRIPP, B., GILLETTE, J. R. & BARTTER, F. C. (1975) Spironolactone and cytochrome P-450: impairment of steroid hydroxylation in the adrenal cortex. *Life Sciences*, **15**, 1639–1648.

MENARD, R. H., GUENTHNER, T. M., KON, H. & GILLETTE, J. R. (1979a) Studies on the destruction of adrenal and testicular cytochrome P-450 by spironolactone. *Journal of Biological Chemistry*, **254**, 1726–1733.

MENARD, R. H., GUENTHNER, T. M., TABURET, A. M., KON, H., POHL, L. R., GILLETTE, J. R., GELBOIN, H. V. & TRAGER, W. F. (1979b) Specifity of the *in vitro* destruction of adrenal and hepatic microsomal steroid hydroxylases by thiosteroids. *Molecular Pharmacology*, **16**, 997–1010.

MILLER, W. R. (1988) Molecular biology of steroid hormone synthesis. *Endocrine Reviews*, **9**, 295–318.

MOCHIZUKI, H., KOMINAMI, S. & TAKEMORI, S. (1988) Examination of differences between benzo(a)pyrene and steroid hydroxylases in guinea pig adrenal microsomes. *Biochimica et Biophysica Acta*, **964**, 83–89.

MOHAMMED, A., HALLBERG, E., RYDSTROM, J. & SLANINA, P. (1985) Toxaphene: accumulation in the adrenal cortex and effect on ACTH-stimulated corticosteroid synthesis in the rat. *Toxicology Letters*, **24**, 137–143.

MURRAY, M. & REIDY, G. F. (1990) Selectivity in the inhibition of mammalian cytochromes P-450 by chemical agents. *Pharmacological Reviews*, **42**, 85–101.

NELSON, D. R., KAMATAKI, T., WAXMAN, D. J., GUENGERICH, F. P., ESTABROOK, R. W., FEYEREISEN, R., GONZALEZ, F. J., COON, M. J., GUNSALUS, I. C., GOTOH, O., OKUDA, K. & NEBERT, D. W. (1993) The P450 superfamily: update on new sequences, gene mapping, accession numbers, early trivial names of enzymes, and nomenclature. *DNA and Cell Biology*, **22**, 1–51.

NESTLER, J. E., MCCLANAHAN, M. A., CLORE, J. N. & BLACKARD, W. G. (1992) Insulin inhibits adrenal 17,20-lyase activity in man. *Journal of Clinical Endocrinology and Metabolism*, **74**, 362–367.

NIKI, E. (1987) Antioxidants in relation to lipid peroxidation. *Chemistry and Physics of Lipids*, **44**, 227–253.

NISHIYAMA, S. & NAKAMURA, K. (1984) Effect of cadmium on plasma aldosterone and serum corticosterone concentrations in male rats. *Toxicology and Applied Pharmacology*, **76**, 420–425.

VON OETTINGEN, W. F. (1955) The halogenated aliphatic, olefinic, cyclic, aromatic and aliphatic–aromatic hydrocarbons including the halogenated insecticides: their toxicity and potential dangers. Public Health Service Publications, Washington, DC.

OTTO, S., MARCUS, C., PIDGEON, C. & JEFCOATE, C. (1991) A novel adrenocorticotropin-inducible cytochrome P450 from rat adrenal microsomes catalyzes polycyclic aromatic hydrocarbon metabolism. *Endocrinology*, **129**, 970–982.

PANESAR, N. S., HERRIES, D. G. & STITCH, S. R. (1979) Effects of cyproterone and cyproterone acetate on the adrenal gland in the rat: studies *in vivo* and *in vitro*. *Journal of Endocrinology*, **80**, 229–238.

PITROLO, D. A., RUMBAUGH, R. C. & COLBY, H. D. (1979) Maturational changes in adrenal xenobiotic metabolism in male and female guinea pigs. *Drug Metabolism and Disposition*, **7**, 52–56.

PREZIOSI, P. & VACCA, M. (1988) Adrenocortical suppression and other endocrine effects of etomidate. *Life Sciences*, **42**, 477–489.

PYYKKÖ, K., TÄHTI, H. & VAPAATALO, H. (1977) Toluene concentrations in various tissues of rats after inhalation and oral administration. *Archives of Toxicology*, **38**, 169–176.

RECKNAGEL, R. O., GLENDE, E. A., DOLAK, J. A. & WALLER, R. L. (1989) Mechanisms of carbon tetrachloride toxicity. *Pharmacology and Therapeutics*, **43**, 139–154.

RIBELIN, W. E. (1984) The effects of drugs and chemicals upon the structure of the adrenal gland. *Fundamental and Applied Toxicology*, **4**, 105–119.

ROBERTSON, W. R., LAMBERT, A., MITCHELL, R., KENDLE, K. & PETROW, V. (1989) The effect of some antiprostatic steroids upon cortisol production by guinea pig adrenal cells stimulated by ACTH. *Biochemical Pharmacology*, **38**, 3669–3671.

SAUNDERS, F. J. & ALBERTI, R. L. (1978) *Aldactone (Spironolactone): A Comprehensive Review*, New York: Searle, Inc.

SEVANIAN, A. & HOCHSTEIN, P. (1985) Mechanisms and consequences of lipid peroxidation in biological systems. *Annual Review of Nutrition*, **5**, 365–390.

SHERRY, J. H., FLOWERS, L., O'DONNELL, J. P., LACAGNIN, L. B. & COLBY, H. D. (1986) Metabolism of spironolactone by adrenocortical and hepatic microsomes: relationship to cytochrome P-450 destruction. *Journal of Pharmacology and Experimental Therapeutics*, **236**, 675–683.

SIMPSON, E. R. & WATERMAN, M. R. (1988) Regulation of the synthesis of steroidogenic enzymes in adrenal cortical cells by ACTH. *Annual Review of Physiology*, **5**, 427–440.

STAATS, D. A., LOHR, D. & COLBY, H. D. (1988a) Relationship between mitochondrial lipid peroxidation and alpha-tocopherol levels in guinea pig adrenal cortex. *Biochimica et Biophysica Acta*, **961**, 279–284.

STAATS, D. A., LOHR, D. & COLBY, H. D. (1988b) Effects of tocopherol depletion on the regional differences in adrenal microsomal lipid peroxidation and steroid metabolism. *Endocrinology*, **123**, 975–980.

STAATS, D. A., LOHR, D. & COLBY, H. D. (1989). α-Tocopherol depletion eliminates the regional differences in adrenal mitochondrial lipid peroxidation. *Molecular and Cellular Endocrinology*, **62**, 189–195.

SUHARA, K., FUJIMURA, Y., SHIROO, M. & KATAGIRI, M. (1984) Multiple catalytic properties of the purified and reconstituted cytochrome P-450 (P-450_{sccII}) system of pig testis microsomes. *Journal of Biological Chemistry*, **259**, 8729–8736.

SUNDSFJORD, J. A., MARTON, P., JORGENSEN, H. & AAKVAAG, A. (1974) Reduced aldosterone secretion during spironolactone treatment in primary aldosteronism: report of a case. *Journal of Clinical Endocrinology and Metabolism*, **39**, 734–739.

SZABO, S. & LIPPE, I. TH. (1989) Adrenal gland: chemically induced structural and functional changes in the cortex. *Toxicologic Pathology*, **17**, 317–329.

SZABO, S., HÜTTER, I., KOVACS, K., HORVATH, E., SZABO, D. & HORNER, H. C. (1980) Pathogenesis of experimental adrenal hemorrhagic necrosis ('apoplexy'). Ultrastructural, biochemical, neuropharmacologic and blood coagulation studies with acrylonitrile in the rat. *Laboratory Investigation*, **42**, 533–546.

TAPPEL, A. L. (1968) Will antioxidant nutrients slow aging processes? *Geriatrics*, **23**, 97–105.

TOWN, C., HENDERSON, L., CHANG, D., MORTILLO, M. & GARLAND, W. (1993)

Distribution of 1-aminobenzotriazole in male rats after administration of an oral dose. *Xenobiotica*, **23**, 383–390.

TREMBLAY, Y., FLEURY, A., BEADOIN, C., VALLÉE, M. & BÉLANGER, A. (1994) Molecular cloning and expression of guinea pig cytochrome P450c17 cDNA (steroid 17α-hydroxylase/17,20 lyase): tissue distribution, regulation and substrate specificity of the expressed enzyme. *DNA and Cell Biology*, **13**, 1199–1212.

TRIBBLE, D. L., AW, T. Y. & JONES, D. P. (1987) The pathophysiological significance of lipid peroxidation in oxidative cell injury. *Hepatology*, **7**, 377–386.

TUCK, M. L., SOWERS, J. R., FITTINGOFF, D. B., FISHER, J. S., BERG, G. J., ASP, N. D. & MAYES, D. M. (1981) Plasma corticosteroid concentrations during spironolactone administration: evidence for adrenal biosynthetic blockade in man. *Journal of Clinical Endocrinology and Metabolism*, **52**, 1057–1061.

VELTMAN, J. C. & MAINES, M. D. (1986) Regulatory effect of copper on rat adrenal cytochrome P-450 and steroid metabolism. *Biochemical Pharmacology*, **35**, 2903–2909.

VERNETTI, L. A., MACDONALD, J. R., WOLFGANG, G. H. I., DOMINICK, M. A. & PEGG, D. G. (1993) ATP depletion is associated with cytotoxicity of a novel lipid regulator in guinea pig adrenocortical cells. *Toxicology and Applied Pharmacology*, **118**, 30–38.

VINSON, G. P. & KENYON, C. J. (1978) Steroidogenesis in the zones of the mammalian adrenal cortex. In CHESTER JONES, I. & HENDERSON, I. W. (Eds), *General, Comparative and Clinical Endocrinology of the Adrenal Cortex*, Vol. 2, pp. 201–264, New York: Academic Press.

WANG, H. P. & KIMURA, T. (1976) Ferrous ion-mediated cytochrome P450 degradation and lipid peroxidation in adrenal cortex mitochondria. *Biochimica et Biphysica Acta*, **423**, 374–381.

WARNER, W., HARRIS, L. S. & CARCHMAN, R. A. (1977) Inhibition of corticosteroidogenesis by delta-9-tetrahydrocannabinol. *Endocrinology*, **101**, 1815–1820.

XU, D., VOIGT, J. M., MICO, B. A., KOMINAMI, S., TAKEMORI, S. & COLBY, H. D. (1994) Inhibition of adrenal cytochromes P450 by 1-aminobenzotriazole *in vitro*. *Biochemical Pharmacology*, **48**, 1421–1426.

YARRINGTON, J. T., LOUDY, D. E., SPRINKLE, D. J., GIBSON, J. P., WRIGHT, C. L. & JOHNSTON, J. O. (1985) Degeneration of the rat and canine adrenal cortex caused by α-(1,4-dioxido-3-methylquinoxalin-2-yl)-N-methylnitrone (DMNM). *Fundamental and Applied Toxicology*, **5**, 370–381.

6

The Adrenal Medulla as a Toxicological Target Organ

MARY J. TUCKER

Zeneca Pharmaceuticals, Alderley Park, Macclesfield

6.1 Anatomy and Physiology

The adrenal medulla occupies the central area of the adrenal glands and comprises approximately 10% of the volume of the gland. Although the general structure is similar in all species, there are variations. In the mouse, rat and hamster the glands are round, while in guinea pig, dog and man they are flat. In the primate species, the marmoset, they are a flat triangular shape, while in the Rhesus monkey the glands are elongated and flat.

In humans the blood supply to the gland includes three arteries, the superior suprarenal artery which comes from a branch of the aorta, the median suprarenal direct from the aorta, and the inferior suprarenal which branches from the renal artery; some of the larger arterioles penetrate the capsule and go direct to the medulla. The sinusoidal spaces of the medulla drain into larger veins, which unite and emerge from the gland as the single suprarenal vein joining the vena cava. In the rat there is a dual blood supply; numerous branches from the aorta penetrate the capsule and form the network of cortical capillaries, while a number of smaller arteries and arterioles pass through the cortex and form a medullary capillary complex. As in humans, a single large vein drains the adrenal gland, the right joins the aorta and the left the renal vein. Lymphatic capillaries, other than those associated with the larger veins, have not been demonstrated. The innervation of the medulla is complex, and although the major component comes from the splanchnic nerve, there are significant contributions from other nerves. The splanchnic nerve fibres which supply the medulla end in claw-like terminations around each cell; the number of synapses/cell (4–5) remains unchanged throughout life (Tomlinson & Coupland 1990). In most species the medulla is sharply demarcated from the cortex but in humans, where the adrenal gland is divided into head, body and tail, the medulla is not present in the tail portion. In the rat and hamster, sections of the medulla frequently show finger-like projections, or small groups, of cortical cells (Figure 6.1), the extent and incidence varies with strain. Some species of mammals, such as the mouse, have a well-developed adrenal medulla at birth while others, including humans, do not. It is thought that in those species with less well-

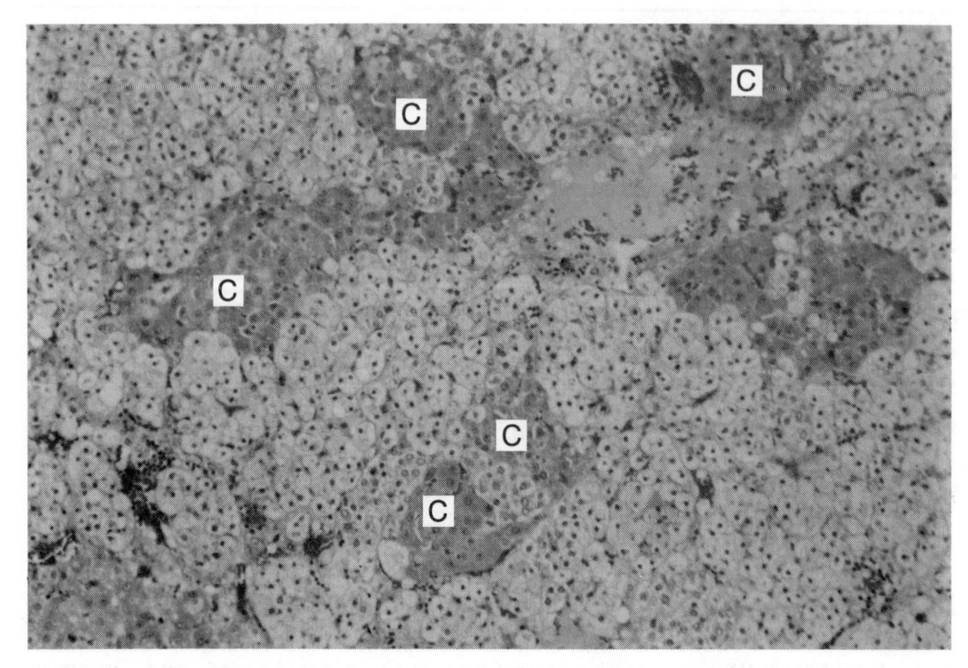

Figure 6.1 Normal rat adrenal medulla. Small groups of cortical cells (C) interspersed between the cells of the medulla. Haematoxylin & eosin (H & E)

developed medullas, the function is, for some time, performed by the aortic paraganglia (the organs of Zuckerlandl). Since species such as the mouse do not possess these paraganglia, it is essential for them that the adrenal medulla is fully functional at birth. The histological appearance of the cells of the medulla is primarily of regular polyhedral chromaffin cells arranged in small groups or cords surrounded by sinusoidal venules (Figure 6.2). The 'glandular' pattern is more clearly-defined in some species, eg the marmoset, when compared with the rat (Figure 6.3). The cytoplasm of the cells is extensive, and in sections stained with haematoxylin and eosin stains more basophilic than cortical cells; nuclei are small, round and contain little chromatin. Mitotic figures are rare in the medullary cells of the adult animal, but in the rat it has been shown that a small proportion of cells do divide throughout life (Malvaldi *et al.* 1968, Coupland & Tomlinson 1989, Jones & Clarke 1993).

Ultrastructural examination of medullary cells shows that the catecholamines, adrenaline and noradrenaline, are present as granules, the latter being more electron-dense. In mature rats and mice they are located in different cells, but in humans, and immature rats, within the same cell. The traditional method for demonstrating catecholamines is the chromaffin reaction; in sections stained with potassium dichlorate, medullary cells stain a brown colour due to the oxidation of stored catecholamines. This is an insensitive method which has been superseded by immunohistochemical techniques which are specific for catecholamines and the synthesizing enzymes of the biosynthetic pathways. These techniques can be used on formalin-fixed, paraffin-embedded tissues (Verhofstad *et al.* 1985). Another component of chromaffin cells, chromogranin A, can also be identified immunohistochemically, and used to distinguish chromaffin cells from sympathetic neurones

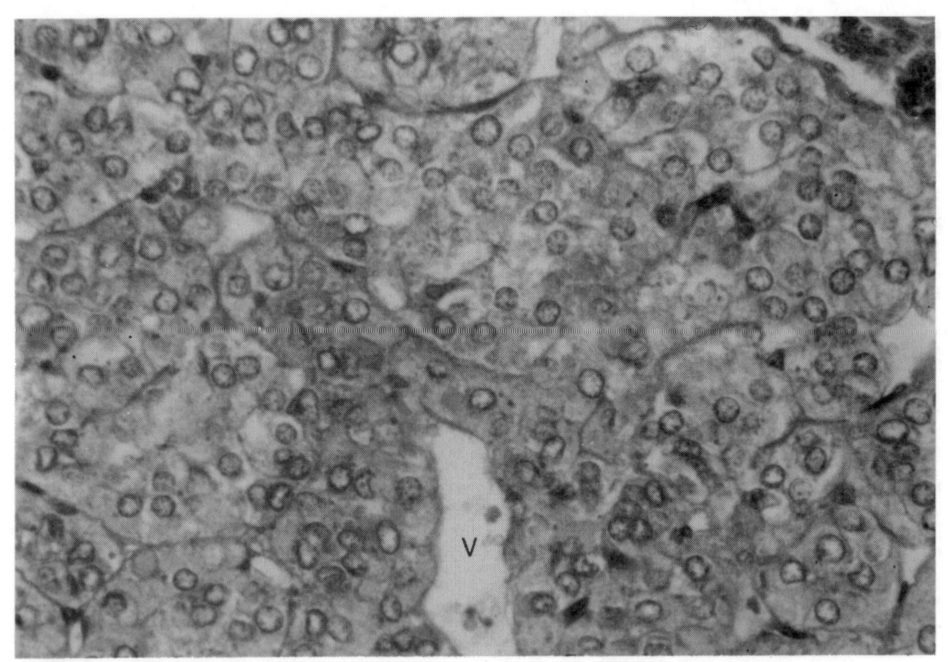

Figure 6.2 Chromaffin cells. Regular polyhedral cells in small groups with extensive pale cytoplasm and round pale nuclei containing little chromatin, and a small sinusoidal venule (V). Haematoxylin & eosin (H & E)

(Fischer-Colbrie *et al.* 1985). In addition to the chromaffin cells, the medulla contains single, or grouped, ganglion cells (Figure 6.4) whose axons synapse around the chromaffin cells. In microscopic sections of the medulla of rodents, ganglion cells are always few in number and may be absent. A third cell type, possibly intermediate between chromaffin cells and ganglion cells, has been designated a small-granule-containing (SGC) cell.

The physiological functions of the medulla are centred around the activity of the chromaffin granules. These complex structures have a variety of functions (Winkler *et al.* 1986); other components of the chromaffin granules include neuropeptides and encephalins. The function and control of these components is not as clearly understood as that of the catecholamines for which the biosynthetic pathways are well known (Hedge *et al.* 1987). The substrates for synthesis are phenylalanine and tyrosine. The former is converted by phenylalanine hydroxylase to tyrosine, which is the initial substrate in the catecholamine pathway. Tyrosine hydroxylase converts tyrosine to dopa, the rate-limiting step in the pathway, and dopa is then converted to dopamine by dopa carboxylase. These stages all occur in the cytosol. Dopamine then enters the chromaffin granules to undergo conversion to noradrenaline by dopamine hydroxylase and re-enters the cytosol for the final conversion to adrenaline by phenylethanolamine-N-methyltransferase (PNMT). It is believed that two types of chromaffin cell exist; one stores and secretes adrenaline, the other noradrenaline. PNMT is found only in the adrenaline-secreting cells thus restricting the production to these cells. Synthesis of PNMT itself may be induced by glucocorticoids, and the sinusoids which drain the adrenal cortex contain high concentrations

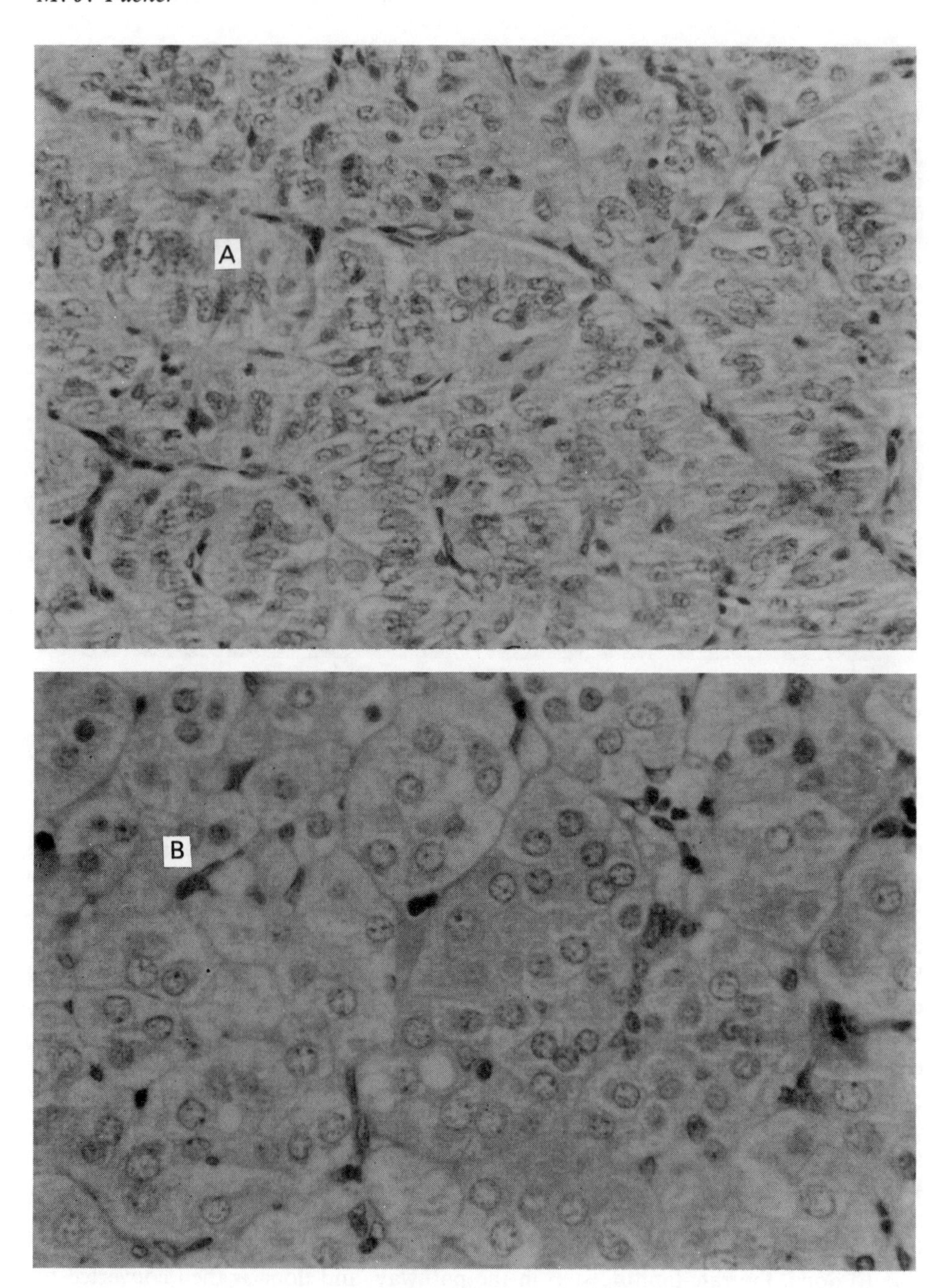

Figure 6.3 Species differences in medullary architecture. Normal adrenal medulla of (A) marmoset (*Callithrix jacchus*) and (B) rat showing differences in cell patterns.

of cortisol, a potent glucocorticoid. Thus the function of the medulla is partly controlled by the adrenal cortex (Neville 1969). In humans $>80\%$ of the catecholamines are stored in the adrenal as adrenaline; in the Long-Evans rat the level is $>90\%$ (Tischler *et al.* 1985). Chromogranin proteins and neuropeptides, in contrast

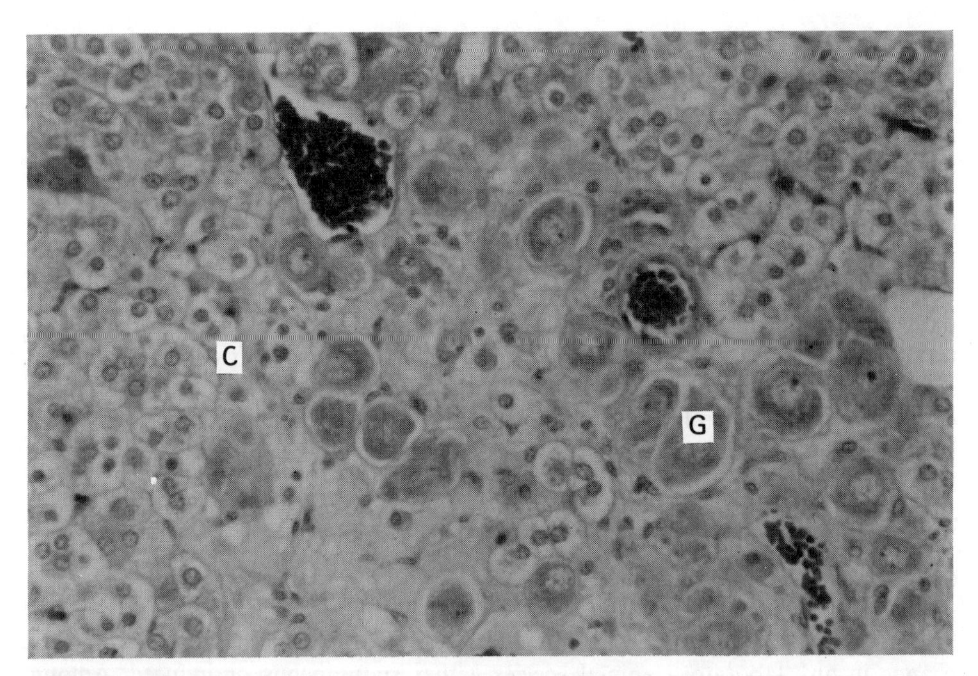

Figure 6.4 Medullary ganglion cells. Groups of large ganglion cells (G) surrounded by chromaffin cells (C) in rat adrenal medulla.

to the catecholamines, are synthesized in the endoplasmic reticulum (ER) and stored in the Golgi apparatus. Medullary cells also contain serotonin and histamine, but it is not known if they are synthesized in the cells or taken up from the circulation.

Chromaffin cell function is regulated by humoral and neurogenic signals; the latter predominate in the adult. The sequence of events which lead to medullary catecholamine secretion is known as stimulus–secretion coupling. Sympathetic stimulation causes release of acetylcholine from the pre-ganglionic nerve synapses at the chromaffin cells which depolarizes the cell membrane and increases calcium permeability. This increased intracellular calcium stimulates the release of catecholamines from the chromaffin granules by exocytosis. The granules move to, and fuse with, the cell membrane, releasing the catecholamines and other products into the extracellular space. A number of additional factors have been identified in the control of catecholamine synthesis. Other neurotransmitters and peptide hormones react with their receptors and pass into the cells by transduction pathways that involve adenosine monophosphate, protein kinase A, phosphatidylinositol and protein kinase C. All of these pathways interact at many levels to regulate hormone synthesis (Tischler & DeLellis 1988). Other factors which may modify function include steroid hormones (Unsicker 1989) and the extracellular matrix (Doupe *et al.* 1985), in addition to growth factors which act upon receptors for protein kinase.

Approximately 50% of the catecholamines secreted by the medulla circulate as free hormone, the remainder is bound to albumin. Catecholamines are cleared from the plasma by enzymatic conversion to other (inactive) products, uptake by other tissues, and by enzymatic degradation and excretion through liver and kidneys. The depletion of circulating catecholamines produces a reflex increase in splanchnic nerve discharge to stimulate catecholamine secretion and biosynthesis of tyrosine hydroxylase and other enzymes. Among the more important events which trigger catecholamine production are hypoglycaemia, hypothermia, hypoxia and stress. Synthesis also increases with age in the rat, where increased levels are related to increased splanchnic nerve discharges observed after the age of 300 days (Ito *et al.* 1986). Differences have been demonstrated between strains in the response to stress, with Wistar-Kyoto rats showing two-fold higher levels of catecholamines, in response to foot shock, than Brown-Norway rats (McMarty & Kopin 1978).

Catecholamines act on two major receptor types which are widely and differentially distributed throughout the tissues: the α and β receptors. In general, α receptor stimulation is excitatory, causing effects such as vascoconstriction and muscle contraction. Activation of β_1 receptors in the heart also has excitatory effects on contractility, heart rate and conduction velocity. β_2 receptors mediate various metabolic processes such as glycogenolysis and lipolysis, arteriolar dilatation, and relaxation of bronchial and gastrointestinal muscle.

6.2 Non-proliferative Conditions

The adrenal medulla does not produce a large range of non-proliferative histological change in any laboratory animal species, either spontaneous or induced. Among reported spontaneous conditions inflammation is rare, although small inflammatory cell infiltrates are not an uncommon finding (Figure 6.5). Necrosis is even less frequently seen. In the mouse, virus infections may produce changes in medullary cells,

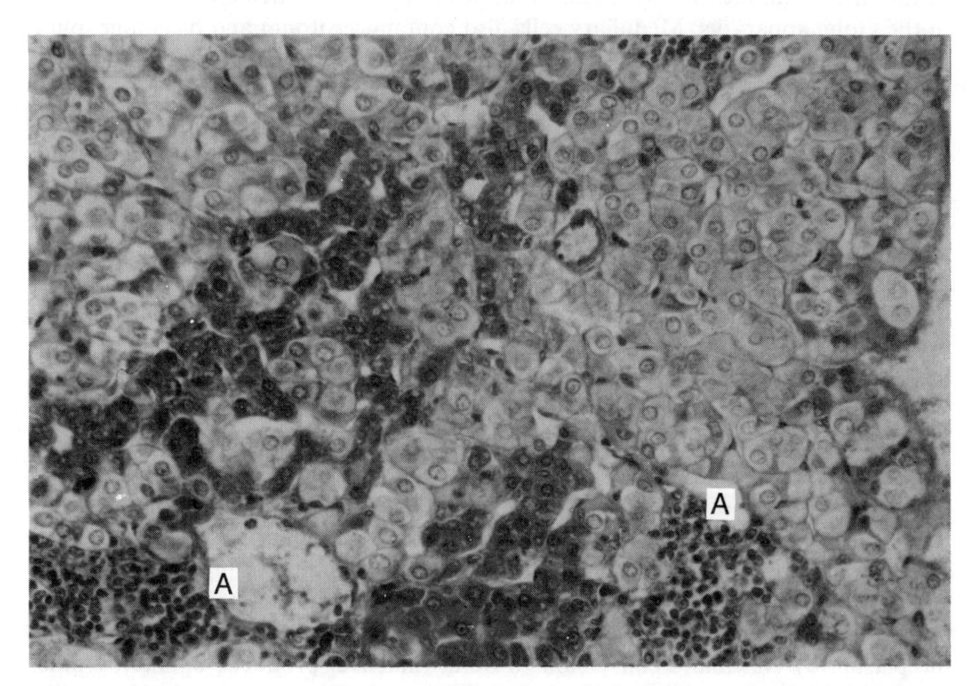

Figure 6.5 Infiltrates in rat medulla. Two small foci of mixed inflammatory cell infiltration (A) in rat adrenal medulla.

although the cortex is more commonly involved. The type and incidence of the response to the virus varies with the strain. Mouse hepatitis virus may produce syncitia in medullary cells and adenovirus, cellular disintegration and inflammation (Barthold 1983). Spontaneous physiological changes have been demonstrated frequently with many of the conditions which fall loosely into the category of stress. An example is that reported by Kvetnansky *et al.* (1971) where catecholamine levels were shown to rise in rats during and after stress induced by immobilization. Relatively little has been published on the toxic effects of chemicals in the adrenal medulla; one reason may be the difficulty of demonstrating changes in medullary function. Clearly a sustained increase in catecholamines may have important effects on many distant organs, but the causal relationship to altered adrenal function may not be recognized in the absence of any obvious morphological change in the adrenal glands. Toxicological pathologists are likely to encounter few non-proliferative changes in the medulla. Focal necrosis (Figure 6.6) has been reported after administration of cysteamine hydrochloride (McComb *et al.* 1981), acrylonitrile (Szabo *et al.* 1980) and pyrazole (Szabo *et al.* 1981). It is not known if the necrosis is a direct effect of these chemicals on the cells. Fatty change in medullary cells (Figure 6.7) has been seen in rats treated with triparanol (Lullman-Rauch & Reil 1974) and tamoxifen (Lullman & Lullman-Rauch 1981), and cynomolgus monkeys treated with a novel anti-neoplastic agent (Gopinath *et al.* 1987). Other histological changes include intracytoplasmic eosinophilic inclusion bodies, which are predominately periodic acid-Schiff positive, and have been reported in rhesus monkeys treated with polychlorinated biphenyls and also by guinea pigs treated with a dioxin derivative (McConnell & Tally 1977). This range of structural changes is small, as might be expected in a tissue of uniform cells and structure.

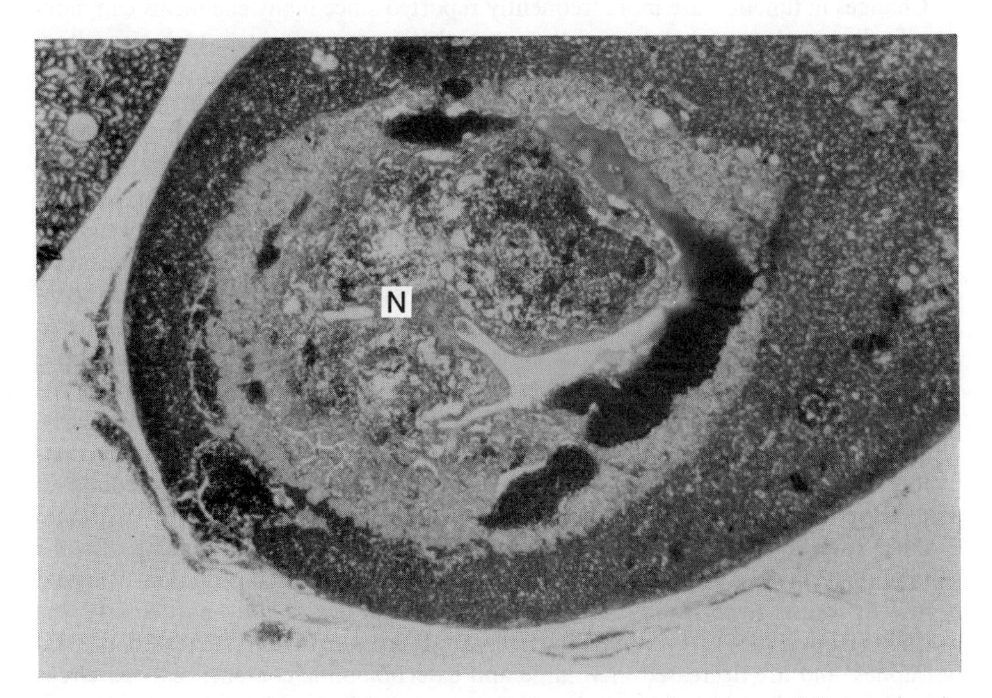

Figure 6.6 Necrosis of adrenal medulla. Large central area of medullary necrosis (N) in rat adrenal medulla.

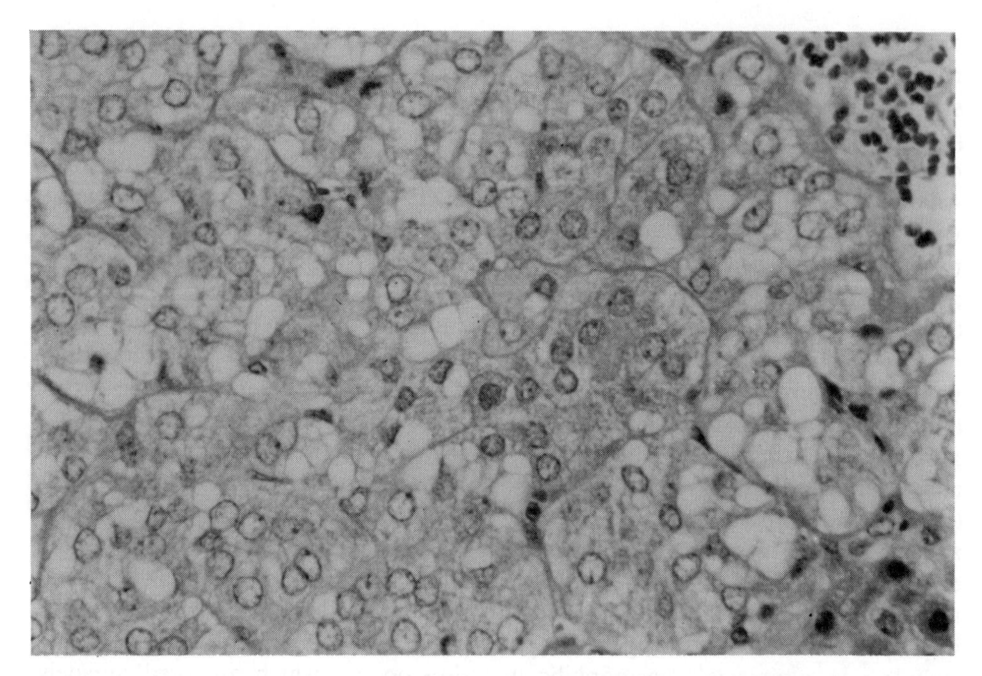

Figure 6.7 Fatty change in adrenal medulla. Fat vacuoles in medullary chromaffin cells rat adrenal medulla.

Changes in function are more frequently reported since many chemicals can, non-specifically, produce stimulation of the sympathetic nervous system with a resultant increase in catecholamine levels. Such effects can be of short duration as other pathways are initiated to counter the increase, and medullary cells will also rapidly return to normal when the cause of the stimulation has been removed. Changes in catecholamine synthesis should result in a change in medullary cell size, but this may be difficult to detect. Adrenal gland weight is not a good indicator of medullary hypertrophy since the medulla contributes such a small proportion to the weight of the whole gland and, because of the presence of groups of cortical cells, accurate enucleation and isolation of the medulla is impossible. Morphometric analysis of individual cell size is difficult as care must be taken with fixation of the tissue to avoid shrinkage of the cells. Also, as the proportion of medullary cells present in a microscopic section varies with the plane of section, it is difficult to standardize the numbers of cells examined, and multiple sections should be analysed. Similarly, although the splanchnic nerve is the chief source of sympathetic stimulation, other factors are involved and it is not possible to surgically denervate the medulla to study the effects of sympathetic stimulation.

Since catecholamines have widespread activity in the body, the toxicity of some xenobiotics can be attributed to effects produced by increased secretion. Thyroid hormones cause hypertrophy of the whole adrenal gland, but particularly the medulla (Hopsu 1960). Noradrenaline cells are more sensitive to increased thyroid hormones, and are decreased in volume and catecholamine content; reduced circulating levels of thyroid hormones have the opposite effect with an increase in noradrenaline and adrenaline secretion. Long-term administration of growth hormone

(GH) to rats causes medullary hypertrophy without increased catecholamine levels (Evans *et al.* 1948). In mice, GH did not affect adrenal weight, but increased catecholamine levels (Hopsu 1960).

Reserpine causes a reflex increase in the activity of cholinergic nerves in the medulla (USDHHR 1982), with a consequent stimulation of adrenal catecholamines. Nicotine is known to directly stimulate nicotinic acetylcholine receptors and has been shown to increase adrenal catecholamine levels in rats (Boelsterli *et al.* 1984), and circulating adrenaline levels in guinea pigs (Hexum & Russet 1987). Insecticides are another class of chemicals which have been shown to affect several endocrine glands including the adrenal medulla. Chlordecane significantly reduces adrenaline levels and increases noradrenaline; the mechanism is thought to be an inhibition of PNMT, with a subsequent decreased adrenaline synthesis. In addition chlordecane may inhibit mitochondrial function with decreased uptake into storage granules (Baggett *et al.* 1980). Another insecticide, 1-methyl-4-phenyl-1,2,3,6-tetrahydropyridine (MPTH), administered to rats decreased adrenaline levels and tyrosine hydroxylase activity (Ambrosio & Mahy 1989).

Administration of sugar alcohols (polyols), such as mannitol and xylitol, to rodents produced hypertrophy of the adrenal medulla (Salsbury 1980, Boelsteri & Zbinden 1985, Bär 1988). These compounds are thought to produce their effects by altering calcium homeostasis; they have been shown to increase calcium absorption from the intestinal tract (Roe & Bär 1985) and thus may increase intracellular calcium and influence catecholamine production. Few other compounds have been reported in the literature as medullary toxins but those that have are reviewed by Ribelin (1984) and Colby & Longhurst (1992).

6.3 Proliferation Conditions

The low incidence of inflammatory and degenerative changes in the adrenal medulla is in marked contrast to the frequency of proliferative lesions seen in the rat. In this respect, the rat differs markedly from humans and other laboratory animal species. Hyperplasia involves both adrenaline- and noradrenaline-producing cells (Tischler *et al.* 1985), but the hyperplastic cells are thought to have normal innervation (Tomlinson & Coupland 1990). Diffuse hyperplasia occurs when the medullary cells are increased without nodular formation or cortical compression (Figure 6.8). It is a bilateral and multicentric condition and varies with strain; the incidence in a range of rat strains has been cited by Tischler & Coupland (1994). In male Wistar rats the number of adrenaline-producing cells increases by 40%, and noradrenaline cells by 60%, in animals aged 2 years. In the Long-Evans rat (which is Wistar-derived) the ratio of medulla to cortex changes with age, and at 2 years is almost double that at 1 year (Tischler *et al.* 1985). The individual cells may be hypertrophied and, ultrastructurally, show increased numbers of lysosomes.

Nodular hyperplasia also occurs frequently as a spontaneous and induced condition of the adrenal medulla. The frequency of the condition is clearly genetically controlled, and those strains prone to develop the condition spontaneously are more likely to do so in response to a variety of stimuli such as hormones and xenobiotics (Boelesterli & Zbinden 1985). Nodules of hyperplastic cells may be located in any part of the medulla, but are often positioned at the juxta-cortical

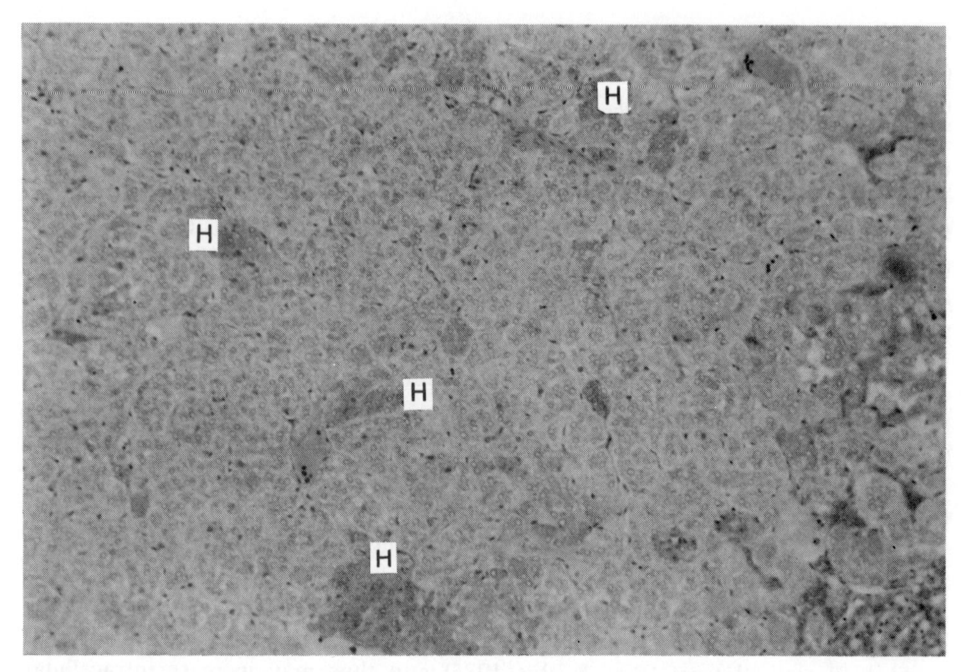

Figure 6.8 Diffuse medullary hyperplasia. Small groups of darker stained hyperplastic cells (H) scattered through rat medulla.

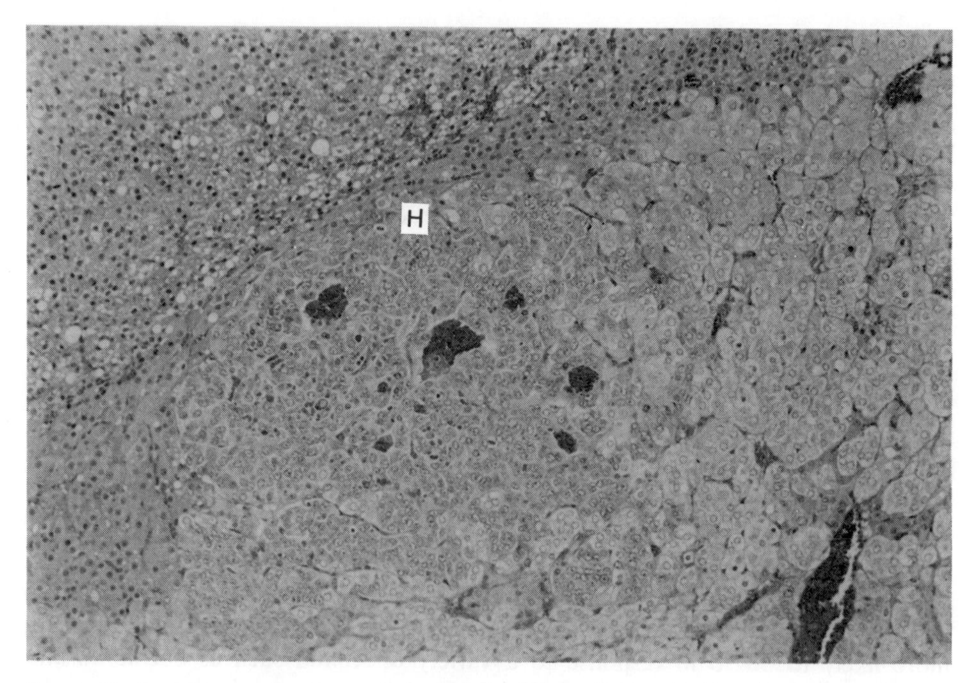

Figure 6.9 Nodular medullary hyperplasia. Nodule of hyperplastic cells (H) at the juxta-cortical region rat medulla.

region of the medulla (Figure 6.9). The cells are usually smaller than normal, polygonal in shape with a high nuclear/cytoplasmic ratio; the cytoplasm is basophilic and the cells may show mitotic figures. This condition has been described in different strains by various authors including Yeakel (1947), Strandberg (1983) and Roe & Bär (1985). The distinction between hyperplasia of medullary cells and neoplasia (phaeochromocytoma) is somewhat arbitrary, with some pathologists ignoring the condition of hyperplasia and defining all proliferative lesions as neoplastic (Hollander & Snell 1976). This has made the published data for the incidences of medullary tumours unreliable. Roe & Bär (1985) have described a five-point grading system for hyperplasia from Grade 1, with only one or two small hyperplastic nodules, to Grade 5, with a hyperplastic nodule occupying at least two-thirds of the medulla. They suggest that a Grade 6 would, therefore, represent a benign phaeochromocytoma. The diagnostic criterion distinguishing a large hyperplastic lesion from a tumour is an absence of compression of adjacent medulla and cortex; tumours, however small, should have a clear margin and show some degree of compression. There is a growing acceptance for this type of progressive development from hyperplasia to neoplasia in the rat adrenal medulla. Hyperplasia and neoplasia are more common in male rats, but the incidence varies with strain. It is apparent that, whatever the initial stimulus, any prolonged or abnormal type of stimulation of catecholamines can cause medullary cell proliferation. Support for this is provided by the work of Greenberg *et al.* (1986) who demonstrated the expression of protooncogenes after increased stimulation.

Polyols added to the diet at concentrations up to 20% increase the incidence of medullary hyperplasia (Roe & Bär 1985). The mechanism, as for the hypertrophy produced by polyols, is thought to be increased calcium absorption. In the gastrointestinal tract calcium is absorbed with monosaccharides. It has been suggested that polyols are poorly absorbed in the upper part of the tract (stomach and duodenum) and breakdown, to the more absorbable monosaccharides, occurs throughout the tract; on account of this more calcium is absorbed than occurs with more digestible carbohydrates. Increased calcium entry into the cells is ultimately responsible for stimulation of catecholamine production and the hypertrophy and hyperplasia of the chromaffin cells. The incidence of spontaneous medullary hyperplasia is also increased in rats with severe chronic progressive nephropathy (CPGN). This degenerative condition impairs the animal's ability to handle calcium overload. It is known that overnutrition, in animals-fed *ad libitum*, increases the incidence of CPGN (Saxton & Kimball 1941, Bras & Ross 1964, Yu *et al.* 1982). In addition, the excessive dietary intake of calcium and vitamin D in animals with impaired renal function has widespread effects, with parathyroid hyperplasia, metastatic calcification and an increased incidence of proliferative lesions in the medulla.

Hormones are another group of stimuli known to influence adrenal activity and may also increase proliferative medullary lesions. A clear relationship exists, in the rat, between anterior pituitary hormones and adrenal medullary hyperplasia, but as with all endocrine functions this is a complex, multifaceted relationship. Medullary hyperplasia and neoplasia may be induced by the administration of GH (Moon *et al.* 1950) and GH is 40% homologous with pituitary prolactin and, in high concentrations, can bind to prolactin receptors. Prolactin release in the rat is associated with stress (Minamitani *et al.* 1987), and is also associated with increased production and secretion of catecholamines. In many strains prolactin-secreting pituitary tumours are common (Lee *et al.* 1982). The pituitary prolactin production is controlled by dopamine which acts as an inhibitor; thus any xenobiotic which inhibits dopamine itself, which includes psychotropic agents, may increase prolactin secretion and consequently medullary hyperplasia and neoplasia.

The most common tumour of the adrenal medulla is the phaeochromocytoma which may be benign or malignant and is usually unilateral. As with all tumours the incidence varies with strain but, as with hyperplasia, phaeochromocytomas are more common in males (Cheng 1980), with some strains such as the Fischer 344 reporting incidences of 17% in males and 3.5% in females (Tischler & DeLellis 1988). Malignant phaeochromocytomas are uncommon and distant metastases rare (Goodman *et al.* 1979). A relationship between the incidence of spontaneous phaeochromocytomas and CPGN was first observed by Gilman *et al.* (1953). The mechanism, as for medullary hyperplasia, is likely to be due to effects on calcium homeostasis. The histological appearance of the tumours is of small cells showing a degree of nuclear pleomorphism and a more intense basophilia. Compression of adjacent tissue is invariably present and there is usually a clearly-defined margin (Figure 6.10). Increasing atypia and mitotic activity may be seen with the larger tumours (Figure 6.11). The appearance of phaeochromocytomas in the rat is different from that of other species including humans, where the tumours retain the pale frothy appearance of the normal medullary cell. In addition, changes in blood pressure were reported to be minimal in rats while they are a common feature of the tumour in man (Ribelin *et al.* 1984). This led to speculation that the rat phaeochromocytoma is different from other species, however more recent ultrastructural and functional studies tend to disprove this. The tumours show chromaffin granules and increased production of noradrenaline only (Bosland & Baer 1984). The reason for this is probably clarified by the work of Tischler *et al.* (1990) who demonstrated that in

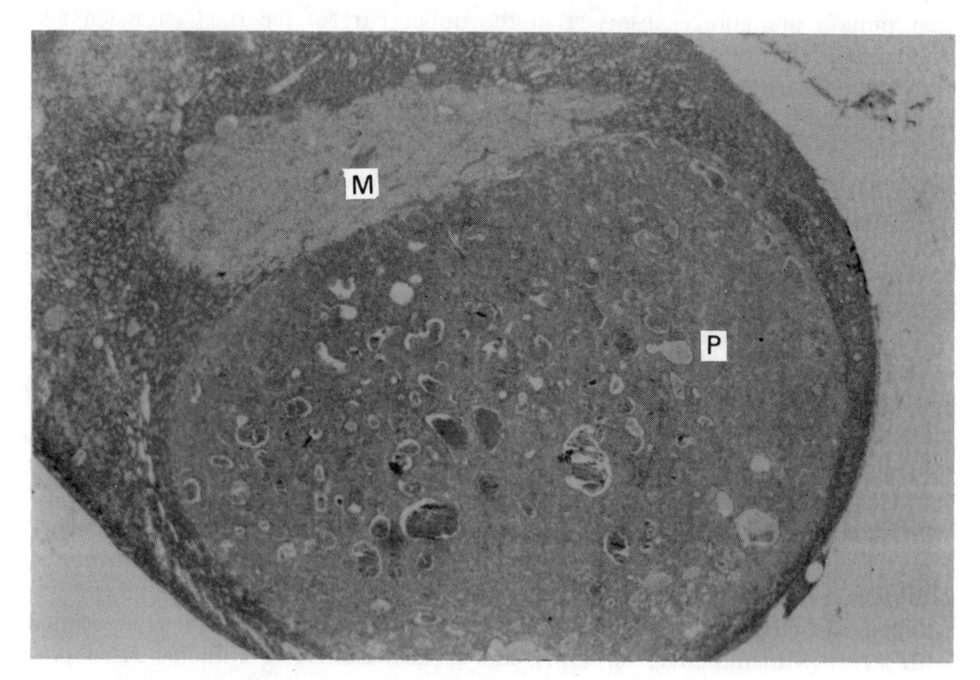

Figure 6.10 Phaeochromocytoma rat adrenal. Large phaeochromocytoma (P) below medulla (M) showing a clearly defined margin and compression of the cortex.

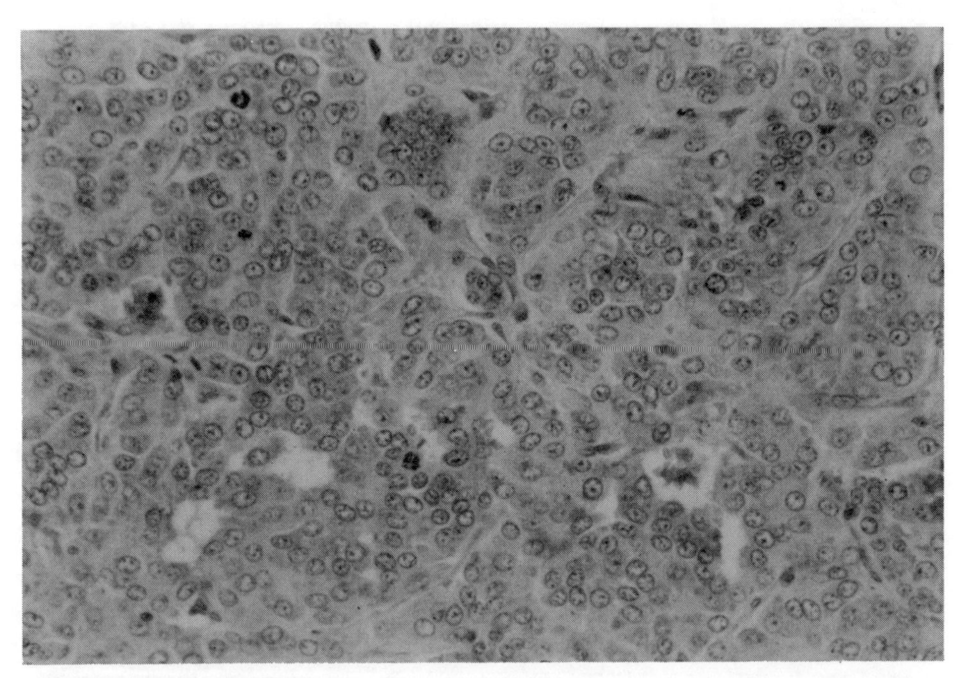

Figure 6.11 Phaeochromocytoma rat adrenal. Phaeochromocytoma showing disorganized architecture, cellular atypia and mitotic figures.

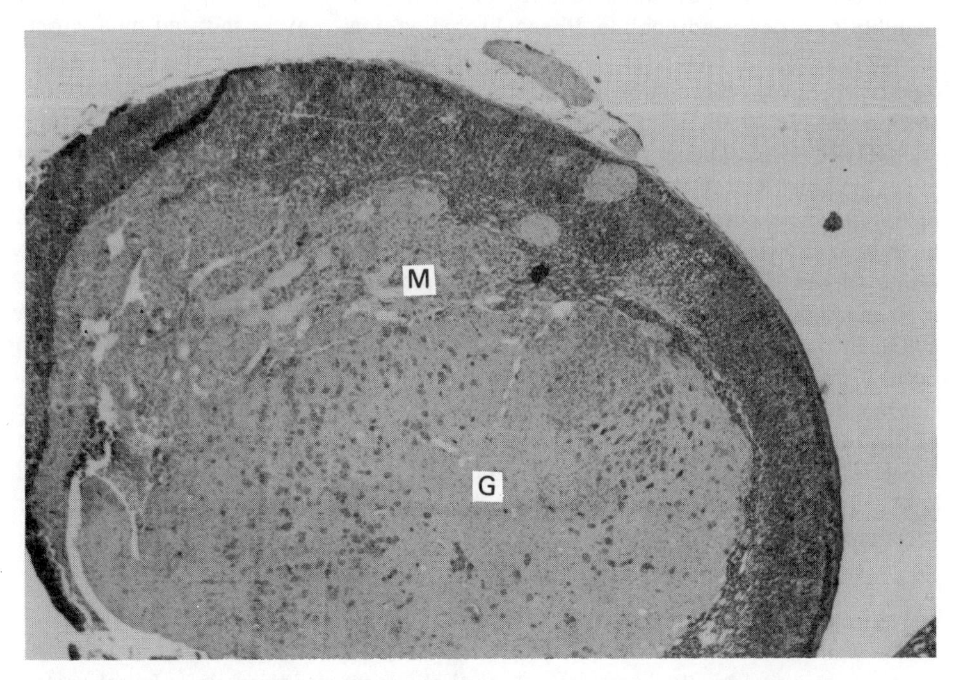

Figure 6.12 Ganglioneuroma rat adrenal. Ganglioneuroma (G) below medulla (M) showing scattered ganglion cells in a neuromatous stroma.

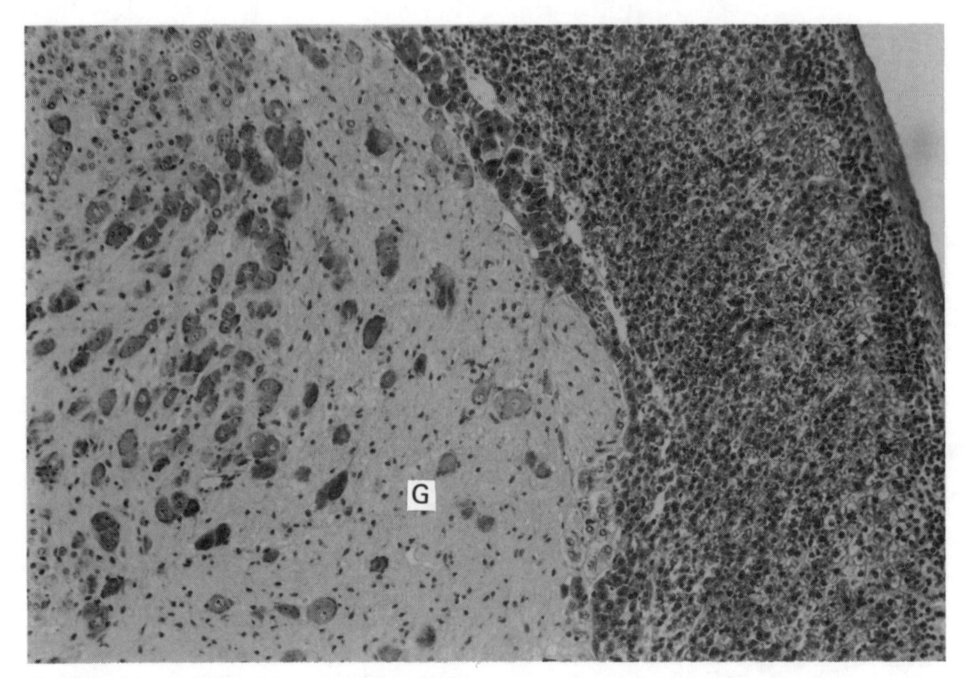

Figure 6.13 Ganglioneuroma rat adrenal. Higher-power view of ganglioneuroma to show scattered ganglion cells and neuromatous stroma.

phaeochromocytomas induced by a phosphodiesterase inhibitor, tyrosine hydoxylase and dopamine hydroxylase could be identified but not PNMT, the enzyme which converts noradrenaline to adrenaline; this is consistent with the tumour's inability to produce adrenaline. Phaeochromocytomas may be induced by a variety of chemicals which have been reviewed by Russfield (1966) and Ribelin (1984). Other agents which have been shown to produce these tumours include growth hormone (Moon *et al.* 1950), nicotine (Eränkö 1955, Boelsterli *et al.* 1984), reserpine (USDHHR 1982, Diener 1988), an anti-inflammatory drug (Mosher & Kircher 1988), polyols (Roe & Bär 1985) and synthetic retinols (Kamm 1982, Kurokawa *et al.* 1985). These compounds, for which the effects on adrenal function are known, have been shown to affect catecholamine production prior to tumour development. Other species show varying incidences of phaeochromocytomas. They are uncommon in mice although they have been induced by polyoma virus (Tischler *et al.* 1993). In the hamster they are among the most common spontaneous tumours to occur (Chvédoff *et al.* 1980).

Ganglioneuroma is an uncommon medullary tumour. The distinguishing feature of this tumour is the presence of ganglion cells (Figures 6.12 and 6.13) and this and other features have been described by Glaister *et al.* (1977) and Reznik & Ward (1983). There are no data on the induction of this type of tumour by any chemical or other stimulus.

6.4 Summary and Conclusions

In comparison with other organs such as the kidney, and even the adrenal cortex, toxic effects in the adrenal medulla are infrequent. This would seem to be an inconsistent observation when one considers the widespread activities of catecholamines and the diverse stimuli which affect them. It is possible that our inability, until recently, to detect functional changes may account for this. As yet little is known of the effects of variables such as age, sex and hormonal status in the normal rat or in those with induced medullary lesions. Even less is known in other common laboratory species such as the mouse and dog. Further investigations of the physiological control and function of the organ should prove fruitful and contribute to our understanding of the medulla and its response to toxins.

References

AMBROSIO, S. & MAHY, N. (1989) Acute peripheral catecholamine changes in rat after MPTP and MPP + treatment. *Revista Espanôla de Fsiologia*, **45**, 157–161.

BÄR, A. (1988) Sugars and adrenomedullary proliferative lesions: the effects of lactose and various polyalcohols. *Journal of the American College of Toxicology*, **7**, 71–81.

BAGGETT, J. McC., THURESON-KLEIN, A. & KLEIN, R. L. (1980) Effects of chlordecone on the adrenal medulla of the rat. *Toxicology and Applied Pharmacology*, **52**, 313–322.

BARTHOLD, S. W. (1983) Adrenal lesions due to infections. In JONES, T. C., MOHR, U. & HUNT, R. D. (Eds), *Endocrine System*, pp. 85–97, Berlin: Springer.

BOELSTERLI, U. A. & ZBINDEN, G. (1985) Early biochemical and morphological changes of the rat adrenal medulla induced by xylitol. *Archives of Toxicology*, **57**, 25–30.

BOELSTERLI, U. A., GRUZ-ORIVE, L. M. & ZBINDEN, G. (1984) Morphometric and biochemical analysis of adrenal medullary hyperplasia induced by nicotine in rats. *Archives of Toxicology*, **56**, 113–116.

BOSLAND, M. C. & BAER, A. (1984) Some functional characteristics of adrenal medullary tumours in aged male Wistar rats. *Veterinary Pathology*, **21**, 129–140.

BRAS, G. & ROSS, M. H. (1964) Kidney disease and nutrition in the rat. *Toxicology and Applied Pharmacology*, **6**, 246–262.

CHENG, L. (1980) Phaeochromocytoma in rats: incidence, etiology, morphology and functional activity. *Journal of Environmental and Pathological Toxicology*, **4**, 219–228.

CHVÉDOFF, M., CLARKE, M. R., IRISARRI, E., FACCINI, J. M. & MONRO, A. M. (1980) Effects of housing conditions on food intake, body weight and spontaneous lesions in mice. A review of the literature and results of an 18 month study. *Food and Cosmetic Toxicology*, **18**, 517–522.

COLBY, H. D. & LONGHURST, P. A. (1992) Toxicology of the adrenal gland. In ATTERWILL, C. K. & FLACK, J. D. (Eds), *Endocrine Toxicology*, pp. 243–281, Cambridge: Cambridge University Press.

COUPLAND, R. E. & TOMLINSON, A. (1989) The development and maturation of the adrenal medullary chromaffin cells of the rat *in vivo*. A descriptive and quantitative study. *International Journal of Developmental Neuroscience*, **7**, 419–438.

DIENER, R. M. (1988) Phaeochromocytomas and reserpine; review of carcinogenicity bioassay. *Journal of the American College of Toxicology*, **7**, 95–109.

DOUPE, A. J., LANDIS, S. C. & PATTERSON, P. H. (1985) Environmental influences in the development of neural crest derivatives; glucocorticoids, growth factors, and chromaffin cell plasticity. *Journal of Neuroscience*, **5**, 2119–2142.

ERÄNKÖ, O. (1955) Nodular hyperplasia and increase noradrenaline content in the adrenal medulla of nicotine-treated rats. *Acta Pathologica Microbiologica*, **36**, 210–218.

EVANS, H. M., SIMPSON, M. E. & LI, C. H. (1948) The gigantism produced in normal rats by injection of pituitary growth hormone. 1. Body growth and organ changes. *Growth*, **12**, 210–218.

FISCHER-COLBRIE, R., LASSMAN, H., HAGN, C. & WINKLER, H. (1985) Immunological studies on the distribution of chromogranin A and B in the endocrine and nervous tissues. *Neuroscience*, **16**, 547–555.

GILMAN, J., GILBERT, C. & SPENCE, I. (1953) Phaeochromocytoma in the rat. Pathogenesis and collateral reactions and its relation to comparable tumours in man. *Cancer*, **6**, 494–511.

GLAISTER, J. R., SAMUELS, D. M. & TUCKER, M. J. (1977) Ganglioneuroma-containing tumours of the adrenal medulla in Alderley Park rats. *Laboratory Animals*, **11**, 35–37.

GOODMAN, D. G., WARD, J. M., SQUIRE, R. A., CHU, K. C. & LINHART, M. S. (1979) Neoplastic and non-neoplastic lesions in ageing F344 rats. *Toxicology and Applied Pharmacology*, **48**, 237–248.

GOPINATH, C., PRENTICE, D. E. & LEWIS, D. J. (1987) The endocrine glands. In *Atlas of Experimental Toxicological Pathology*, pp. 108–109, Lancaster: MTP Press Ltd.

GREENBERG, M. E., ZOFF, E. B. & GREENE, L. A. (1986) Stimulation of neuronal acetylcholine receptors induces rapid gene transcription. *Science*, **234**, 80–83.

HEDGE, G. A., COLBY, H. D. & GOODMAN, R. L. (1987) *Clinical Endocrine Physiology*, Philadelphia: W. B. Saunders.

HEXUM, T. D. & RUSSET, L. R. (1987) Plasma encephalin-like peptide response to chronic nicotine infusion in guinea pig. *Brain Research*, **406**, 370–372.

HOLLANDER, C. F. & SNELL, K. C. (1976) Tumours of the adrenal gland. In TURUSOV, V. S. (Ed.), *Pathology of Tumours in Laboratory Animals*, Vol. 1, *Tumours of the Rat*, Part 2, pp. 273–293, Lyon: IARC Scientific Publication No. 6.

HOPSU, V. (1960) Effects of experimental alterations of thyroid function on the adrenal medulla of the mouse. *Acta Endocrinologica Supplementum*, **48**, 1–87.

ITO, I., SATO, A., SATO, Y. & SUZUKI, H. (1986) Increase in adrenal catecholamine secretion and adrenal sympathetic nerve unitary activities with aging in rats. *Neuroscience Letters*, **69**, 263–268.

JONES, H. B. & CLARKE, N. A. B. (1993) Assessment of the influence of subacute phenobarbitone administration on multi-tissue cell proliferation in the rat using bromodeoxyuridine immunocytochemistry. *Archives of Toxicology*, **67**, 622–628.

KAMM, J. J. (1982) Toxicology, carcinogenicity and teratogenicity of some orally administered retinoids. *Journal of the American Academy of Dermatology*, **6**, 652–659.

KUROKAWA, Y., HAYASHI, Y., MAEKAWA, A., TAKAHASHI, M. & KUKUBO, T. (1985) High incidences of phaechromocytoma after long term administration of retinal acetate to F344/Du Crj rats. *Journal of the National Cancer Institute*, **74**, 715–723.

KVETNANSKY, R., WEISE, V. K., GEWIRTZ, G. P. & KOPIN, I. J. (1971) Synthesis of catecholamines in rats during and after immobilization stress. *Endocrinology*, **89**, 46–49.

LEE, A. K., DELELLIS, R. A., BLOUNT, M., NUNNEMACHER, G. & WOLFE, H. J. (1982) Pituitary proliferative lesions in ageing male Long-Evans rats. A model of mixed multiple endocrine neoplasia syndrome. *Laboratory Investigation*, **46**, 595–682.

LULLMAN, H. & LULLMAN-RAUCH, R. (1981) Tamoxifen-induced generalized lipidosis in rats subchronically treated with high doses. *Toxicology and Applied Pharmacology*, **61**, 138–146.

LULLMAN-RAUCH, R. & REIL, G. H. (1974) Cholophentermine-induced lipidosis-like utrastructural alterations in lungs and adrenal glands of several species. *Toxicology and Applied Pharmacology*, **30**, 408–421.

MALVALDI, G., MENCACCI, P. & VIOLA-MAGNI, M. P. (1968) Mitoses in the adrenal medullary cells. *Experientia*, **24**, 475–477.

MCCARTY, R. & KOPIN, I. J. (1978) Sympatho-adrenal medullary activity and behaviour during exposure to footshock stress: a comparison of seven rat strains. *Physiology and Behaviour*, **21**, 567–572.

MCCOMB, D. J., KOVACS, K., HOFNER, H. C., GALLAHER, G. T., SCHWEDES, U., USADEL, K. H. & SZABO, S. (1981) Cysteamine-induced adrenal necrosis in rats. *Experimental and Molecular Pathology*, **35**, 422–434.

McConnell, E. E. & Tally, F. A. (1977) Intracytoplasmic hyaline globules in the adrenal medulla of laboratory animals. *Veterinary Pathology*, **14**, 435–440.

Minamitani, N., Minamitani, T., Lechman, R. M., Bollinger-Gruber, J. & Reichlin, S. (1987) Paraventricular nucleus mediates prolactin secretory responses to restraint, stress, ether stress, and 5-hydroxy-L-tryptophan injection in rat. *Endocrinology*, **120**, 860–867.

Moon, H. D., Simpson, M. E., Li, C. H. & Evans, H. M. (1950) Neoplasms in rats treated with pituitary growth hormone. II. Adrenal glands. *Cancer Research*, **10**, 364–370.

Mosher, A. H. & Kircher, C. H. (1988) Proliferative lesions of the adrenal medulla in rats treated with zomperic sodium. *Journal of the American College of Toxicology*, **7**, 83–91.

Neville, A. M. (1969) The adrenal medulla. In Symington, T. (Ed.), *Functional Pathology of the Adrenal Gland*, Part II, pp. 219–224, Edinburgh: Churchill Livingstone.

Reznik, G. & Ward, J. M. (1983) Neuroblastoma adrenal rat. In Jones, T. C., Mohr, U. & Hunt, R. D. (Eds), *Endocrine System*, pp. 35–37, Berlin: Springer.

Ribelin, W. E. (1984) The effects of drugs and other chemicals upon the structure of the adrenal gland. *Fundamental and Applied Toxicology*, **4**, 105–119.

Ribelin, W. E., Roloff, M. V. & Houser, R. M. (1984) Minimally functional rat adrenal medullary phaeochromocytomas. *Veterinary Pathology*, **21**, 281–285.

Roe, F. J. C. & Bär, A. (1985) Enzootic and epizootic adrenal medullary proliferative disease of rats: influence of dietary factors which affect calcium absorption. *Human Toxicology*, **4**, 27–52.

Russfield, A. B. (1966) Tumours of endocrine glands and secondary sex organs. PHS Publication 1332, US Department of Health, Education and Welfare, Washington DC.

Salsbury, D. (1980) The effects of life-time feeding studies on patterns of senile lesions in rats and mice. *Drug and Chemical Toxicology*, **3**, 1–33.

Saxton, J. A. & Kimball, G. A. (1941) Relation to nephrosis and other diseases of albino rats to age and to modification of diet. *Archives of Pathology*, **32**, 951–965.

Strandberg, J. D. (1983) Hyperplasia adrenal medulla, rat. In Jones, T. C., Mohr, U. & Hunt, R. D. (Eds), *Endocrine System*, pp. 18–22, Berlin: Springer.

Szabo, S.,Q Hütter, I., Kovacs, K., Howarth, E., Szabo, D. & Horner, H. C. (1980) Pathogenesis of experimental adrenal haemorrhagic necrosis ('apoplexy'). Ultrastructural, biochemical, neuropharmacologic, and blood coagulation studies with acrylonitrile in the rat. *Laboratory Investigation*, **42**, 533–546.

Szabo, S., McComb, D. J., Kovacs, K. & Hütter, I. (1981) Adrenocortical haemorrhagic necrosis. *Archives of Pathology and Laboratory Medicine*, **105**, 536–539.

Tischler, A. S. & Coupland, R. E. (1994) Changes in structure and function of the adrenal medulla. In Mohr, U., Dungworth, D. L. & Capen, C. C. (Eds), *Pathobiology of the Ageing Rat*, Vol. 2, pp. 245–268, Washington DC: ILSI Press.

Tischler, A. S. & DeLellis, R. A. (1988) The rat adrenal medulla. II. Proliferative lesions. *Journal of the American College of Toxicology*, **7**, 23–41.

Tischler, A. S., DeLellis, R. A., Perlman, R. L., Allen, J. M., Costopoulos, D., Lee, Y. C., Nunnemacher, G., Wolfe, H. J. & Bloom, S. R. (1985) Spontaneous proliferative lesions of the adrenal medulla in aging Long-Evans rats. Comparison to PC12 cells, small granule-containing cells and human medullary hyperplasia. *Laboratory Investigation*, **53**, 486–498.

Tischler, A. S., Ruzicka, L. A., Vanpelt, C. S. & Sandusky, G. E. (1990) Catecholamine-synthesizing enzymes and chromogranin proteins in drug-induced proliferative lesions in the rat adrenal medulla. *Laboratory Investigation*, **63**, 44–51.

Tischler, A. S., Freund, R., Carroll, J., Cahill, A. L., Perlman, R. L., Alroy, J. & Riseberg, J. C. (1993) Polyoma-induced neoplasms of mouse adrenal medulla. *Laboratory Investigation*, **68**, 541–549.

Tomlinson, A. & Coupland, R. E. (1990) The innervation of the adrenal gland. IV. Innervation of the rat adrenal medulla from birth to old age. A descriptive and quantitative morphometric and histochemical study of the innervation of chromaffin cells and adrenal medullary neurones in Wistar rats. *Journal of Anatomy*, **169**, 209–236.

UNITED STATES DEPARTMENT OF HEALTH AND HUMAN RESOURCES (1982) *Third Annual Report on Carcinogenesis*, Reserpine, p. 257.

UNSICKER, K. (1989) Development and plasticity of sympathoadrenal cells. *International Journal of Developmental Neuroscience*, **7**, 63–73.

VERHOFSTAD, A. A. J., COUPLAND, R. E. & PARKES, T. L. (1985) Immunohistochemical and biochemical study on the development of the noradrenaline-storing cells of the adrenal medulla of the rat. *Cell and Tissue Research*, **242**, 233–243.

WINKLER, H., APPS, D. K. & FISCHER-COLBRIE, R. (1986) The molecular function of adrenal chromaffin granules: established facts and unresolved topics. *Neuroscience*, **18**, 261–290.

YEAKEL, E. M. (1947) Medullary hyperplasia of the adrenal gland in aged Wistar rat and grey Norway rats. *Archives of Pathology*, **44**, 71–77.

YU, B. P., MASORO, E. J., MURATA, I., BERTRANSA, H. A. & LYND, F. T. (1982) Life span study of SPF Fischer 344 male rats fed *ad libitum* or restricted diets: longevity, growth, lean body mass and disease. *Journal of Gerontology*, **37**, 130–141.

SECTION FOUR

Adrenocortical and Glucocorticosteroid Modulation of Toxicity

7

Glucocorticosteroid Modulation of Toxicity

PHILIP W. HARVEY

AgrEvo UK Limited, Chesterford Park, Saffron Walden

7.1 An Introduction to Hormonal Modulation of Toxicity

Hormones of a particular endocrine axis have been used experimentally and clinically to modulate toxic insult against a component gland within the same endocrine axis. For example, hormones from the hypothalamo-pituitary-gonadal axis have been used to ameliorate testicular toxicity. Glode *et al.* (1981) reported that a GnRH analogue (D-Leu^6 GnRH) could protect against cyclophosphamide-induced testicular damage in mice. There have been similar reports in rat and dog and this has led to the suggestion that the use of GnRH analogues may be useful in a clinical setting for the protection of the testes in cancer chemotherapy (reviewed in Garside & Harvey 1992). Sex steroids are also known to protect the testes from toxic insult: in the mouse, both progesterone (Wolkowski-Tyl & Preston 1979) and oestrogen (Gunn *et al.* 1965) are protective against the necrotizing effects of cadmium to the testes. Other structurally-related steroids, such as the natural and synthetic glucocorticoids corticosterone and dexamethasone, are without such effect, and the necrotizing potential of cadmium to the testes and other organs and tissues is not modulated by these compounds, at least in the mouse (Wolkowski-Tyl & Preston 1979). In the rat, however, progesterone does not protect the testes from cadmium, indeed the reverse is true since progesterone has been reported to significantly exacerbate and enhance cadmium toxicity, especially to the liver (Shiraishi *et al.* 1993). Thus, different steroids with structural similarity but differing pharmacological activity can modulate toxicity in different ways and this is species-dependent.

Whilst it can be appreciated how hormones within an axis can modulate toxicity to a constituent gland where endocrine mechanisms are operative (for example, the use of hormones to protect the testes from cytotoxic insult has the rationale that quiescent or hormonally suppressed testes are less vulnerable to toxic insult, eg Morris & Shalet 1990, Garside & Harvey 1992), it is less clear how hormones alter toxicity in non-endocrine target organs. For example, it is not currently known how progesterone (a steroid with relatively few pharmacological actions) pre-treatment can exacerbate the general toxicity and hepatoxicity of cadmium in rats (Shiraishi *et al.* 1993). However, the literature related to hormone interactions in toxicity is broad, and that devoted to the glucocorticosteroids and their role in modulating

Table 7.1 Summary of some toxic interactions between glucocorticosteroids and metallic compounds, drugs, chemicals and natural toxins

Compound	Glucocorticosteroid	Nature of toxic interaction	Reference
(a) Metal compounds			
Cadmium chloride	Corticosterone, dexamethasone	No effect on necrotizing potential in testes and other organs (progesterone was protective)	Wolkowski-Tyl & Preston (1979)
Cisplatin	Methylprednisolone	Protection against nephrotoxicity	Koikawa *et al.* (1993)
Cisplatin	Triamcinolone	Protection against nephrotoxicity	Reznick & Gambarayan (1994)
(b) Drugs and pharmacological agents			
Methotrexate	Corticosterone, dexamethasone	Cortiscosterone ameliorates haematoligic, liver and kidney toxicity	English *et al.* (1987)
Paracetamol	Corticosterone, cortisol	Cortisol pre-treatment caused marked reduction of hepatic reduced glutathione following paracetamol treatment	Harvey *et al.* (1993)
Paracetamol	Dexamethasone	Reduction of hepatic reduced glutathione content and enhanced hepatotoxicity	Madhu *et al.* (1992)
Cephaloridine	Methylprednisolone	Protection against nephrotoxicity	Harvey *et al.* (1995)
Cisplatin	Methylprednisolone	Protection against nephrotoxicity	Koikawa *et al.* (1993)
Cisplatin	Triamcinolone	Protection against nephrotoxicity	Reznick & Gambarayan (1994)
Adrenalin	9α-Fluorocortisol	Potassium-displacing cardiotoxic effectiveness of adrenalin markedly intensified	Kimura *et al.* (1968)
Isoproterenol	Desoxycorticosterone acetate, hydrocortisone	Cardiotoxicity enormously potentiated (mediated via mineralocorticoid and blood electrolyte disturbance). Hydrocortisone less effective than desoxycorticosterone acetate.	Guideri *et al.* (1974)
3-Acetylpyridine	Corticosterone	Exacerbation of toxicity to hippocampal neurons *in vivo* and *in vitro*	Sapolsky (1985, 1986b) Sapolsky *et al.* (1988)

(c) Chemicals			
Carbon tetrachloride	Corticosterone	Exacerbation of hepatotoxicity	Lloyd & Franklin (1991)
Carbon tetrachloride	Hydrocortisone	Protection against hepatotoxicity	Sudhir & Budhiraja (1992)
Perfluoroctanoic acid	Corticosterone	Hepatomegaly mediated by corticosterone	Thotassery *et al.* (1992)
Ozone	Hydrocortisone	No effect on pulmonary oedema	Giri *et al.* (1975)
Tetraethylammonium-bromide	Hydrocortisone, dexamethasone triamcinolone	Glucocorticoids all diminished toxicity by increasing excretion	Kourounakis & Selye (1976)
Dextran	Corticosterone	Amelioration of dextran-induced shock	Kogure *et al.* (1986)
Dimethylaminoazobenzene	Hydrocortisone, corticosterone	Hydrocortisone enhances appearance of hepatic tumours (corticosterone without effect)	Lacassagne *et al.* (1966)
2,3,7,8-Tetrachlorodibenzo-*p*-dioxin	Corticosterone	Hypophysectomy aggravates toxicity, corticosterone restores tolerance	Gorski *et al.* (1988)
Ammonia	Hydrocortisone hydrochloride	Adrenalectomy induces toxic hypersensitivity, hydrocortisone hydrochloride restores tolerance	Lee *et al.* (1977)
Paraquat	Corticosterone	Exacerbation of toxicity to hippocampal neurons *in vitro*	Sapolsky *et al.* (1988)
(d) Natural toxins			
Kainic acid	Corticosterone	Exacerbation of hippocampal neurotoxicity *in vivo* and *in vitro*	Sapolsky (1986a,b) Sapolsky *et al.* (1988) Elliott & Sapolsky (1992)
Aflatoxin B1	Hydrocortisone	Marked increase of hepatotoxicity and mortalities	Chentanez *et al.* (1988)
Naja nigricollis venom	Hydrocortisone	Amelioration of hepatotoxicity but exacerbation of nephrotoxicity	Mohamed *et al.* (1974)
Trichothecene mycotoxins	Dexamethasone, predinisolone, hydrocortisone	Glucocorticoids reduce general toxicity, lethality, pulmonary oedema and gastrointestinal effects	Tremel *et al.* (1985) Mutoh *et al.* (1988) Hunder *et al.* (1994)
Vitamin D_2	Corticosterone, dexamethasone	Enhancement of toxicity	Kunitomo *et al.* (1989)

toxicological responses in non-endocrine tissues such as brain, liver and kidney is relatively advanced, where the pharmacological basis for the interaction with a toxicant is known. The glucocorticosteroids have an enormous range of fundamental metabolic, physiological and pharmacological actions (eg effects on DNA, RNA and protein synthesis, carbohydrate and protein metabolism, anti-inflammatory, weak mineralocorticoid and immunosuppressive actions and enzyme induction – see earlier chapters). Therefore, more than any other class of steroids, the glucocorticosteroids have the potential for interacting with toxicants within the context of a whole-body response.

The literature on glucocorticosteroid modulation of toxicity has partly developed from the need to better explore how, and which, factors in the endocrine stress response influence toxic responses. For reviews and discussion of stress and toxicity see Vogel (1987, 1993) and Harvey (1994). Indeed, early research was conducted by Selye, for example, on the influence of steroids and stress on the toxicity and disposition of tetraethylammonium bromide (Kourounakis & Selye 1976). The widespread clinical use of the glucocorticosteroids and the real potential for polypharmacy and drug interactions has elevated this literature from a position of research interest, to one of clear clinical relevance. Furthermore, as the glucocorticosteroids have been shown in certain incidences to prevent toxicity (methylprednisolone protects against cisplatin nephrotoxicity in the rat, eg Koikawa *et al.* 1993) there is again the possibility of major clinical applications where drug toxicity can be modified by the use of glucocorticosteroids.

This chapter reviews the literature relating to glucocorticosteroid modulation of experimental toxicity. Studies in this area conducted on laboratory rodents typically involve the administration of natural or synthetic glucocorticosteroids with a toxicant, and the effects this treatment has on a target organ is then studied with respect to the effects of the toxicant and steroid alone. Target organs most frequently studied are the brain, liver and kidney and this review is consequently organized by target organ. A summary of glucocorticosteroid modulation of toxicity is given in Table 7.1 and this is organised by chemical. Steroids most frequently employed include corticosterone, cortisol (hydrocortisone), prednisolone, methylprednisolone, dexamethasone and less frequently triamcinolone and 9α-fluorocortisol and interactions have been reported with natural toxins (eg kainic acid, aflatoxin B1, trichothecene mycotoxins and vitamin D), drugs and other pharmacologically active agents (eg methotrexate, paracetamol, cisplatin, cephaloridine, isoproterenol and adrenalin) and chemicals (eg carbon tetrachloride, paraquat, tetraethylammonium bromide, dimethylaminoazobenzene and 2,3,7,8-tetrachlorodibenzo-*p*-dioxin). The literature specifically related to glucocorticosteroid interactions with natural toxins has been previously reviewed elsewhere (Harvey *et al.* 1994).

7.2 The Brain and Neurotoxity

Glucocorticoids in general exert catabolic and, in excess, degenerative actions throughout the body, including the brain and especially the hippocampus (McEwen *et al.* 1986, Miller 1992). The hippocampus is rich in corticosteroid receptors and this fact has allowed Sapolsky and co-workers to progress a programme of research investigating interactions of corticosterone with neurotoxicants in the hippocampus.

Sapolsky (1985) investigated synergy between corticosterone (administered in dose levels to rats producing a prolonged blood concentration similar to that which occurs physiologically in the stress response) and 3-acetylpyridine. 3-Acetylpyridine (a neurotoxic antimetabolite which inhibits ATP synthesis) was microinfused into Ammons horn where it destroyed dentate gyrus neurons. Rats that had been pre-treated with corticosterone (as recently as 24 h before 3-acetylpyridine treatment) showed markedly greater volume (five-fold increase) of hippocampal damage compared with controls treated with 3-acetylpyridine only. Conversely, adrenalectomy reduced the degree of damage of 3-acetylpyridine treatment. Employing a similar experimental protocol, Sapolsky (1986a) examined potential synergy between corticosterone and a second neurotoxic agent, kainic acid (an exitotoxin operating on glutamate sensitive neurons). Rats were adrenalectomized or treated with 10 mg/day corticosterone subcutaneously and then had a microinfusion of kainic acid into the hippocampus. As in the previous study, corticosterone pre-treatment resulted in a marked increase in the volume of hippocampal damage following microinfusion of kainic acid, whereas adrenalectomy reduced the degree of damage, compared with controls treated with kainic acid only. Thus, the neurotoxicity of several agents can be exacerbated by corticosterone manipulation. Sapolsky (1986a) proposed that glucocorticoid exposure appeared to damage the hippocampus by inducing a state of 'metabolic vulnerability' in hippocampal neurons, which resulted in impaired capacity to survive a toxic insult.

In subsequent work, the mechanism of glucocorticoid-induced neuronal vulnerability to toxic insult was investigated and Sapolsky (1986b) reported that administration of brain fuels such as mannose, fructose, glucose or β-hydroxybutyrate reduced hippocampal damage following the co-administration of corticosterone with either kainic acid or 3-acetylpyridine. The critical feature of exacerbation of neurotoxicant activity by corticosterone appears to revolve around glucose ultilization, which is a fundamental peripheral action of glucocorticoids. A second mechanism may involve neuronal calcium regulation and Elliott & Sapolsky (1992) have shown that in cultured hippocampal neurons, corticosterone resulted in a large increase (23-fold) of neuronal calcium following kainic acid treatment compared with control cultures. It is worth noting that *in vitro*, corticosterone-induced reductions of viability of hippocampal neurons subjected to toxic insult from kainic acid, 3-acetylpyridine or paraquat (a superoxide radical generator) can be clearly demonstrated using extremely low concentrations of corticosterone (10^{-9} M) (Sapolsky *et al.* 1988). Presumably, the sensitivity of the hippocampus to corticosterone exacerbation of neurotoxic insult is a function of corticosteroid receptor concentration, and this may also account for the finding that relatively acute duration changes in corticosterone exposure are effective in altering neuronal response to toxic insult (Sapolsky 1985).

7.3 The Liver and Hepatotoxicity

As the major site of metabolism, the liver is a target for glucocorticoids where they modulate carbohydrate, protein and lipid metabolism, and induce glycogen deposition and gluconeogenesis (eg Haynes & Murad 1991). Glucocorticoids are also inducers of various cytochrome P450 isoenzymes (eg Ortiz de Montellano 1986, Ruckpaul & Rein 1989, Timbrell 1991) but may inhibit other enzymes including aryl hydrocarbon hydroxylase (Bogdanffy *et al.* 1986). In specific studies in rats, the detoxification of benzo(a)pyrene by ethanol is mediated by corticosterone (Bogdanffy *et al.* 1984), and corticosterone has also been shown to mediate ethanol induction of liver tryptophan oxygenase (Moerland *et al.* 1985). *In vitro*, corticosterone, hydroxycortisone and dexamethasone increase the activity of γ-glutamyltransferase in a highly differentiated hepatoma (Fao) cell line (Barouki *et al.* 1982). Additionally, hepatomegaly caused by perfluorooctanoic acid (a non-metabolizable peroxisome proliferator) has been shown to be dependent on corticosterone (Thotassery *et al.* 1992). Thus, glucocorticoids, and specifically corticosterone, exert specific metabolic-toxicokinetic, in addition to general biochemical, effects in the liver.

Lloyd & Franklin (1991) examined the modulation of carbon tetrachloride hepatotoxicity by corticosterone pre-treatments and adrenalectomy in rats. They found that when rats were treated with corticosterone (20–40 mg/kg intraperitoneally) and then treated with carbon tetrachloride (1 ml/kg) 30 min later, hepatotoxicity (measured by blood enzyme markers such as alanine aminotransferase and also by histopathological examination of livers) was increased compared with controls treated with carbon tetrachloride only. Adrenalectomy afforded some protection in carbon tetrachloride hepatotoxic challenge. The exacerbation of carbon tetrachloride hepatotoxicity by corticosterone is consistent with the concept of corticosterone-induced 'metabolic vulnerability' to toxic insult.

Harvey *et al.* (1993) investigated the effect of corticosterone, cortisol (hydrocortisone) or pregnenolone 16α-carbonitrile pre-treatments on subsequent hepatic response to paracetamol in rats. In this study, rats were pre-treated for 6 days with 20 mg/kg/day corticosterone or cortisol or pregnenolone 16α-carbonitrile, given subcutaneously, followed by a single dose of paracetamol orally (1600 mg/kg) on Day 6. Paracetamol markedly increased absolute and relative liver weights and, interestingly, both corticosterone and cortisol pre-treatment attenuated this response. Levels of reduced glutathione in the liver were measured and cortisol pre-treatment was shown to produce extreme depletion 48 hr after paracetamol, which was much more marked than that occurring in rats given paracetamol only or with corticosterone pre-treatment. Pregnenolone 16α-carbonitrile protected against paracetamol-induced glutathione depletion (which is consistent with a previous report in hamsters by Madhu & Klaasen 1991). The dose of paracetamol markedly increased blood concentrations of aspartate aminotransferase, alanine aminotransferase and sorbitol dehydrogenase (which are markers for hepatotoxicity) and whilst the highest concentrations overall were detected in the corticosterone pre-treatment group 24 hr after paracetamol, these failed to achieve statistical significance compared with results from rats given paracetamol only. Thus hepatoxicity was enhanced rather than strongly exacerbated. Madhu *et al.* (1992) report similar findings in the mouse. In their study, very high dose dexamethasone pre-treatments (75 mg/kg given intraperitoneally for 4 days) enhanced paracetamol (500 mg/kg intraperitoneally) hepatotoxicity and of particular note was that dexamethasone treatment with paracetamol reduced liver glutathione content, which is consistent with the effect of cortisol reported by Harvey *et al.* (1993) in the rat.

The results from these paracetamol studies broadly agreed with Lloyd & Franklin's (1991) work on carbon tetrachloride in that corticosterone, cortisol and dexamethasone predisposed the liver to toxic insult. However, whilst corticosterone in the rat (Harvey *et al.* 1993) and dexamethasone in the mouse (Madhu *et al.* 1992) moderately enhance paracetamol liver toxicity, cortisol appears to possess the most potent action as this compound promoted an almost complete paracetamol-induced depletion of liver glutathione in rats at equivalent, or lower dose levels than the other two steroids (Harvey *et al.* 1993). As glutathione depletion is a prerequisite factor in paracetamol hepatoxicity, with the benzoquinone imine metabolite being the proximal cause of liver damage, the depletion caused by cortisol pre-treatment is significant and warrants further study.

In fact, cortisol has been previously demonstrated to markedly exacerbate the hepatotoxicity of aflatoxin B1 (Chentanez *et al.* 1988). In this study, rats were treated for 7 days with cortisol in the dose range 1–10 mg/kg and then treated with 3 mg/kg aflatoxin B1. The cortisol pre-treatment group showed marked exacerbation of aflatoxin toxicity, compared with aflatoxin B1-treated controls, as evidenced by increased mortalities, increased activity of plasma glutamic pyruvic and oxaloacetic transaminases and increased plasma triglycerides (it should be noted that glucocorticoids increase the rate of carboxylation of pyruvate to oxaloacetate in liver mitochondria; Haynes & Murad 1991. Chentanez *et al.* (1988) also reported a dose-related increase in hepatic necrosis in cortisol pre-treated rats subsequently treated with aflatoxin B1, and the mechanism of action was attributed to increased aniline hydroxylase activity and 2,3-epoxide formation. This in turn facilitated binding to hepatic DNA and proteins and lipid peroxide formation.

Although the consensus of reports suggest that pre-treatment regimes of natural glucocorticoids such as corticosterone and cortisol exacerbate hepatotoxic insult from a variety of chemicals, there are reports of protective effects. Sudhir & Budhiraja (1992) have reported that hydrocortisone (cortisol) has a protective effect against carbon tetrachloride hepatotoxicity in rats. Similarly, in a histopathological study of changes caused by the venom of the Egyptian snake, *Naja nigricollis*, Mohamed *et al.* (1974) found that the co-administration of hydrocortisone ameliorated venom-induced hepatotoxicity. In this study, mice were injected with venom (0.2 mg/kg) alone or in combination with hydrocortisone (1.0 mg/kg) intraperitoneally. The livers of animals that had also received hydrocortisone showed less vacuolation and diminished cellular infiltration of the portal tracts compared with livers from mice treated with venom only. Interestingly, Mohamed *et al.* (1973) reported that whilst hydrocortisone protected the liver from venom toxicity, this was not the case in the kidney (see next section). Indeed, it is apparent from the studies reviewed that there is no consistency as to whether the natural or synthetic glucocorticosteroids protected from or exacerbated hepatotoxicity. This draws attention to the dangers of over generalizing the effects of glucocorticoids and of ignoring the specific mechanism of action of the co-administered toxicant.

7.4 The Kidney and Nephrotoxicity

The natural glucocorticosteroids, particularly corticosterone, possess weak mineralocorticoid activity and promote water and sodium retention in the nephron. Although the relative potency of the natural glucocorticoids is much lower than that of aldosterone, their higher blood concentration allows compounds such as corticosterone and cortisol to exert biologically relevant effects on the kidney (eg Bowman & Rand 1984). The concentration of solutions containing excreted toxicants, and the subsequent re-absorption of these solutions by co-administered corticosteroid

Figure 7.1 Outer cortex of kidney sampled 96 hr after treatment [haematoxylin & eosin (H&E) × 220]. (a) Section from a rat treated with cephaloridine only, showing severe toxic nephrosis characterized by coagulative necrosis of the convoluted proximal tubules and granular cytoplasm. NT, Necrotic tubule. (b) Section from a rat treated with cephaloridine and methylprednisolone, showing only slight evidence of toxic nephrosis and clearly indicating [when compared with (a)] a protective effect of methylprednisolone against cephaloridine-induced nephrotoxicity. Note also the granular appearance to the cytoplasm. (c) Section from a rat treated with methylprednisolone only, demonstrating the lack of adverse effects on the kidney of the treatment regime employed with this steroid. This is indistinguishable from controls. From Harvey *et al.* (1995).

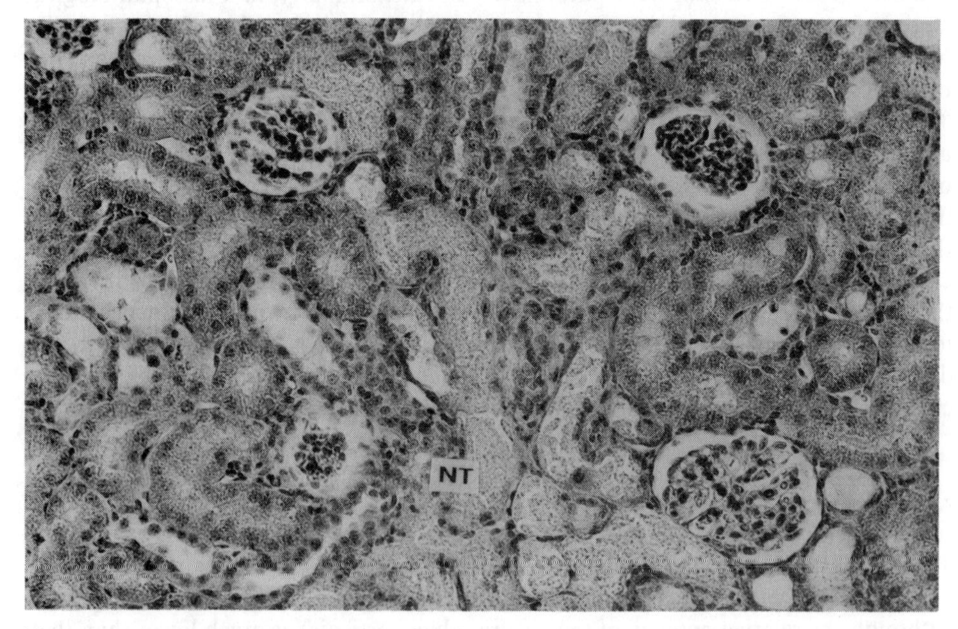

(a)

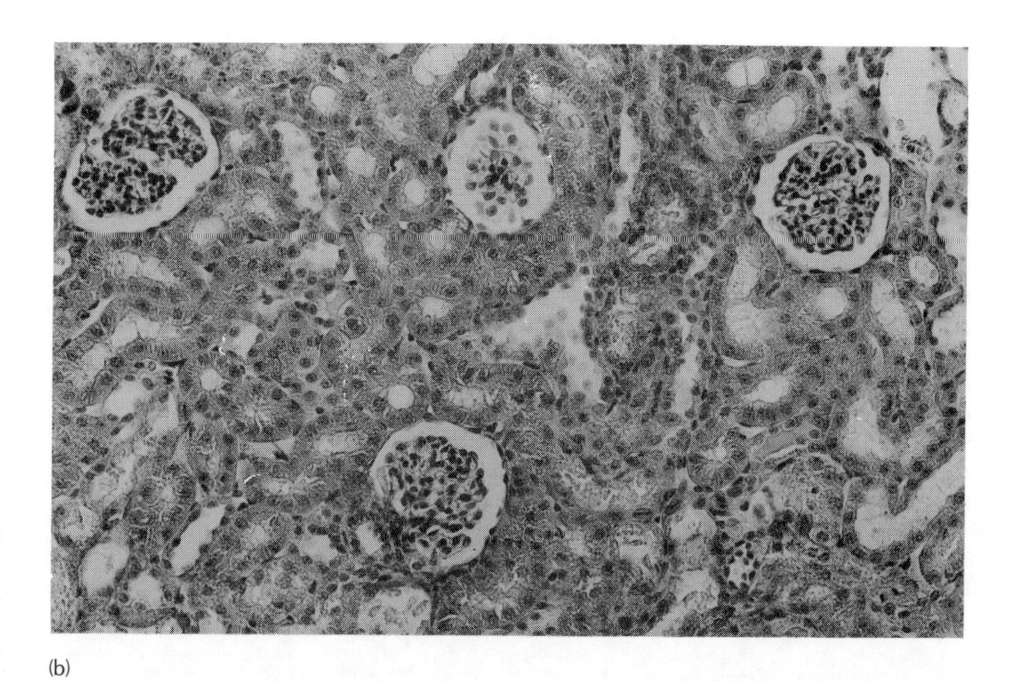

(b)

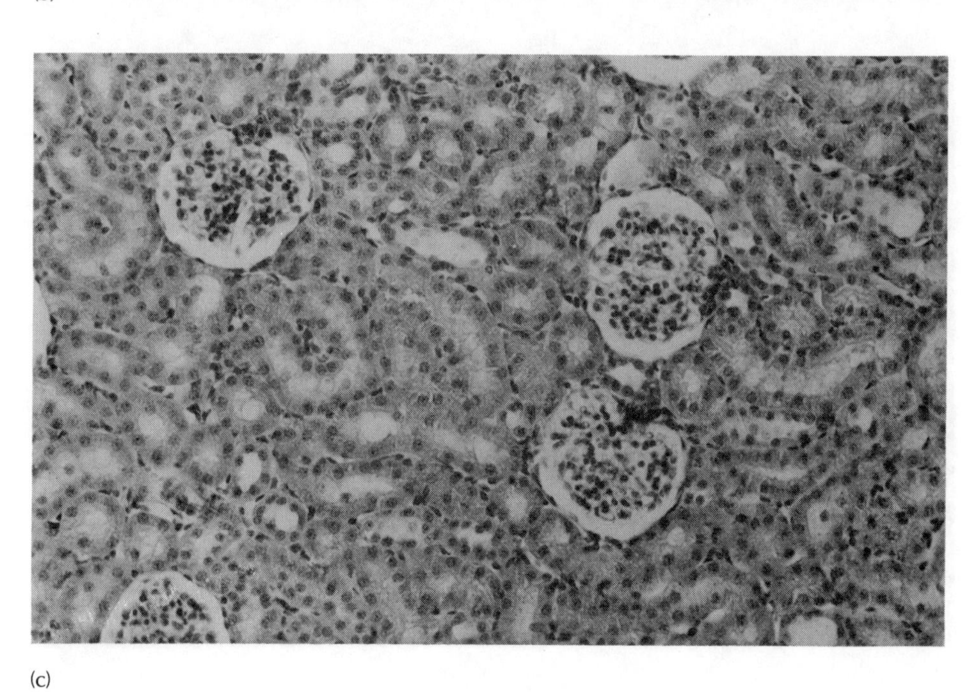

(c)

Figure 7.2 Inner cortex of kidney sampled 96 hr after treatment [haematoxylin & eosin (H & E) × 220]. (a) Section from a rat treated with cephaloridine only, showing severe toxic nephrosis characterized by gross distension of the terminal pars recta by exfoliated cellular debris. EC, Exfoliated cellular debris. Note the granular appearance of the cytoplasm. (b) Section from a rat treated with cephaloridine and methylprednisolone, showing only slight damage (note also proteinaceous fluid-filled tubule and the granular cytoplasm) with resolution through cellular regeneration (note mitotic figures) and clearly indicating [when compared with (a)] a protective effect of methylprednisolone against cephaloridine-induced nephrotoxicity. FT, Proteinaceous fluid-filled tubule; arrows mark mitotic figures. (c) Section from a rat treated with methylprednisolone only, demonstrating the lack of adverse effects on the kidney of the treatment regime employed with this steroid. This is indistinguishable from controls. From Harvey *et al.* (1995).

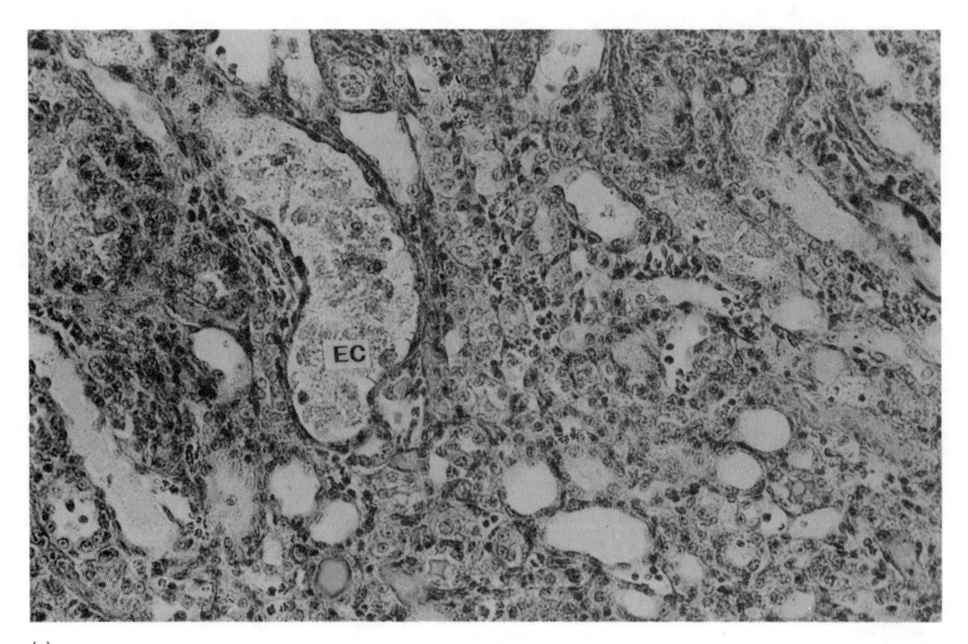

(a)

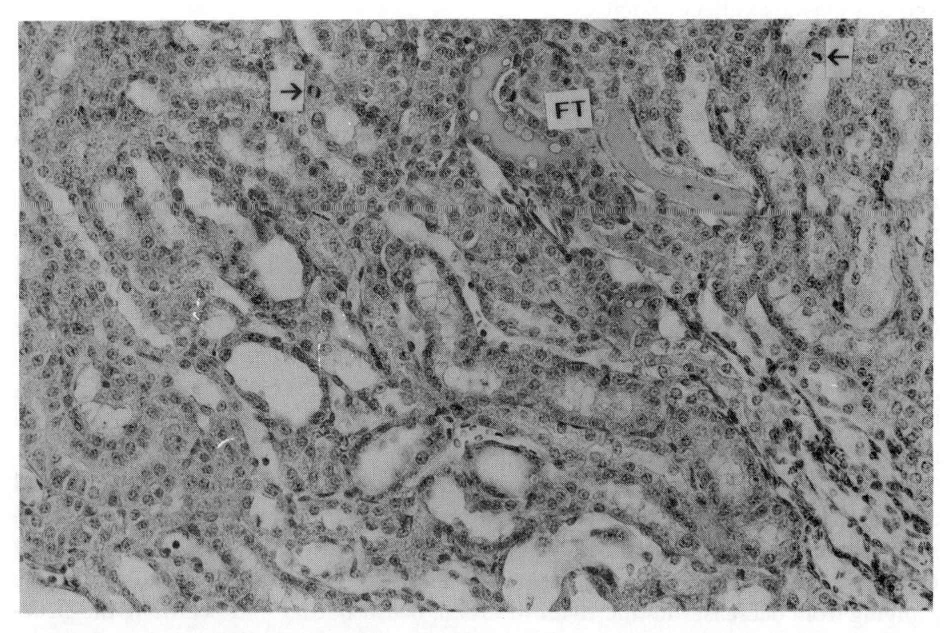

(b)

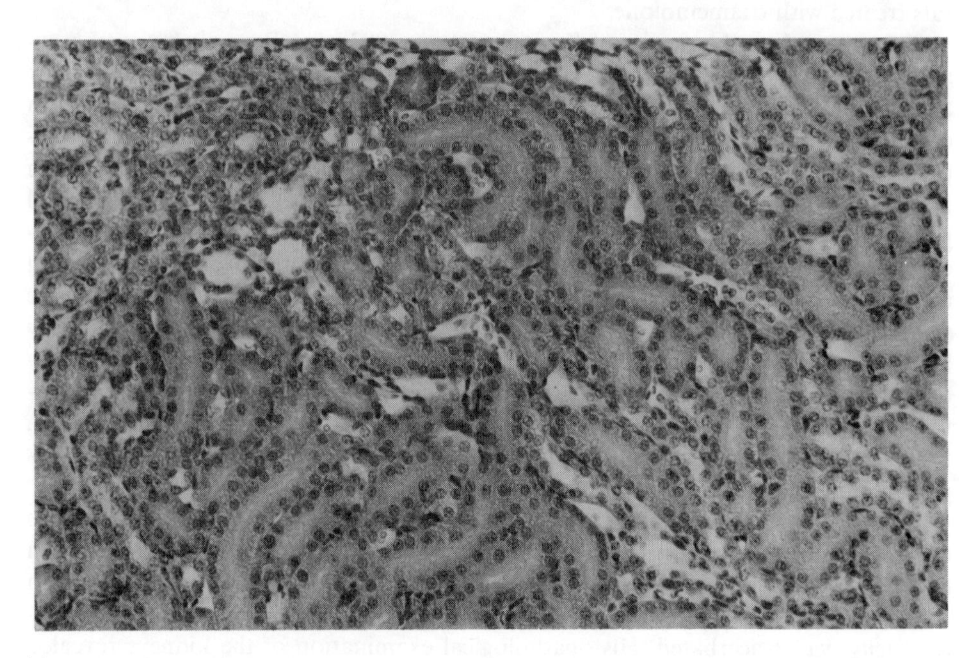

(c)

action, may be a mechanism contributing to the nephrotoxic profiles of toxicants. Additionally, the kidney, as with other organs and tissues, is a parimary target for the other general effects of glucocorticoids (eg metabolic effects, enzyme induction) and this may also affect toxicity.

There have been several recent studies of glucocorticoid toxicity interactions in the kidney. Koikawa *et al.* (1993) have recently reported a protective 'prophylactic' effect of methylprednisolone on cisplatin (an alkylating agent used in cancer chemotherapy)-induced nephrotoxicity in the rat. In this study, rats were injected subcutaneously with methylprednisolone in the dose range 10–50 mg/kg up to 4 hr before cisplatin injection (6.5 mg/kg administered intravenously) and methylprednisolone pre-treatment significantly reduced blood urea nitrogen and creatinine levels compared with rats treated with cisplatin alone. Koikawa *et al.* (1993) suggest that the protective effects of methylprednisolone may be related to the stabilization of lysosomal membranes, although there may be other factors involved.

The protective effect of glucocorticoid pre-treatment on cisplatin-induced nephrotoxicity in the rat has been confirmed. Reznick & Gambarayan (1994) treated rats intraperitoneally with a single dose of cisplatin (7.5 mg/kg) with or without a single intramuscular dose of triamcinolone (4 mg/kg administered 6 hr before cisplatin). These results showed that cisplatin resulted in marked elevations of serum urea and creatinine on the third day after treatment but that pre-treatment with triamcinolone significantly reduced these parameters. Histopathology confirmed that whilst cisplatin caused extensive lesions of the S3 segment of the proximal tubule (ranging from decreases in microvilli to loss of brush border and necrosis), the structure of the renal cortex was normal, with only slight damage to renal medulla, in rats treated with triamcinolone.

Taken together, the results of Koikawa *et al.* (1993) and Reznick & Gambarayan's (1994) studies suggest that glucocorticoids protect in cisplatin-induced kidney toxicity and this has implications for the potential clinical use of these steroids in cisplatin cancer chemotherapy. The use of two different steroids, methylprednisolone and triamcinolone, points to a common pharmacological mechanism of drug interaction. Harvey *et al.* (1995) investigated whether glucocorticoids could protect in the nephrotoxicity of other agents. Cephaloridine was administered subcutaneously to rats (3750 mg/kg) with or without methylprednisolone (100 mg/kg administered 2 hr before cephaloridine). Results showed that methylprednisolone significantly protected against cephaloridine-induced nephrotoxicity as evidenced by kidney weights, blood urea and creatinine and histopathology. Indeed, histopathology revealed marked toxic nephrosis and coagulative necrosis of the proximal convoluted tubules in the outer cortex and gross distension by exfoliated cell debris in the medulla of cephaloridine-treated rats, whereas those treated with methylprednisolone showed only slight nephrosis [see Figures 7.1(a)–(c) and 7.2(a)–(c)].

Despite the evidence that glucocorticoids protect against nephrotoxicity there is one report of exacerbation. Mohamed *et al.* (1974) studied the interaction between hydrocortisone and the venom from the snake *Naja nigricollis* in mice. Although hydrocortisone had a protective effect in venom-induced hepatotoxicity, nephrotoxicity was exacerbated. Histopathological examination of the kidneys revealed that those from mice treated with hydrocortisone and venom were more severely affected (showing a greater degree of 'round cell infiltration' and cytoplasmic vacuolation) than those from mice treated with venom only.

7.5 The Heart and Cardiotoxicity

Interactions between glucocorticoids and selective cardioactive compounds have been known for many years. The potentiation of catecholamine cardiotoxicity by adrenal corticoids is well documented (eg Kimura *et al.* 1968 and references therein), and one mechanism is by corticosteroids altering electrolytes. In the study by Kimura *et al.*, (1968), the potassium-displacing cardiotoxic effectiveness of adrenalin was markedly intensified by pre-treatments with 9α-fluorocortisol in rats and cats. However, in a more recent study corticosterone has been shown to enhance the cardiotoxicity of adrenalin and induce hypersensitivity in rats by inhibiting the biosynthesis of prostaglandins (Rascher *et al.* 1979).

Similarly, the cardiotoxicity of isoproterenol (a β-adrenergic agonist) has been shown to be enormously potentiated in rats pre-treated with a 25 mg pellet implant of desoxycorticosterone acetate (Guideri *et al.* 1974). In this study, desoxycorticosterone resulted in greatly increased sensitivity to isoproterenol and produced severe arrhythmias, ventricular fibrillation and death at a dose which was otherwise without any effect on the heart. Guideri *et al.* (1974) also found that hydrocortisone (cortisol) was less effective than desoxycorticosterone acetate in sensitizing the myocardium to the effects of isoproterenol. Given that electrolyte disturbances, and prostaglandin synthesis inhibition, are recognized mechanisms in the potentiation of the toxicity of cardiostimulants (ie catecholamines, β-adrenergic agonists) differences in these properties of corticosteroids could account for their differences in potentiating cardiotoxic effects.

The studies previously mentioned have tended to examine the effects of glucocorticoid pre-treatments or concurrent treatment with cardioactive compounds in intact animals *in vivo*. Different effects of glucocorticoids can be produced when these experimental situations are varied. For example, methylprednisolone and hydrocortisone have been reported to protect the ischaemic myocardium and reduce infarct size (largely by direct coronary vasodilation and possibly anti-inflammatory actions and lysosomal membrane stabilization) in isolated perfused hearts, or following coronary arterial ligation *in vivo*, in rats, cats and dogs (Brachfeld 1976 and references therein). It is possible that chemicals causing or exacerbating ischaemia may have their adverse effects similarly modified. Overall the effects of the co-administration of corticosteroids will depend on when they are given in relation to toxic insult, in addition to their specific properties (ie the steroids' potential for causing electrolyte imbalance or anti-inflammatory effects) and the mechanism of action of the cardiotoxic agents.

7.6 General Toxicity

There are reports of glucocorticoid toxicity interactions which are general in nature or single reports of an effect in an organ. English *et al.* (1987) have reported that the general toxicity of methotrexate (a folic acid antagonist used in cancer chemotherapy) varies in the rat according to time of day and endogenous corticosterone rhythm. Abolishing endogenous corticosterone secretion increased toxicity (using haematologic, liver and kidney markers) and interestingly the method chosen was the administration of dexamethasone (a synthetic glucocorticoid used here to suppress endogenous corticosteroid secretion). This apparently ambiguous finding suggests that corticosterone specifically (rather than simply a glucocorticoid) alters methotrexate toxicity and the mechanism probably revolves around differences in the properties of the two steroids, perhaps involving mineralocorticoid actions.

Further studies have shown that the toxicity of a diverse range of chemicals is modified by glucocorticoids. For example, hydrocortisone pre-treatment does not affect pulmonary oedema induced in rats by ozone exposure (Giri *et al.* 1975) but hydrocortisone, dexamethasone or triamcinolone pre-treatments are effective in diminishing the toxicity of tetraethylammonium bromide by increasing its urinary excretion (Kourounakis & Selye 1976). Both corticosterone and dexamethasone are reported to enhance the toxicity of vitamin D_2 in rats (Kunitomo *et al.* 1989) and corticosterone significantly ameliorates dextran-induced shock in rats (Kogure *et al.* 1986). The administration of cortisol to rats has been shown to enhance the appearance of liver tumours induced by dimethylaminoazobenzene, whereas corticosterone was without effect (Lacassagne *et al.* 1966). The lethality of T-2 toxin, a trichothecene mycotoxin, is reduced by the co-administration of dexamethasone (Tremel *et al.* 1985). In this study, an intravenous dose of 0.75 mg/kg T-2 toxin killed two-thirds of rats, but administration of dexamethasone (1.6 mg/kg intravenously) shortly before or after the T-2 toxin resulted in a four-fold reduction in lethality, and particularly reduced lung oedema and diarrhoea (presumably indicative of gastrointestinal toxicity). This finding has also been confirmed in mice, where both prednisolone and hydrocortisone pre-treatments reduced T-2 toxin lethality (Mutoh *et al.* 1988).

Several studies have involved surgical removal of either the pituitary (resulting in adrenocortical underactivity) or adrenal (removal of endogenous corticosterone). Hypophysectomy has been reported to significantly aggravate the toxicity of 2,3,7,8-tetrachlorodibenzo-*p*-dioxin in rats and normal tolerance to this toxicant is restored if corticosterone is given (Gorski *et al.* 1988). Adrenalectomy induces hypersensitivity to ammonia intoxication in the rat and administration of hydrocortisone hydrochloride restores normal sensitivity (Lee *et al.* 1977). It is apparent that in both of these studies, the removal of endogenous corticosterone enhances toxicity, but it should be noted that both procedures result in significant alteration of endocrine status involving hormonal systems other than endogenous glucocorticosteroids.

7.7 Mechanisms of Glucocorticosteroid Modulation of Toxicity

It is clear from the literature reviewed that there are multiple mechanisms involved in glucocorticoid modulation of toxicity and this is not least evident from findings that glucocorticoids can both enhance or protect against toxicity. Indeed, it has been previously noted that the direction of glucocorticoid modulation of toxic response depends on the pharmacological and toxicological properties of the chemical toxicant, the pharmacological properties of the glucocorticosteroid, the target organ, the temporal relationship of the steroid–toxicant co-administration regime, in addition to strain, sex, species and, in certain instances, age considerations (Harvey *et al.* 1994).

The diverse range of pharmacological, physiological and metabolic actions of the glucocorticosteroids certainly allows equally diverse modes of interaction with toxicants, and it is worthwhile briefly reviewing examples of mechanisms of glucocorticoid modulation of toxicity as suggested in the literature. Sapolsky (1986a) suggested that the mechanism by which corticosterone exacerbated kainic acid hippocampal neurotoxicity involved glucocorticoid-induced metabolic vulnerability. The glucocorticoids are well known to exert catabolic and, in excess, degenerative actions throughout the body, including the brain and especially the hippocampus which is rich in corticosterone receptors (McEwen *et al.* 1986, Miler 1992). Corticosterone is therefore suggested to weaken the tolerance of hippocampal neurons to survive an additional toxic challenge (Sapolsky 1986a). Sapolsky's hypothesis of metabolic vulnerability lends itself to extrapolation to other organs and tissues as the catabolic and degenerative metabolic actions of glucocorticoids occur generally throughout the body. There are several lines of evidence in support of a metabolic mechanism of glucocorticoid-induced sensitivity to neurotoxicity. Firstly, the use of metabolic fuels such as mannose, fructose, glucose and 3-hydroxybutyrate can ameliorate hippocampal damage following the co-administration of corticosterone with either kainic acid or 3-acetylpyridine (Sapolsky 1986b). Second, 3-acetylpyridine is a neurotoxic antimetabolite that inhibits ATP synthesis and kainic acid is an excitory neurotoxin which by its nature increases metabolic demand and both, therefore, induce toxicity by metabolic mechanisms. The primary role of the glucocorticoids is the regulation of glucose metabolism and, as such, the potential for interaction with metabolically active neurotoxicants is clear. Elliott & Sapolsky (1992) have also shown that in cultured hippocampal neurons, administration of corticosterone results in a large increase of neuronal calcium following kainic acid treatment, compared with control cultures, and whilst this may be an energetic effect dependent on neuronal glucose transport it may also implicate other mechanisms operative in glucocorticoid modulation of neurotoxicity. Interestingly, it has been reported that dexamethasone enhances vitamin D_2 toxicity, and the mechanism of this was that the steroid substantially increased hypercalcaemia and tissue calcification above that which normally occurs with vitamin D_2 intoxication (Kunitomo *et al.* 1989).

In the liver, the exacerbation of carbon tetrachloride hepatoxicity by corticosterone has been attributed to increased mono-oxygenase activity and the increased generation of free-radical and reactive metabolites (Lloyd & Franklin 1991). Similarly, the enhancement of paracetamol toxicity by corticosterone, cortisol and dexamethasone (Madhu *et al.* 1992, Harvey *et al.* 1993) has been associated with marked reduction of liver glutathione therefore allowing more free reactive metabolite to induce cytotoxicity through free-radical and covalent binding mechanisms. In the kidney, Koikowa *et al.* (1993) attributed the protection against cisplatin-induced nephrotoxicity by methylprednisolone to stabilization of the lysosomal membrane. In contrast, Reznick & Gambarayan (1994) suggest that glucocorticoids, specifically triamcinolone, protect the kidney from cisplatin-induced nephrotoxicity by suppressing inflammatory responses and local tissue swelling resulting from cisplatin. Anti-inflammatory actions of glucocorticoids may also be the mechanism of modulation of trichothecene mycotoxicity in reducing pulmonary oedema and local gastrointestinal effects typical of these toxins (Tremel *et al.* 1985, Mutoh *et al.* 1988, Hunder *et al.* 1994).

One generic mechanism of glucocorticoid influence on cellular toxicity concerns the role of these steroids in the regulation of P-glycoprotein and multi-drug resistance (mdr) gene expression. The multi-drug resistance phenotype is associated with the over-expression of a plasma transmembrane phosphoglycoprotein (P-glycoprotein). P-Glycoprotein is thought to act as an ATP-dependent efflux pump, the function of which is to pump foreign chemicals out of the cell. The over-expression of P-glycoprotein confers multi-drug resistance which is a particular problem in cancer chemotherapy. Glucocorticoids have been shown to alter the activity of this system. Fardel *et al.* (1993) analysed doxorubicin P-glycoprotein-mediated efflux in cultured rat hepatocytes also treated with dexamethasone and reported that intracellular doxorubicin concentrations were higher in dexamethasone-treated cells. Thus, doxorubicin was actively being pumped out of cells not exposed to dexamethasone, and this correlated with P-glycoprotein induction. The use of mdr gene probes showed that dexamethasone modulated P-glycoprotein induction through mdr 1 gene regulation. Although the relevance of the mdr-P-glycoprotein-efflux pump is obviously most apparent in the transport of cytotoxic anticancer drugs (Fardel *et al.* 1993) and conferring multi-drug resistance to cancer cells, it may also be a potentially important aspect of modulation of general cytotoxicity in certain cells exposed to toxic insult. The fact that dexamethasone can regulate this system not only implicates glucocorticoids in the modulation of cytotoxicity at a molecular level, but may also have clinical applications in blocking multi-drug resistance in cancer.

Although glucocorticoids are often suggested to alter toxicity by inducing various metabolic enzymes and thereby potentially altering toxicant metabolism, disposition and excretion (eg Szabo 1979, Kourounakis & Selye 1976), it is an important point that many studies employing co-administration regimes of a glucocorticoid with a toxicant are single-dose designs, allowing little opportunity for significant enzyme induction to occur. This is further emphasized by the use of natural steroids in the physiological dose range (eg Lloyd & Franklin 1991) and, where steroid administration is at the same time, slightly before or slightly after toxicant administration. The use of relatively high doses of synthetic steroids in long-lasting depot regimes could afford opportunity for significant enzyme induction, where blood concentrations of the steroid remain elevated for prolonged periods of time, but this must be coincident with toxicant exposure. Therefore, glucocorticosteroid interactions with drugs and chemicals, and the subsequent modulation of toxic respones, can occur through a variety of mechanisms, several of which may be concurrently operative, such that it is not possible to readily predict the direction of effect. Finally, the nature of the interaction is as much a function of the mechanism of action of the toxicant as it is the steroid, and is also dependent on the target tissue and general factors such as the temporal relationships of the co-administration regime.

7.8 General Conclusions

The general implications arising from this review of glucocorticoid modulation of toxicity fall broadly into two areas. The first concerns regulatory toxicology of new medicinal and chemical compounds, where regulatory guidelines stipulate that the highest dose level tested should be the maximum tolerated dose (MTD). The MTD is therefore a clear toxic insult and challenge which is expected to produce toxicity and morbidity, is likely to be noxious, and can be considered stressful to tolerate. Indeed, Miller (1992) has recognized that the MTD is stressful and the consequences of stress–toxicity interactions, involving adrenocortical products, has been discussed in Chapter 1 of this volume and by Vogel (1987, 1993) and Harvey (1994). The main consideration is to what extent is toxicity at the MTD due to glucocorticosteroid interactions or due to the chemical toxicant alone. This is more than an academic argument since species differences in adrenocortical activation, endocrinology and toxicological profile of drugs and chemicals must be fully understood before extrapolation can be made to humans. The literature on glucocorticosteroid administration with drugs and chemicals that examines interactions and the modulation of toxicity is not, however, confounded by considerations of stress, and information may be gained concerning the use of excessive dose levels on toxicity profiles.

The second implication arising from glucocorticosteroid modulation of toxicity concerns clinical pharmacology and toxicology. The finding in animal models that glucocorticosteroids can ameliorate toxicity is worthy of investigation in terms of clinical application. One important area is the use of glucocorticosteroids to ameliorate general non-target cytotoxicity resulting from aggressive cancer chemotherapy regimes. The clinical efficacy of cisplatin is tempered by toxicity to the kidney and there are at least two reports in rats where methylprednisolone and triamcinolone are shown both biochemically and histopathologically to protect the kidneys from cisplatin toxicity (Koikawa *et al.* 1993, Reznick & Gambarayan 1994). Glucocorticosteroids are frequent additions to multi-agent cancer chemotherapeutic regimes for certain conditions but are not necessarily included to block side-effect toxicity (or indeed for their potential in modulating the mdr-P-glycoprotein-efflux pump system which modulates the development of the cellular multi-drug resistance phenotype (Fardel *et al.* 1993) as previously discussed). Conversely, animal models have also shown that glucocorticosteroids can exacerbate toxicity and, whilst these studies often employ excessive dose levels of steroid and drug (eg paracetamol) in an acute exposure regime, consideration should be given to the clinical situation where lower therapeutic dose levels are used for longer periods of time. The chronic use of anti-inflammatory steroids and analgesics containing paracetamol, for example, is extremely common creating a potential for drug interaction and adverse reactions. An additional consideration is the tolerance of drugs in patients with adrenal dysfunction syndromes such as Cushing's syndrome or Addison's disease, where drug toxicity may occur as a result of altered endogenous glucocorticosteroid status.

Finally, glucocorticosteroids can both enhance and protect against experimentally-induced toxic insult, and the mechanisms of action depend on the pharmacology of the steroid, the pharmacology of the toxic agent, the target organ, the temporal relationship of the co-administration regime, in addition to species, sex and age considerations. That glucocorticosteroids modulate toxicity implicates the adrenal gland as a potential endogenous modulator of toxicity and in the whole-body response to chemical insult.

References

Barouki, R., Chobert, M. N., Billon, M. C., Finidori, J., Tsapis, R. & Hanoune, J. (1982) Glucocorticoid hormones increase the activity of gamma-glutamyltransferase in a highly differentiated hepatoma cell line. *Biochimica et Biophysica Acta*, **721**, 11–21.

Bogdanffy, M. S., Schatz, R. A. & Brown, D. R. (1984) Adrenal mediation of ethanol's inhibition of benzo(a)pyrene metabolism. *Journal of Toxicology and Environmental Health*, **13**, 799–810.

BOGDANFFY, M. S., ROBERTS, A. E., SCHATZ, R. A. & BROWN, D. R. (1986) Regioselective inhibition of benzo(a)pyrene metabolism by corticosterone in comparison with metyrapone and alpha-naphthoflavone. *Toxicology Letters*, **31**, 57–64.

BOWMAN, W. C. & RAND, M. J. (1984) *Textbook of Pharmacology*, Oxford: Blackwell.

BRACHFELD, N. (1976) Metabolic evaluation of agents designed to protect the ischemic myocardium and to reduce infarct size. *American Journal of Cardiology*, **37**, 528–532.

CHENTANEZ, T., PATRADILOK, P., GLINSUKON, T. & PIYACHATURAWAT, P. (1988) Effects of cortisol pre-treatment on the acute hepatotoxicity of aflatoxin B1. *Toxicology Letters*, **42**, 237–248.

ELLIOTT, E. M. & SAPOSKY, R. M. (1992) Corticosterone enhances kainic acid-induced calcium elevation in cultured hippocampal neurons. *Journal of Neurochemistry*, **59**, 1033–1040.

ENGLISH, J., AHERNE, G. W., ARENDT, J. & MARKS, V. (1987) The effects of abolition of the endogenous corticosteroid rhythm on the circadian variation in methotrexate toxicity in the rat. *Cancer Chemotherapy and Pharmacology*, **19**, 287–290.

FARDEL, O., LECUREUR, V. & GUILLOUZO, A. (1993) Regulation of P-glycoprotein expression in cultured rat hepatocytes. *FEBS Letters*, **327**, 189–193.

GARSIDE, D. & HARVEY, P. W. (1992) Endocrine toxicology of the male reproductive system. In ATTERWILL, C. K. & FLACK, J. D. (Eds), *Endocrine Toxicology*, pp. 285–312, Cambridge: Cambridge University Press.

GIRI, S. N., BENSON, J., SIEGEL, D. M., RICE, S. A. & SCHIEDT, M. (1975) Effects of pre-treatment with anti-inflammatory drugs on ozone-induced lung damage in rats. *Proceedings of the Society for Experimental Biology and Medicine*, **150**, 810–814.

GLODE, L. M., ROBINSON, J. & GOULD, S. F. (1981) Protection from cyclophosphamide-induced testicular damage with an analogue of gonadotrophin releasing hormone. *Lancet*, **1**, 1132–1134.

GORSKI, J. R., LEBOFSKY, M. & ROZMAN, K. (1988) Corticosterone decreases toxicity of 2,3,7,8-tetrachlorobenzo-p-dioxin (TCDD) in hypophysectomized rat. *Journal of Toxicology and Environmental Health*, **24**, 349–360.

GUIDERI, G., BARLETTA, M. A. & LEHR, D. (1974) Extraordinary potentiation of isoproterenol cardiotoxicity by corticoid pre-treatment. *Cardiovascular Research*, **8**, 775–786.

GUNN, S. A., GOULD, T. C. & ANDERSON, W. A. D. (1965) Protective effect of estrogen against vascular damage to the testis caused by cadmium. *Proceedings of the Society for Experimental Biology and Medicine*, **119**, 901–905.

HARVEY, P. W. (1994) Stress and toxicity. *Human & Experimental Toxicology*, **13**, 275–276.

HARVEY, P. W., ROUTH, M. R., REES, S. J., HEALING, G., RUSH, K. C., PURDY, K., EVERETT, D. J. & COCKBURN, A. (1993) Steroid pre-treatments and subsequent hepatic response to paracetamol in rats: do glucocorticoids modulate liver toxicity? *Medical Science Research*, **21**, 165–167.

HARVEY, P. W., HEALING, G., REES, S. J., EVERETT, D. J. & COCKBURN, A. (1994) Glucocorticosteroid interactions with natural toxins; a mini review. *Natural Toxins*, **2**, 341–346.

HARVEY, P. W., HEALING, G., MAJOR, I. R., MCFARLANE, M., PURDY, K. A., OLATUNDE, O., GARCIA CONESA, M. T., EVERETT, D. J. & COCKBURN, A. (1995) Glucocorticoid amelioration of nephrotoxicity: a study of cephaloridine-methylprednisolone interaction in the rat. *Human & Experimental Toxicology*, **14**, 554–561.

HAYNES, R. C. & MURAD, F. (1991) Adrenocorticotropic hormone; adrenocortical steroids and their synthetic analogs; inhibitors of adrenocortical steroid biosynthesis. In GOODMAN GILMAN, A., RALL, T. W., NIES, A. E. & TAYLOR, P. (Eds), *Goodman and Gilmans – the Pharmacological Basis of Therapeutics*, pp. 1459–1489, 7th Edn, New York: Macmillan.

HUNDER, G., FICHTL, B. & FORTH, W. (1994) Influence of glucocorticoids and activated charcoal on the lethality of rats after acute poisoning with T-2 toxin, diacetoxystirpenol or roridin A. *Natural Toxins*, **2**, 120–123.

KIMURA, H., BAJUSZ, E., HERRLICH, H. C. & RABB, W. (1968) Pre-necrotic epinephrine and corticoid-induced myocardial electrolyte shifts. Histochemical and chemical studies. *Cardiology*, **53**, 175–191.

KOGURE, K., ISHIZAKI, M. & NEMOTO, M. (1986) Antishock effects of corticosterone on dextran-induced shock in rats. *American Journal of Physiology, Endocrinology and Metabolism*, **251**, 569–575.

KOIKAWA, Y., UOZUMI, J., VEDA, T., YASUMASU, T. & KUMAZAWA, J. (1993) Prophylactic effect of methylprednisolone against cisplatin-induced nephrotoxicity in rats. *Toxicology Letters*, **66**, 281–286.

KOUROUNAKIS, P. & SELYE, H. (1976) Influence of steroids and stress on toxicity and disposition of tetraethylammonium bromide. *Journal of Pharmacological Science*, **65**, 1838–1840.

KUNITOMO, M., FUTAGAWA, Y., TANAKA, Y., YAMAGUCHI, Y. & BANDO, Y. (1989) Cholesterol reduces and corticosteroids enhance the toxicity of vitamin D in rats. *Japanese Journal of Pharmacology*, **49**, 381–388.

LACASSAGNE, A., JAYLE, M. F., HURST, L. & PASQUALINI, J. R. (1966) Influence of various corticosteroids on the production of hepatic cancer by p-dimethylaminoazobenzene in rats. *Comptes Rendus Hebdomadaires des Seances de l'Academie des Sciences, Serie D: Sciences Naturelles*, **262**, 2117–2119.

LEE, H. W., KIM, S. & PAIK, W. K. (1977) Effect of hydrocortisone on ammonia intoxication in the adrenalectomized rat. *Acta Endocrinologica*, **84**, 789–794.

LLOYD, S. A. & FRANKLIN, M. R. (1991) Modulation of carbon tetrachloride hepatotoxicity and xenobiotic-metabolizing enzymes by corticosterone pre-treatment, adrenalectomy and sham surgery. *Toxicology Letters*, **55**, 65–75.

MADHU, C. & KLAASEN, C. D. (1991) Protective effect of pregnenolone-16α-carbonitrile on acetaminophen-induced hepatotoxicity in hamsters. *Fundamental and Applied Toxicology*, **109**, 305–313.

MADHU, C., MAZIASZ, T. & KLAASEN, C. D. (1992) Effect of pregnenolone-16α-carbonitrile and dexamethasone on acetaminophen-induced hepatotoxicity in mice. *Toxicology and Applied Pharmacology*, **115**, 191–198.

MCEWEN, B. S., DEKLOET, E. R. & ROSTENE, W. (1986) Adrenal steroid receptors and actions in the nervous system. *Physiological Reviews*, **66**, 1121–1188.

MILLER, D. B. (1992) Caveats in hazard assessment – stress and neurotoxicity. In ISAACSON, R. L. & JENSEN, K. F. (Eds), *The Vulnerable Brain and Environmental Risks*, Vol. 1, *Malnutrition and Hazard Assessment*, pp. 239–266, New York: Plenum.

MOERLAND, J., STOWELL, L. & GJERDE, H. (1985) Ethanol increases rat liver tryptophan oxygenase: evidence for corticosterone mediation. *Alcohol*, **2**, 255–259.

MOHAMED, A. H., NAWAR, N. N. & MOHAMED, F. A. (1974) Influence of hydrocortisone on the microscopic changes produced by *Naja nigricollis* venom in kidney, liver and spleen. *Toxicon*, **12**, 45–48.

MORRIS, I. D. & SHALET, S. M. (1990) Protection of gonad function from cytotoxic chemotherapy and irradiation. *Baillieres Clinical Endocrinology and Metabolism*, **4**, 97–118.

MUTOH, A., ISHII, K. & VENO, Y. (1988) Effect of radioprotective compounds and anti-inflammatory agents on the acute toxicity of trichothecenes. *Toxicology Letters*, **40**, 165–174.

ORTIZ DE MONTELLANO, P. R. (1986) *Cytochrome P-450 – Structure, Mechanism and Biochemistry*, New York: Plenum.

RASCHER, W., DIETZ, R. & SCHOEMIG, A. (1979) Modulation of sympathetic vascular tone by prostaglandins in corticosterone-induced hypertension in rats. *Clinical Science*, **57**(Suppl. 5), 235–237.

REZNICK, L. V. & GAMBARAYAN, S. P. (1994) Protective effect of anti-inflammatory corticosteroid triamcinolone in cisplatin nephrotoxicity. *Renal Physiology and Biochemistry*, **17**, 50–56.

RUCKPAUL, K. & REIN, H. (1989) *Basis and Mechanisms of Regulation of Cytochrome P-450*, Vol. 1, *Frontiers in Biotransformation*, London: Taylor & Francis.

SAPOLSKY, R. M. (1985) Glucocorticoid toxicity in the hippocampus: Temporal aspects of neuronal vulnerability. *Brain Research*, **359**, 300–305.

SAPOLSKY, R. M. (1986a) Glucocorticoid toxicity in the hippocampus. Temporal aspects of synergy with kainic acid. *Neuroendocrinology*, **43**, 440–444.

SAPOLSKY, R. M. (1986b) Glucocorticoid toxicity in the hippocampus: Reversal by supplementation with brain fuels. *Journal of Neuroscience*, **6**, 2240–2244.

SAPOLSKY, R. M., PACKAN, D. R. & VALE, W. W. (1988) Glucocorticoid toxicity in the hippocampus: *in vitro* demonstration. *Brain Research*, **453**, 367–371.

SHIRAISHI, N., BARTER, R. A., UNO, H. & WAALKES, M. P. (1993) Effect of progesterone pre-treatment on cadmium toxicity in the male Fischer (F344/NCr) rat. *Toxicology and Applied Pharmacology*, **118**, 113–118.

SUDHIR, S. & BUDHIRAJA, R. D. (1992) Comparison of the protective effect of withaferin-'A' and hydrocortisone against CCl_4–induced hepatotoxicity in rats. *Indian Journal of Physiology and Pharmacology*, **36**, 127–129.

SZABO, S. (1979) Effect of ACTH, corticoids and PCN (pregnenolone-16α-carbonitrile) on drug response and disposition. *Acta Endocrinologica*, **230**, supplementum, I–III.

THOTTASSERY, J., WINBERG, L., YOUSSEF, J., CUNNINGHAM, M. & BADR, M. (1992) Regulation of perfluorooctanoic acid-induced peroxisomal enzyme activities and hepatocellular growth by adrenal hormones. *Hepatology*, **15**, 316–322.

TIMBRELL, J. A. (1991) *Principles of Biochemical Toxicology*, London: Taylor & Francis.

TREMEL, H., STRUGALA, G., FORTH, W. & FICHTL, B. (1985) Dexamethasone decreases lethality of rats in acute poisoning with T-2 toxin. *Archives of Toxicology*, **57**, 75–75.

VOGEL, W. H. (1987) Stress – the neglected variable in experimental pharmacology and toxicology. *Trends in Pharmacological Sciences*, **8**, 35–38.

VOGEL, W. H. (1993) The effect of stress on toxicological investigations. *Human & Experimental Toxicology*, **12**, 265–271.

WOLKOWSKI-TYL, R. & PRESTON, S. F. (1979) The interaction of cadmium binding proteins (Cd-bp) and progesterone in cadmium-induced tissue and embryo toxicity. *Teratology*, **20**, 341–352.

8

Glucocorticosteroids

Deleterious effects on neurons and potential exacerbation of environmental neurotoxicity

LAURA J. MCINTOSH and ROBERT M. SAPOLSKY

Stanford University, California

8.1 Introduction

Glucocorticosteroids (GCs) are hormones secreted by the adrenals to co-ordinate the body's response to a multitude of stressors. This endocrine response can be evoked by physical stressors, such as strenuous exercise, and also psychological stressors, such as the worry caused in response to life events and difficulties. Secretion of GCs serves to mobilize energy from storage sites and divert it primarily to cardiovascular, pulmonary, sensory and muscle tissues – places where the need for oxygen and glucose is urgent in order to respond to the perceived demand for action. To help provide this energy, GCs suppress systems that are not crucial in this perceived emergency, including digestion, growth, reproduction and immunity. Although these physiological responses are essential for responding to an acute stressor, the identical responses are also triggered by chronic stressors, and therefore serve as the basis of pathology to the systems involved. When exposure to GCs is prolonged (due to stress, pathologic hypersecretion as in Cushing's syndrome, or exogenous administration of synthetic GCs such as prednisone, hydrocortisone or dexamethasone), myopathy, fatigue, hypertension, ulcers and increased risk for adult-onset diabetes and neuron death are a few of the noted consequences (Munck *et al.* 1984). This chapter is concerned with the latter consequence, namely neuron death. The purpose of this chapter is not only to give a short review of the current literature, but also to draw attention to the ramifications of GC effects to neurotoxicology.

Who are the individuals likely to be affected by the deleterious effects of GCs on neurons? Any person with chronically high levels of GCs, particularly if that individual has a concurrent neurological insult such as a stroke, seizure or hypoglycaemic episode. High levels of GCs can be iatrogenic or a function of individual circumstance. Sixteen million prescriptions for GCs (including corticosteroids such as prednisone or hydrocortisone) are written annually in the US to control autoimmunity or as an anti-inflammatory agent (Eufemio 1990). Another population recognized to be at risk for stress-related injury are those individuals who reach old age. Not only do older organisms have higher basal levels of circulating GCs, but they have a decreased capacity to respond appropriately to stressors. This means that homeostasis can be disrupted with a lesser stressor, or that the ability to terminate the stress response and re-establish homeostasis is diminished. Finally, those persons whose lives contain a succession of unrelieved physical or psychological stressors can be included in the group with an elevated stress response and who are at risk for stress-related disease and neuron death [for a review see Sapolsky (1992)].

8.2 Glucocorticoid Neurotoxicity and Neurodegeneration

The negative effects of GC overexposure include damage to the brain. Pharmacologic GC concentrations are particularly damaging to the brain region involved in learning and memory, the hippocampus (Aus der Muhlen & Ockenfels 1979). The hippocampus appears to be targeted as a result of its high concentrations of corticosteroid receptors (McEwen *et al.* 1986). Decreasing GC exposure over an individual's lifespan (by surgical or behavioural means) prevents the hippocampal neuron loss and cognitive impairments typical of aging (Landfield *et al.* 1981, Meaney *et al*, 1988, 1991). Conversely, experimentally induced stress over the course of a few weeks causes atrophy of dendritic processes, which is thought to be the first step in subsequent neuron loss (Woolley *et al.* 1990, Watanabe *et al.* 1992, Magarinos & McEwen, 1993). Months of stress, or pharmacologic elevation of GCs into the stress range, may serve to accelerate senescent pyramidal neuron loss (Sapolsky *et al.* 1985, Kerr *et al.* 1991, Mizoguchi *et al.* 1992, Talmi *et al.* 1993, Levy *et al.* 1994). Furthermore, the neurotoxicity of both GCs and stress increases with age (Kerr *et al.* 1991, Levy *et al.* 1994). Recognizing species differences, it is important to demonstrate the relevance of these studies to humans. Sustained stress will also preferentially damage the primate hippocampus (Uno *et al.* 1989, 1991), and microimplantation of GC-secreting pellets into the primate hippocampus for 1 year causes mild but demonstrable damage, as compared to the contralateral hippocampus exposed to cholesterol (Sapolsky *et al.* 1990). Finally, in humans who oversecrete GCs (Cushing's syndrome), increasing concentrations of circulating GCs are associated with selective hippocampal atrophy (Starkman *et al.* 1992).

The demonstration that GCs could damage neurons during ageing suggested that behavioural manipulations which caused long-term diminution of GC concentrations might be protective. In support of this, we demonstrated that aged rats that had been neonatally handled (a phenomenon long known to lower adult GC concentrations) were spared the hippocampal neuron loss, neuroendocrine dysfunctions and cognitive deficits typical of old age (Meaney *et al.* 1988, 1991).

It was then reported that GCs and stress were not only capable of damaging the neurons outright, but were capable of 'endangering' them. This means that under circumstances where GC exposure is not yet sufficient to kill neurons, their ability to survive other, coincident challenges is nevertheless compromised. As a result, stress and/or GC concentrations in the stress range exacerbate the toxicities of various neurological and neurotoxic insults, and removing the capacity for GC production (adrenalectomy) reduces damage. This has been shown for epileptic seizures and excitotoxin exposure in two areas of the brain, the striatum and hippocampus (Sapolsky 1985a,b, Theoret *et al.* 1985, Stein & Sapolsky 1988, Uhler *et al.* 1994,

Supko & Johnston 1994, Smith-Swintosky 1995); moreover, stress itself will worsen seizure damage to the hippocampus (Stein-Behrens *et al.* 1994b). Furthermore, hypoxic-ischemic damage to other brain regions such as the cortex, caudate and hippocampus is worsened by GCs (Sapolsky & Pulsinelli 1985, Koide *et al.* 1986, Hall 1990, Miller & Davis 1991). GCs will also exacerbate the toxicity of hypoglycemia or exposure to antimetabolites (Sapolsky 1985a,b, 1986, Bennett *et al.* 1993), cholinergic or serotenergic neurotoxins (Johnson *et al.* 1989, Amoroso *et al.* 1993, Hortnagl *et al.* 1993) and oxygen radical generators (Sapolsky *et al.* 1988, McIntosh & Sapolsky, 1994) in the hippocampus. As an additional demonstration of the endangering effects of GCs, inhibition of adrenal steroidogenesis with the drug metyrapone following seizures or hypoxia-ischemia reduces hippocampal damage (Stein & Sapolsky 1988, Morse & Davis 1990, Miller & Davis 1991, Smith-Swintosky 1995).

GCs also impair the capacity of cultured neurons and glia to survive *in vitro* versions of these insults (Sapolsky *et al.* 1988, Virgin *et al.* 1991, Mizoguchi *et al.* 1992, Tombaugh *et al.* 1992), indicating that the GC effect is not a function of peripheral physiological changes. The endangerment is not caused by non-GC steroids, and is mediated by the type II GC receptor (Packan & Sapolsky, 1990, Tombaugh & Sapolsky 1993). As further support for the receptor specificity, the neuronal endangerment is also caused by the type II receptor ligands dexamethasone, RU28362 or methylprednisolone (Koide *et al.* 1986, Hall 1990, Mizoguchi *et al.* 1992, Hortnagl *et al.* 1993).

Thus GCs can be endangering.[1] However, it is important to emphasize some caveats and controversies that have arisen concerning GC neuroendangerment.

(i) Obviously not all insults to the brain are exacerbated by GCs. Examples of non-endangerment include trimethyllead and trimethyltin in the hippocampus (Sapolsky, Krigman & Theoret unpublished data, O'Callaghan personal communication), excitotoxins in the hypothalamus or cerebellum (Sapolsky 1985a), and cholinergic toxins in the septum or parietal cortex (Amoroso *et al.* 1993, Hortnagl *et al.* 1993). In general, it is the regions of the brain with higher concentrations of GC receptors – the hippocampus, cortex, striatum – which are most vulnerable to the endangerment.

(ii) One report suggests that the GC endangerment is due to the increase in body temperature brought on by steroids, a well-known route by which hypoxic-ischemic damage is worsened (Morse & Davis 1990). However, a number of other investigators have reported that GCs augment hypoxic-ischemic damage even after controlling body temperature (Sapolsky & Pulsinelli 1985, Koide *et al.* 1986, Miller & Davis 1991).

(iii) One study suggested that GC toxicity and endangerment, both *in vivo* and *in vitro*, occur only when accompanied by minimal testosterone concentrations (Mizoguchi *et al.* 1992). The authors showed that stress was neurotoxic in the hippocampus only in castrated (male) rats, and that testosterone supplementation prevents GC endangerment of cultured hippocampal neurons. This is intriguing, given the authors' suggestion that the typical decline in testosterone concentrations with age in rats would explain why the neurotoxic potential of stress becomes more

[1] Throughout this chapter, GC 'toxicity' refers to the capacity of the steroid to damage neurons, whereas 'endangerment' refers to GC augmentation of the toxicity of another insult.

pronounced with age (Kerr *et al.* 1991, Levy *et al.* 1994). As arguments against the conclusions of this single study, the literature above demonstrating GC toxicity and endangerment has been carried out with females or gonadally-intact males.

(iv) There are some inconsistencies as to the robustness of GC endangerment in the perinatal brain. For example, GCs do not exacerbate excitotoxin damage in the neonatal hippocampus (Barks *et al.* 1991), yet do so in the neonatal striatum (Supko & Johnston 1994). As another example, one study indicated that GCs increased mortality due to infarction in neonatal rats (Altman *et al.* 1984), while other studies have failed to replicate that finding and have shown, in fact, that GCs decrease ischemic damage (Barks *et al.* 1991, Chumas *et al.* 1993, Tuor *et al.* 1993). Finally, *in vitro* studies using fetal tissue have shown GC endangerment (Sapolsky *et al.* 1988, Virgin *et al.* 1991, Mizoguchi *et al.* 1992, Tombaugh *et al.* 1992). This lack of complete agreement has prompted us to try to evenly balance our own studies in this area between *in vitro* approaches, with the advantage afforded by numerous reductive techniques only applicable to monolayer cultures, and *in vivo* approaches in adults, with the advantage afforded by testing the physiological relevance of *in vitro* observations.

In recent years, a number of groups have sought to understand the mechanisms underlying GC endangerment. Apoptotic internucleosomal cleavage of DNA, the mechanism by which GCs lyse lymphocytes, is not likely to be the answer (Masters *et al.* 1989). Instead, the deficit may lie with a reduced ability to meet cellular energy demand. The insults exacerbated by GCs are energetic in nature, in so far as they either disrupt energy delivery/production or pathologically elevate energy demand. All cause profound drops in ATP and phosphocreatine levels, and their toxicities can all be buffered with energy supplementation (Auer & Siesjo 1988, Beal 1992). As the strongest evidence of the energetic nature of the GC endangerment, it is prevented by coincident supplementation of rats or cultures with excess energy (glucose, mannose or ketones) (Sapolsky 1986a, Sapolsky *et al.* 1988, Tombaugh *et al.* 1992).

Recent work has indicated a likely mechanism by which GCs induce this metabolic endangerment. GCs inhibit glucose transport in tissues such as adipocytes and fibroblasts, as part of a strategy to divert energy to muscles exercising during a stressor (Munck 1971). This effect involves both translocation of pre-existing glucose transport molecules from the cell membrane to intracellular storage sites, as well as inhibition of transcription of the glucose transporter gene (Horner *et al.* 1987, Garvey *et al.* 1989). Similarly, GCs inhibit local cerebral glucose utilization in numerous brain regions (Kadekaro *et al.* 1988, Freo *et al.* 1992) and inhibit glucose transport in cultured neurons and glia (Horner *et al.* 1990, Virgin *et al.* 1991). As with the phenomenon of GC endangerment, the inhibition of glucose transport is not triggered by non-GC steroids, and is type II receptor-mediated (Horner *et al.* 1990).

The inhibition of glucose transport by GCs is of the order of 25–30% in the brain, versus 70% in peripheral tissues. Such mild inhibition is unlikely to kill neurons outright, and GCs do not depress basal ATP concentrations. However, GCs accelerate the decline in ATP concentrations in cultured hippocampal neurons and glia following ischemia or aglycemia (Tombaugh & Sapolsky 1992, Lawrence & Sapolsky 1994). Similarly, GCs by themselves do not affect cytochrome oxidase activity in hippocampal somas, but exacerbate the decline in activity induced by sodium azide (Bennett *et al.* 1993).

What are the consequences of this metabolic endangerment? Which of the numerous derangements in a neuron during a neurological insult are worsened by GCs and hasten neuronal death? One feature GCs appear to impair in neurons is the ability to contain glutamate and calcium cascades.

There is considerable evidence that various neurological insults result in an accumulation of excitatory amino acid (EAA) neurotransmitters, such as glutamate, in brain synapses. This causes longlasting depolarization of the post-synaptic neuron, with subsequent mobilization of cytosolic calcium, and delayed neuron death. This cascade now constitutes a central dogma of cellular neuropathology. Of great relevance to GC endangerment, the toxicity of this cascade is extremely sensitive to energy availability (Beal 1992). A paucity of energy enhances EAA accumulation, through membrane depolarization and increased vesicular release, and also by the reversal of sodium gradient-dependent EAA re-uptake pumps. EAA uptake by both neurons and glia is disrupted by the aberrant ionic gradients caused by energy failure. Thus, more EAAs are released into the synapse and persist longer. This, in turn, mobilizes more cytosolic calcium through a number of different routes. Cytosolic calcium will persist longer in energy deficient cells because of the costs of calcium sequestration and efflux. As the most explicit demonstration of the energy-dependency of this cascade, glucose availability will determine whether glutamate functions as a neurotransmitter or a neurotoxin (Novelli *et al.* 1988, Cox *et al.* 1989, Lysko *et al.* 1989, Zeevalk & Nicklas 1991, Simpson & Isacson 1993).

One may then speculate that GCs, by disrupting neuronal energetics, exacerbate the toxicity of these varied insults by exacerbating steps in the EAA/calcium cascade. As indirect evidence for this scenario, phenytoin, which inhibits EAA release, protects against GC- and stress-induced hippocampal damage (Watanabe *et al.* 1992), while antagonism of the NMDA receptor blocks GC endangerment (Armanini *et al.* 1990), and GC-induced atrophy of dendritic processes (reviewed in McEwen 1992). Furthermore, stress and GCs cause an EAA-dependent increase in hippocampal metabolism (Krugers *et al.* 1992).

More direct evidence also suggests that GCs exacerbate this EAA/calcium cascade. Elevated GC levels increase the magnitude of seizure-induced glutamate and aspartate accumulation in the hippocampus (Stein-Behrens *et al.* 1992). This is physiologically relevant: an increase in GC concentrations from the circadian trough to the peak doubles glutamate accumulation, while a rise into the stress range causes a four-fold increase (Stein-Behrens *et al.* 1994a). As would be expected, stress itself enhances extracellular EAA concentrations in the cortex and hippocampus (Moghaddam 1993) in a GC-dependent manner (Lowy *et al.* 1993, Moghaddam *et al.* 1994). GCs also exacerbate anoxia-induced EAA accumulation in hippocampal cultures, and this is due to inhibition of EAA uptake, rather than enhancement of release (Virgin *et al.* 1992, Chou *et al.* 1994).

Given these GC effects upon EAA trafficking, it is not surprising that GCs elevate resting levels of free cytosolic calcium in cultured hippocampal neurons (Elliott & Sapolsky 1993) and augment the magnitude and duration of calcium mobilization following EAA stimulation (Elliott & Sapolsky 1992, 1993). In support of this, GCs enhance calcium currents in hippocampal neurons (Kerr *et al.* 1992). GCs appear to enhance calcium mobilization more by impairing the removal of free calcium from the cytosolic compartment (either by binding, sequestering or efflux) than by enhancing calcium entry (Elliott & Sapolsky 1993). This is similar to the response seen with glucose deprivation (de Haas Johnson, Brooke & Sapolsky unpublished data).

Given the effects of GCs on calcium concentrations, it is not surprising that GCs exacerbate calcium-dependent degeneration following an insult. In both the cortex and the hippocampus, GCs augment kainate-induced, calcium-dependent proteolysis of the cytoskeletal protein spectrin, and accumulation of the tau protein (Elliott *et al.* 1993, Lowy *et al.* 1994). Increments in GC exposure in the physiological range augment these degenerative endpoints, as does stress itself (Stein-Behrens *et al.* 1994b), while blockade of GC secretion with metyrapone decreases such degeneration (Smith-Swintosky *et al.* 1995).

These GC effects upon the EAA/calcium cascade appear to arise from the disruptive effects of GCs upon energetics. Supplementation with excess sugars reverses the GC effects upon extracellular EAA accumulation (Stein-Behrens *et al.* 1992), free cytosolic calcium mobilization (Elliott & Sapolsky 1992) and cytoskeletal pathology (Elliott *et al.* 1993).

In recent years it has become clear that the generation of reactive oxygen species (ROS) is probably the single most damaging and important consequence of calcium excess (reviewed in Coyle & Puttfarcken 1993). Antioxidants rescue neurons from various insults, including EAAs, thereby underscoring the power of oxygen radicals as inducers of cell death. Depletion of ascorbic acid and glutathione, two cytosolic antioxidant molecules, correlates with onset of oxidative damage in blood endothelium, (Frei *et al.* 1989, Mak *et al.* 1992), and use of an oxygen radical trapping compound protects against EAA toxicity in neuronal culture (Lafon-Cazal *et al.* 1993) and *in vivo* (Carney & Floyd 1991). Protection is also demonstrated by lipophilic antioxidants such as idebenone and α-tocopherol which attenuate EAA damage in neuronal cultures (Miyamoto *et al.* 1990, Schubert *et al.* 1992) and during ischemia (Yamamoto *et al.* 1983). Enzymatic antioxidants such as superoxide dismutase (SOD) and catalase also protect against ischemic injury *in vivo* (Liu *et al.* 1989, Truelove *et al.* 1994) and *in vitro* (Rosenbaum *et al.* 1994).

If GCs increase intracellular calcium concentrations during insults, they should also increase the resulting generation of reactive oxygen species (ROS) and the extent of oxidative damage.

The little that is known about interactions between GCs and oxygen radicals suggests that GCs may be important regulators of oxygen radical processes. GCs exacerbate the neurotoxicity of the ROS generators paraquat and adriamycin in hippocampal cultures (Sapolsky *et al.* 1988, McIntosh & Sapolsky 1994). Recently, Liu *et al.* (1994) indicated that immobilization stress produced increased markers for lipid peroxidation and decreased radical scavenging activity in plasma, and that treatment with glutathione attenuated those changes. Since damage for ROS may result from increased ROS production and/or from decreased antioxidant protection within cells, we have analysed the effects of GCs on the activity of antioxidant enzymes in various brain regions, and observed that the hormones can inhibit the activity of catalase and glutathione peroxidase in a tissue-specific manner. It is unclear at the present time whether the effect of GCs on ROS-mediated neuron death is due to increased production of ROS or a lowering of antioxidant defenses.

Thus, our working model is that GCs disrupt glucose transport in neurons and glia, causing an energetic vulnerability. This then impairs the capacity of the neuron to survive any coincident insult which carries an energetic cost. Such insults can include those activating the EAA/calcium/ROS cascade, as described earlier, or neurologic insults which, while energetic in nature, may not involve that particular cascade (Hortnagl *et al.* 1993).

8.3 Possible Relevance of GCs to Environmental Neurotoxins

It is within the framework of energy vulnerability, glutamate accumulation, calcium dysregulation and ROS generation that it seems plausible for GCs to exacerbate the action of neurotoxicants that also impact those systems. Candidate neurotoxicants include toluene, manganese and methylmercury, and it is worth reviewing the features of their toxicities.

Methylmercury is a heavy metal which can damage cortical and cerebellar neurons when present in the low micromolar range (Sarafian & Verity 1991, Atchison & Hare 1994, Bondy 1994). Its environmental relevance is derived from its release by grain, chemical, mining and lumber industries, its presence in incinerator waste, and in leachate from landfills.

The mechanisms underlying methylmercury neurotoxicity are complex, but ultimately converge in the generation of ROS. Initially, this organometal causes the release of calcium into the cytosolic compartment via an IP3-sensitive pool in the smooth endoplasmic reticulum (ER) (Atchison & Hare 1994). As a result, even relatively low concentrations of the compound can collapse transmembrane potentials in mitochondria (Kauppinen *et al.* 1989, Bondy & McKee 1991). This will dramatically worsen the cytosolic accumulation of calcium since depolarized mitochondria can no longer function as a high capacity site for calcium sequestration. When cytosolic calcium levels are extremely high, the mitochondrial uptake pumps for calcium will reverse their direction, leading to calcium efflux into the cytoplasm (reviewed in Sapolsky 1992, Chapter 10), and attendant protease activation and ROS generation. Finally, at higher methylmercury concentrations, synaptic and even axonal membranes will depolarize (Shrivastav *et al.* 1976, Kauppinen *et al.* 1989), and allow influx of extracellular calcium through voltage-gated channels. Not surprisingly, given these effects, methylmercury disrupts oxidative phosphorylation and protein synthesis (Verity *et al.* 1975) and suppresses ATP concentrations (Sarafian *et al.* 1989).

One of the main consequences of pathologic elevation of free cytosolic calcium in a neuron is generation of ROS. This can arise from calcium-dependent activation of phospholipase A_2 and the arachidonic acid route of ROS generation, or from calcium-dependent generation of xanthine oxidase. Moreover, direct disruption of oxidative phosphorylation and mitochondrial potentials is another potent route for ROS generation. Commensurate with that, methylmercury generates ROS and causes lipid peroxidation (LeBel *et al.* 1990, 1992, Sarafian & Verity 1991), and may suppress antioxidant activity within brain tissue (Yee & Choi 1994). As correlative evidence suggesting that such ROS are important, their generation is more pronounced in brain tissue destined to be damaged by systemic methylmercury exposure (LeBel *et al.* 1992). More directly, the toxicity can be significantly decreased with vitamin E (Chang *et al.* 1978, Shukla *et al.* 1988). Furthermore, the ROS generation is related to release of iron, as the neurotoxicity can also be prevented by iron chelation (Kauppinen *et al.* 1989, LeBel *et al.* 1992).

The second compound to consider is *toluene*, an organic solvent used in a wide range of industrial procedures, and which has demonstrated neurotoxic properties throughout the brain. Many of its mechanisms of action are similar to those of methylmercury, in that it also generates ROS (Mattia *et al.* 1991), collapses mitochondrial membrane potentials (Bondy 1991), and mobilizes free cytosolic calcium concentrations (Bondy 1991). Unlike methylmercury, however, toluene does not depolarize synaptic membranes. Whether it causes the lipid peroxidation predicted to result from high intracellular calcium is not certain (Bondy 1991, Mattia *et al.* 1991). As further potential sources for oxidative damage, toluene reduces glutathione levels (Mattia *et al.* 1991) and stimulates phospholipase activity, a likely source of arachidonic acid release (LeBel & Schatz 1990).

Finally, *manganese* has long been recognized for its capacity to damage the basal ganglia and striatum. Its preferential accumulation in those regions probably arises from their high iron-binding capacity (Hill & Switzer 1984). Manganese compounds generate various reactive chemical species, including hydrogen peroxide and glutathionyl radical. On account of its concentration in the basal ganglia system, manganese is also able to generate the cyclized *o*-quinone form of dopamine (Donaldson *et al.* 1982, Graham 1984, Halliwell 1984, Segura-Aguilar & Lind 1989, Shi 1990), a route of ROS production which is thought to be particularly damaging (Donaldson *et al.* 1981, Bondy 1994). Low micromolar concentrations of manganese cause substantial ATP depletion and calcium accumulation in the striatum (Brouillet *et al.* 1993) and, to a lesser extent, the hippocampus (Sloot *et al.* 1994).

The environmental relevance of manganese derives from the exposure to it by miners and steelworkers (Barbeau *et al.* 1976) in whom extrapyramidal motor disorders resembling Parkinsonism arise more often than in the general population. In addition, methylcyclopentadienyl manganese tricarbonyl (MMT), a manganese derivative added to gasoline to replace lead as an anti-knocking agent, disrupts monoamine levels in multiple brain regions, including the hippocampus (Komura & Sakamoto 1994). MTT has been used in Canada since 1977 and has repeatedly been considered for introduction in the US (Lynam *et al.* 1994). Studies in Canada show increased manganese concentrations in air corresponding to traffic density (Loranger *et al.* 1994), and in food plants grown on soils exposed to high MMT emissions (Brault *et al.* 1994). Garage mechanics, taxi drivers, and persons in congested urban areas are considered particularly at risk from MMT toxicity, although the levels determined during occupational exposure studies meet currently acceptable standards (Abbott 1987, Lynam *et al.* 1994, Zayed *et al.* 1994).

At present, very little is known about whether GCs or stress interact with the toxicity of these compounds. However, it is quite plausible that they do, for the following reasons.

(i) All of these compounds can damage areas of the brain demonstrated to be vulnerable to GC endangerment and are regions that contain demonstrable concentrations of GC receptors (McEwen *et al.* 1986).

(ii) Each compound has been shown to generate ROS.

(iii) At least some of their mechanisms of action are ones shown to be sensitive to exacerbation by GCs.

(iv) GCs might exacerbate oxidative insults through an additional, little-recognized route. Ascorbic acid appears to be transported into cells via glucose transporters (Padh *et al.* 1985, Washko & Levine 1992). As discussed earlier, GCs decrease the numbers of such transporters on the cell surface in tissues throughout the body (a mechanism likely to underly its inhibition of glucose uptake in various tissues, including the brain), potentially reducing this route of antioxidant uptake and thereby reducing cellular antioxidant defenses against these compounds.

(v) Rat intestinal epithelial cells show repressed levels of the mitochondrial form of SOD, Mn-SOD, after GC treatment. Moreover, in the lung, GCs may attenuate developmental increases in antioxidant enzyme level in a dose-dependent fashion

(Asayama *et al.* 1992), although the opposite has also been observed (Frank *et al.* 1985, Langley *et al.* 1993).

It should also be emphasized that some GC actions suggest *protection* from oxidative insults, aside from a possible tissue-specific increase in antioxidant enzymes, as outlined below.

(i) GCs are well-known for their capacity to inhibit the synthesis of prostaglandins, a potent source of oxygen radicals. This occurs via inhibition of phospholipase A_2 and may thus potentially decrease some of the routes of toluene-mediated damage. However, it is far from conclusive whether GCs inhibit prostaglandin synthesis in the brain (Jane *et al.* 1982, Shapira *et al.* 1988).

(ii) GCs decrease spinal cord damage after injury. This has been interpreted as arising from GCs, because of their steroidal structure, protecting against peroxidative attack secondarily to their altering of membrane fluidity (cf. Hall 1990). However, this is a 'megadose' phenomenon, requiring vastly supraphysiological GC concentrations.

(iii) In contrast to the situation just described in intestinal epilethial and lung cells, GCs appear to increase the activity of SOD, catalase and glutathione peroxidase (GSPx) in glomerular epithelial cells, thereby decreasing the degree of cell damage during exposure to oxygen radical generating systems (Yoshioka *et al.* 1994).

(iv) In neonatal animals, GCs produced precocious gut closure that reduced absorption and retention of mercury and manganese given in the diet (Kargacin & Landeka 1990).

One study (Mattia *et al.* 1991) unexpectedly provides a hint that there may indeed be a damaging synergy occurring between at least one of these neurotoxicants and GCs. The authors demonstrated that toluene generated ROS both *in vivo* and *in vitro* preparations of brain tissue. They then reported that the *in vivo* generation was inhibited with the drug metyrapone, which was described as a 'mixed function oxidase inhibitor'. The authors were apparently unaware of the primary clinical use of metyrapone as a steroidogenesis inhibitor which selectively blocks adrenal release of GCs (Cheng *et al.* 1974, Haynes & Murad 1985). The authors' *in vitro* data with metyrapone were not as clearcut as were their *in vivo* data. This suggests that the protective effects of metyrapone may have primarily been via inhibition of GC secretion. Exposure to these toxins is a major stressor, stimulating considerable GC secretion. For example, toluene exposure raises GC concentrations from the 1–10 μg/dl range to 25–35 μg/dl in mice over the course of weeks (Anderson *et al.* 1983, Hsieh *et al.* 1991), levels similar to those seen in response to substantial stressors. To appreciate this, such GC concentrations are sufficient to quadruple seizure-induced glutamate accumulation in the hippocampus (Stein-Behrens *et al.* 1994a). The metyrapone finding suggests that the normative and considerable GC stress-response that accompanies exposure to toluene is enough to augment its generation of ROS. These findings echo reports that metyrapone will decrease seizure-induced neuron death (Stein & Sapolsky 1988), seizure-induced cytoskeletal degeneration (Smith-Swintosky *et al.* 1995) in addition to ischemia-induced damage to the hippocampus (Morse & Davis 1990).

8.4 Conclusions

The literature reviewed in this chapter demonstrates that glucocorticoids can exert a number of deleterious effects upon neurons, compromising their ability to survive a wide variety of neurologic insults. These findings suggest that glucocorticoids might exacerbate the damaging effects of any of a number of environmental neurotoxicants, including those discussed. This remains a largely unexplored area. The possibility that elevated GC concentrations affect populations who contact neurotoxicants either occupationally (especially manufacturing and farm workers) or in the environment in general (through food, water and airborne sources) needs to be examined. If studies suggest an interaction between GCs and these compounds, such findings might ultimately be of clinical relevance for three reasons.

(i) Synthetic GCs (such as prednisone, hydrocortisone, dexamethasone or triamcinalone) are prescribed in vast quantities to control asthma, arthritis, autoimmune disease and other disorders. As noted earlier, approximately 16 million such perscriptions are written annually in the US (Eufemio 1990). Potentially, populations with known exposure to these compounds might have to be viewed as contraindicated for GC perscriptions.

(ii) Acute exposure to particularly high concentrations of some neurotoxic compounds might well take place in the context of a workplace accident that would be viewed as stressful (ie would be accompanied by elevated GC secretion). This might impact the extent of neuropathology caused by the exposure.

(iii) Exposure to the compounds themselves have either been documented to be, or are likely to be, direct stimulants of GC secretion.

Finally, separate from the possibility of studies that stimulate inquiry into the epidemiology of GC effects upon industrial exposure to these compounds, such studies would be of heuristic value for understanding the mechanisms of neurotoxicity of these compounds. If GCs augment neuron death caused by one of these compounds while, for example, failing to augment a particular cellular effect of that compound, that diminishes the importance of that effect in the process of the neuron death. A similar insight would be gained if, conversely, GCs augment a particular cellular effect of a compound without augmenting its overall toxicity. This is critical because it is not always clear if the parameter being measured is related to the actual cause, rather than being merely a correlate, of the neuron death. Thus, continued study may cast light on the basic mechanisms of neuron death, in addition to aiding the understanding of potential modulators of certain environmental neurotoxicants.

References

ABBOTT, P. I. (1987) Methylcyclopentadienyl manganese tricarbonyl (MMT) in petrol: the toxicological issues. *The Science of the Total Environment*, **67**, 247–251.

ALTMAN, D., YOUNG, R. & YAGEL, S. Effects of dexamethasone in hypoxic-ischemic brain injury in the neonatal rat. *Biology of the Neonate*, **46**, 149–157.

AMAROSO, D., TAMER A., WULFERT, E. & HANIN, I. H. (1993) Long-term exposure to corticosterone aggravates AF64A-induced cholinotoxicity in rat hippocampus. *Society for Neuroscience Abstracts*, **19**, 770.2.

ANDERSSON K., NILSEN O., TOFTGARD, R., ENEROTH, P., GUSTAFSON, J., BATTISTINI, N. & AGNATI L. (1983) Increased amine turnover in several noradrenergic terminal systems and changes in prolactin secretion in the male rat by exposure to various concentrations of toluene. *Neurotoxicology*, **4**, 43–51.

ARMANINI, M., HUTCHINS, C., STEIN, B. & SAPOLSKY, R. (1990) Glucocorticoid endangerment of hippocampal neurons is NMDA-receptor mediated. *Brain Research*, **532**, 7–15.

ASAYAMA, K., HAYASHIBE, H., DOBASHI, K., UCHIDA, N. & KATO, K. (1992) Effect of dexamethasone on antioxidant enzymes in fetal rat lungs and kidneys. *Biology of the Neonate*, **62**, 136–143.

ATCHISON W. & HARE, M. (1994) Mechanisms of methylmercury-induced neurotoxicity. *FASEB Journal*, **8**, 622–628.

AUER, R. & SIESJO, B. (1988) Biological differences between ischemia, hypoglycemia and epilepsy. *Annals of Neurology*, **24**, 699–721.

AUS DER MUHLEN K. & OCKENFELS H. (1979) Morphologische Veranderungen im Diencephalon und Telencephalon nach Storngen des Regelkreises Adenohypophyse-Nebennierenrinde III. Ergebnisse beim Meerschweinchen nach Verabreichung von Cortison und Hydrocortison. *Zeitschrift fur Zellforsch*, **93**, 126–133.

BARBEAU, A., INOUE, N. & CLOUTIER, T. (1976) The role of manganese in dystonia. *Advances in Neurology*, **14**, 339–345.

BARKS, J., POST, M. & TUOR, U. (1991) Dexamethasone prevents hypoxic-ischemic brain damage in the neonatal rat. *Pediatric Research*, **29**, 558–565.

BEAL, M. (1992) Does impairment of energy metabolism result in excitotoxic neuronal death in neurodegenerative illnesses? *Annals of Neurology*, **31**, 119–130.

BENNETT, M., LEHMAN, J., MLADY, G. & ROSE, G. (1993) Synergy between corticosterone and sodium azide in producing a place learning deficit in the Morris water maze. *Society for Neuroscience Abstracts*, **19**, 78.12.

BONDY, S. (1994) Induction of oxidative stress in the brain by neurotoxic agents. In CHANG, L. (Ed.), *Principles of Neurotoxicology. Neurological Disease and Therapy*, Vol. 26, pp. 563–582, New York, Marcel Dekker.

BONDY, S. C. & MCKEE, M. (1991) Disruption of the potential across the synaptosomal plasma membrane and mitochondria by neurotoxic agents. *Toxicology Letters*, **58**, 13–21.

BRAULT, N., LORANGER, S., COURCHESNE, F., KENNEDY, G. & ZAYED, J. (1994) Bioaccumulation of manganese by plants: influence of MMT as a gasoline additive. *The Science of the Total Environment*, **153**, 77–85.

BROUILLET, E., SHINOBU, L., MCGARVEY, U., HOCHBERG, F. & BEAL, M. (1993) Manganese injection into the rat striatum produces excitotoxic lesions by impairing energy metabolism. *Experimental Neurology*, **120**, 89–96.

CARNEY, J. M. & FLOYD, R. A. (1991) Protection against oxidative damage to CNS by a phenyl-tert-butyl nitrone (PBN) and other spin-trapping agents: a novel series of nonlipid free radical scavengers. *Journal of Molecular Neuroscience*, **3**, 47–55.

CHANG, L., GILBERT, M. & SPRECHER, J. (1978) Modification of methylmercury neurotoxicity by vitamin E. *Environmental Research*, **17**, 356–362.

CHENG, S., HARDING, B. & CARBALLEIRA, A. (1974) Effects of metyrapone on pregnenolone biosynthesis and on cholesterol-cytochrome P-450 interaction in the adrenal. *Endocrinology*, **94**, 1451–1458.

CHOU, Y., LIN, W. & SAPOLSKY, R. (1994) Glucocorticoids exacerbate cyanide-induced aspartate accumulation in hippocampal cultures. *Brain Research*, **654**, 8–13.

CHUMAS, P., DEL BIGIO, M., DRAKE, J. & TUOR, U. (1993) A comparison of the protective effect of dexamethasone to other potential prophylactic agents in a neonatal rat model of cerebral hypoxia-ischemia. *Journal of Neurosurgery*, **79**, 414–419.

COX, J., LYSKO, P. & HENNEBERRY, R. (1989) Excitatory amino acid neurotoxicity at the NMDA receptor in cultured neurons: role of the voltage-dependent magnesium block. *Brain Research*, **499**, 267–273.

COYLE, J. & PUTTFARCKEN, P. (1993) Oxidative stress, glutamate and neurodegenerative disorders. *Science*, **262**, 684–690.

DONALDSON, J., LABELLA, F. & GESSA, D. (1981) Enhanced autoxidation of daopmine as a possible basis of manganese neurotoxicity. *Neurotoxicology*, **2**, 53–60.

DONALDSON, J., MCGREGOR, D. & LABELLA, F. (1982) Manganese neurotoxicity: a model for free radical mediated neurodegeneration? *Canadian Journal of Physiology and Pharmacology*, **60**, 1398–1405.

ELLIOTT, E. & SAPOLSKY, R. (1992) Corticosterone enhances kainic acid-induced calcium mobilization in cultured hippocampal neurons. *Journal of Neurochemistry*, **59**, 1033–1039.

ELLIOTT, E. & SAPOLSKY, R. (1993) Corticosterone impairs hippocampal neuronal calcium regulation: possible mediating mechanisms. *Brain Research*, **602**, 84–90.

ELLIOTT, E., MATTSON, M., VANDERKLISH, P. K., LYNCH, G., CHANG, I. & SAPOLSKY, R. (1993) Corticosterone exacerbates kainate-induced alterations in hippocampal tau immunoreactivity and spectrin proteolysis *in vivo*. *Journal of Neurochemistry*, **61**, 57–63.

EUFEMIO, M. A. (1990) Advances in the therapy of osteoporosis – Part V. Steroid-induced osteoporosis. *Geriatric – Medicine Today*, **9**, 41–54.

FRANK, L., LEWIS, P. L. & SOSENKO, I. R. S. (1985) Dexamethasone stimulation of fetal rat lung antioxidant enzyme activity in parallel with surfactant stimulation. *Pediatrics*, **75**, 569–576.

FREI, B., ENGLAND, L. & AMES, B. N. (1989) Ascorbate is an outstanding antioxidant in human blood plasma. *Proceedings of the National Academy of Sciences., USA*, **86**, 6377–6381.

FREO, U., HOLLOWAY, H., KALOGERAS, K., RAPOPORT, S. & SONCRANT, T. (1992) Adrenalectomy or metyrapone-pre-treatment abolishes cerebral metabolic responses to the serotonin agonist 1-(2,5-dimethoxy-4-iodophenyl)-2-aminopropane (DOI) in the hippocampus. *Brain Research*, **586**, 256–264.

GARVEY, W., HUECKSTEADT, T., LIMA, F. & BIRNBAUM, M. (1989) Expression of a glucose transporter gene cloned from brain in cellular models of insulin resistance: dexamethasone decreases transporter mRNA in primary cultured adipocytes. *Molecular Endocrinology*, **3**, 1132–1138.

GRAHAM, D. (1984) Catecholamine toxicity: a proposal for the molecular pathogenesis of manganese neurotoxicity and Parkinson's disease. *Neurotoxicology*, **5**, 83–88.

HALL, E. (1990) Steroids and neuronal destruction or stabilization. *Steroids and Neuronal Activity*, Ciba Foundation Symposium Vol. 153, pp. 206–215, Chichester: John Wiley.

HALLIWELL, B. (1984) Manganese ions, oxidation reactions and the superoxide radical. *Neurotoxicology*, **5**, 113–120.

HAYNES, R. & MURAD, F. (1985) Adrenocorticotropic hormone; adrenocortical steroids and their synthetic analogs; inhibitors of adrenocortical steroid biosynthesis. In GILMAN, A., GOODMAN, L., RALL, T. & MURAD, F. (Eds), *The Pharmacological Basis of Therapeutics*, 7th Edn, pp. 1459–1470, New York: Macmillan.

HILL, J. & SWITZER, R. (1984) The regional distribution and cellular localization of iron in the rat brain. *Neuroscience*, **II**, 595–606.

HORNER, H., MUNCK, A. & LIENHARD, G. (1987) Dexamethasone causes translocation of glucose transporters from the plasma membrane to an intracellular site in human fibroblasts. *Journal of Biological Chemistry*, **262**, 17696–17703.

HORNER, H., PACKAN, D. & SAPOLSKY, R. (1990) Glucocorticoids inhibit glucose transport in cultured hippocampal neurons and glia. *Neuroendocrinology*, **52**, 57–63.

HORTNAGL, H., BERGER, M., HAVELEC, L. & HORNYKIEWICZ, O. (1993) Role of glucocorticoids in the cholinergic degeneration in rat hippocampus induced by ethylcholine aziridinium (AF64A). *Journal of Neuroscience*, **13**, 2939–2945

HSIEH, G., SHARMA, R. & PARKER, R. (1991) Hypothalamic-pituitary-adrenocortical axis activity and immune function after oral exposure to benzene and toluene. *Immunopharmacology*, **21**, 23–30.

JANE, J. A., RIMEL, R., POBERESKIN, L. H., TYSON, G. W., STEWARD, O. & GENNARELLI, T. A. (1982) Outcome and pathology of head injury. In GROSSMAN, R. G.

& GILDENBERG, P. (Eds), *Head Injury: Basic and Clinical Aspects*, pp. 229–237, New York: Raven Press.

JOHNSON, M., STONE, D., BUSH, L., HANSON, G. & GIBB, J. (1989) Glucocorticoids and 3,4-methylenedioxymethamphetamine (MDMA)-induced neurotoxicity, *European Journal of Pharmacology*, **161**, 181–187.

KADEKARO, M., MASANORI, I. & GROSS, P. (1988) Local cerebral glucose utilization is increased in acutely adrenalectomized rats. *Neuroendocrinology*, **47**, 329–335.

KARGACIN, B. & LANDEKA, M. (1990) Effect of glucocorticoids on metal retention in rats. *Bulletin of Environmental Contamination and Toxicology*, **45**, 655–633.

KAUPPINEN, R., KOMULAINEN, H. & TAIPALE, H. (1989) Cellular mechanisms underlying the increase in cytosolic free calcium concentration induced by methylmercury in cerebrocortical synaptosomes from guinea pig. *Journal of Pharmacology and Experimental Therapeutics*, **248**, 1248–1253.

KERR, D., CAMPBELL, L., APPLEGATE, M., BRODISH, A. & LANDFIELD, P. (1991) Chronic stress-induced acceleration of electrophysiologic and morphometric biomarkers of hippocampal aging. *Journal of Neuroscience*, **II**, 1316–1321.

KERR, D., CAMPBELL, L., THIBAULT, O. & LANDFIELD, P. (1992) Hippocampal glucocorticoid receptor activation enhances voltage-dependent calcium conductances: relevance to brain aging. *Proceedings of the National Academy of Sciences, USA*, **89**, 8527–8531.

KOIDE, T., WIELOCH, T. & SIESJO, B. (1986) Chronic dexamethasone pre-treatment aggravates ischemic neuronal necrosis. *Journal of Cerebral Blood Flow and Metabolism*, **6**, 395–408.

KOMURA, J. & SAKAMOTO, M. (1994) Chronic oral administration of methylcyclopentadienyl manganese tricarbonyl altered brain biogenic amines in the mouse: comparison with inorganic manganese. *Toxicology Letters*, **73**, 65–72.

KRUGERS, H., JAARSMA, D. & KORF, J. (1992) Rat hippocampal lactate efflux during electroconvulsive shock or stress is differentially dependent on entorhinal cortex and adrenal integrity. *Journal of Neurochemistry*, **58**, 826–833.

LAFON-CAZAL, M., PIETRI, S., CULCASI, M. & BOCKAERT, J. (1993) NMDA-dependent superoxide production and neurotoxicity. *Nature*, **364**, 535–537.

LANDFIELD, P., BASKIN, R. & PITLER, T. (1981) Brain-aging correlates; retardation by hormonal–pharmacological treatments. *Science*, **214**, 581–585.

LANGLEY, S. C., RICKETT, G. W. M., HUNT, A., KELLY, F. J., POSTLE, A. D. & YORK, D. A. (1983) Effects of the glucocorticoid agonist, RU28362, and the antagonist RU486 on lung phosphatidylcholine and antioxidant enzyme development in the genetically obese Zucker rat. *Biochemical Pharmacology*, **45**, 543–550.

LAWRENCE, M. & SAPOLSKY, R. (1994) Glucocorticoids accelerate ATP loss in cultured hippocampal neurons. *Brain Research*, **646**, 303–308.

LEBEL, C. & SCHATZ, R. (1990) Altered synaptosomal phospholipid metabolism after toluene: possible relationship with membrane fluidity, sodium, potassium-ATPase and phospholipid methylation. *Journal of Pharmacology and Experimental Therapeutics*, **253**, 1189–1195.

LEBEL, C., ALI, S., MCKEE, M. & BONDY, S. (1990) Organometal-induced increases in oxygen reactive species: the potential of 2′-7′-dichlorofluorescin diacetate as an index of neurotoxic damage. *Toxicology and Applied Pharmacology*, **104**, 17–23.

LEBEL, C., ALI, S. & BONDY, S. (1992) Deferoxamine inhibits methyl-mercury induced increases in reactive oxygen species formation in rat brain. *Toxicology and Applied Pharmacology*, **112**, 161–167.

LEVY, A., DACHIR, S., ARBEL, I. & KADAR, T. (1994) Aging, stress and cognitive function. In *Annals of the New York Academy of Sciences*, **717**, 79–88.

LIU, J., WANG, X. & MORI, A. (1994) Immobilization stress-induced antioxidant defense changes in rat plasma: effect of treatment with reduced glutathione. *International Journal of Biochemistry*, **26**, 511–518.

LIU, T. H., BECKMAN, J. S., FREEMAN, B. A., HOGAN, E. L. & HSU, C. Y. (1989) Polyethylene glucol-conjugated superoxide dismutase and catalase reduce ischemic brain injury. *American Journal of Physiology*, **256**, H589–H594.

LORANGER, S., ZAYED, J. & FORGET, E. (1994) Manganese contamination in Montreal in relation with traffic density. *Water, Air and Soil Pollution*, **74**, 385–391.

LOWY, M., GAULT, L. & YAMAMOTO, B. (1993) Adrenalectomy attenuates stress-induced elevations in extracellular glutamate concentrations in the hippocampus. *Journal of Neurochemistry*, **61**, 1957–1963.

LOWY, M., WITTENBERG, L. & NOVOTNEY, S. (1994) Adrenalectomy attenuates kainic acid-induced spectrin proteolysis and heat shock protein 70 induction in hippocampus and cortex. *Journal of Neurochemistry*, **63**, 886–893.

LYNAM, D. R., PFEIFER, G. D., FORT, B. F., TER HARR, G. L. & HOLLRAH, D. P. (1994) Atmospheric exposure to manganese from use of methylcyclopentadienyl manganese tricarbonyl (MMT) performance additive. *The Science of the Total Environment*, **146/7**, 103–108.

LYSKO, P., COX, J., ALESSANDRA, M. & HENNEBERRY, R. (1989) Excitatory amino acid neurotoxicity in at the NMDA receptor in cultured neurons: pharmacological characterization. *Brain Research*, **449**, 258–264.

MAGARINOS, A. & MCEWEN, B. (1993) Blockade of glucocorticoid synthesis by cyanoketone prevents chronic stress-induced dendritic atrophy of hippocampal neurons in the rat. *Society for Neuroscience Abstracts*, **19**, 71.2.

MAK, I. T., BOEHME, P. & WEGLICKI, W. B. (1992) Antioxidant effects of calcium channel blockers against free radical injury in endothelial cells. Correlation of protection with preservation of glutathione levels. *Circulation Research*, **70**, 1099–1105.

MASTERS, J., FINCH, C. & SAPOLSKY, R. (1989) Glucocorticoid endangerment of hippocampal neurons does not involve DNA cleavage. *Endocrinology*, **124**, 3083–3088.

MATTIA, C., LEBEL, C. & BONDY, S. (1991) Effect of toluene and its metabolites on cerebral oxygen radical formation. *Biochemical Pharmacology*, **42**, 879–886.

MCEWEN, B. (1992) Re-examination of the glucocorticoid hypothesis of stress and aging. *Progress in Brain Research*, **93**, 365–383.

MCEWEN, B., DE KLOET, E. & ROSTENE, W. (1986) Adrenal steroid receptors and actions in the nervous system. *Physiological Reviews*, **66**, 1121–1187.

MCINTOSH, L. J. & SAPOLSKY, R. M. (1994) Glucocorticoids exacerbate the *in vitro* neurotoxicity of the oxygen radical generator adriamycin. *Society for Neuroscience Abstracts*, **20**, 1660.

MEANEY, M., AITKEN, D., BHATNAGER, S., VAN BERKEL, C. & SAPOLSKY, R. (1988) Effects of neonatal handling on age-related impairments associated with the hippocampus. *Science*, **239**, 766–769.

MEANEY, M., AITKEN, D. & SAPOLSKY, R. (1991) Postnatal handling attenuates neuroendocrine, anatomical and cognitive dysfunctions associated with aging in female rats. *Neurobiology of Aging*, **12**, 31–40.

MILLER, G. & DAVIS, J. (1991) Post-ischemic surge in corticosteroids aggravates ischemic damage to gerbil CAI pyramidal cells. *Society for Neuroscience Abstracts*, **17**, 302.4.

MIYAMOTO, M. & COYLE, J. T. (1990) Idebenone attenuates neuronal degeneration induced by intrastriatal injection of excitotoxins. *Experimental Neurology*, **108**, 38–45.

MIZOGUCHI, K., KUNISHITA, T., CHUI, D. & TABIRA, T. (1992) Stress induces neuronal death in the hippocampus of castrated rats. *Neuroscience Letters*, **138**, 157–163.

MOGHADDAM, B. (1993) Stress preferentially increases extraneuronal levels of excitatory amino acids in the prefrontal cortex: comparison to hippocampus and basal ganglia. *Journal of Neurochemistry*, **60**, 1650–1657.

MOGHADDAM, B., STEIN-BEHRENS, B. & SAPOLSKY, R. (1994) Role of endogenous glucocorticoid secretion in stress-induced augmentation of glutamate accumulation in the hippocampus. *Brain Research*, **665**, 251–256.

MORSE, J. & DAVIS, J. (1990) Regulation of ischemic hippocampal damage in the gerbil: adrenalectomy alters the rate of CAI cell disappearance. *Experimental Neurology*, **110**, 86–94.

MUNCK, A. (1971) Glucocorticoid inhibition of glucose uptake by peripheral tissues. Old and new evidence, molecular mechanisms and physiological significance. *Perspectives in Biology and Medicine*, **14**, 265–283.

MUNCK, A., GUYRE, P. & HOLBROOK, N. (1984) Physiological actions of glucocorticoids in stress and their relation to pharmacological actions. *Endocrine Reviews*, **5**, 25–43.

NOVELLI, A., REILLY, J., LYSKO, P. & HENNEBERRY, R. (1988) Glutamate becomes neurotoxic via the NMDA receptor when intracellular energy levels are reduced. *Brain Research*, **45**, 205–211.

PACKAN, D. & SAPOLSKY, R. (1990) Glucocorticoid endangerment of the hippocampus: tissue, steroid and receptor specificity. *Neuroendocrinology*, **51**, 613–620.

PADH, H., SUBRAMONIAM, A. & ALEO, J. (1985) Glucose inhibits cellular ascorbic acid uptake by fibroblasts *in vitro*. *Cell Biology International Reports*, **9**, 531–536.

ROSENBAUM, D. M., KALBERG, J. & KESSLER, J. A. (1994) Superoxide dismutase ameliorates neuronal death from hypoxia in culture. *Stroke*, **25**, 857–862.

SAPOLSKY, R. (1985a) A mechanism for glucocorticoid toxicity in the hippocampus; increased neuronal vulnerability to metabolic insults. *Journal of Neuroscience*, **5**, 1228–1234.

SAPOLSKY, R. (1985b) Glucocorticoid toxicity in the hippocampus: temporal aspects of neuronal vulnerability. *Brain Research*, **339**, 300–307.

SAPOLSKY, R. (1986) Glucocorticoid toxicity in the hippocampus: reversal by supplementation with brain fuels. *Journal of Neuroscience*, **6**, 2240–2248.

SAPOLSKY, R. (1992) *Stress, the Aging Brain and the Mechanisms of Neuron Death*, Cambridge, MA: MIT Press.

SAPOLSKY, R. & PULSINELLI, W. (1985) Glucocorticoids potentiate ischemic injury to neurons: therapeutic implications. *Science*, **229**, 1397–1401.

SAPOLSKY, R., KREY, L. & MCEWEN, B. (1985) Prolonged glucocorticoid exposure reduces hippocampal neuron number; implications for aging. *Journal of Neuroscience*, **5**, 1221–1227.

SAPOLSKY, R., PACKAN, D. & VALE, W. (1988) Glucocorticoid toxicity in the hippocampus: *in vitro* demonstration. *Brain Research*, **453**, 367–373.

SAPOLSKY, R., UNO, H., REBERT, C. & FINCH, C. (1990) Hippocampal damage associated with prolonged glucocorticoid exposure in primates. *Journal of Neuroscience*, **10**, 2897–2903.

SARAFIAN, T., HAGLER, J., VARTAVERIAN, L. & VERITY, A. (1989) Rapid cell death induced by methyl mercury in suspension of cerebellar granule neurons. *Journal of Neuropathology and Experimental Neurology*, **48**, 1–11.

SARAFIAN, T. & VERITY, M. (1991) Oxidative mechanisms underlying methyl mercury neurotoxicity. *International Journal of Developmental Neuroscience*, **9**, 147–156.

SCHUBERT, D., KIMURA, H. & MAHER, P. (1992) Growth factors and vitamin E modify neuronal glutamate toxicity. *Proceedings of the National Academy of Sciences, USA*, **89**, 8264–8269.

SEGURA-AGUILAR, J. & LIND, C. (1989) On the mechanism of the manganese-induced neurotoxicity of dopamine: prevention of quinone-derived oxygen toxicity by DT diaphorase and superoxide dismutase. *Chemico-Biological Interactions*, **72**, 309–316.

SHAPIRA, Y., DAVIDSON, E., WEIDENFELD, Y., COTEV, S. & SHOHAMI, E. (1988) Dexamethasone and indomethacin do not affect brain edema following head injury in rats. *Journal of Cerebral Blood Flow and Metabolism*, **8**, 395–402.

SHI, X. (1990) The glutathionyl radical formation in the reaction between manganese and glutathione and its neurotoxic implications. *Medical Hypotheses*, **33**, 83–95.

Shrivastav, B., Brodwick, B. & Narahashi, T. (1976) Methylmercury: effects on electrical properties of squid axon membranes. *Life Sciences*, **18**, 1077–1083.

Shukla, G., Srivastava, R. & Chandra, S. (1988) Prevention of cadmium-induced effects on regional glutathione status of rat brain by vitamin E. *Journal of Applied Toxicology*, **8**, 355–360.

Simpson, J. & Isacson, O. (1993) Mitochondrial impairment reduces the threshold for *in vivo* NMDA-mediated neuronal death in the striatum. *Experimental Neurology*, **21**, 57–63.

Sloot, W., van der Sluijs-Gelling, A. & Gramsbergen, J. (1994) Selective lesions by manganese and extensive damage by iron after injection into rat striatum or hippocampus. *Journal of Neurochemistry*, **62**, 205–212.

Smith-Swintosky, V., Pettigrew, L., Sapolsky, R., Phares, C., Craddock, S., Brooke, S. & Mattson, M. (1995) Metyrapone, an inhibitor of glucocorticoid production, reduces brain injury induced by focal and global ischemia and seizures. *Journal of Cerebral Blood Flow and Metabolism* (in press).

Starkman, M., Gebarski, S., Berent, S. & Schteingart, D. (1992) Hippocampal formation volume, memory dysfunction and cortisol levels in patients with Cushing's syndrome. *Biological Psychiatry*, **32**, 756–764.

Stein, B. & Sapolsky, R. (1988) Chemical adrenalectomy reduces hippocampal damage induced by kainic acid. *Brain Research*, **437**, 175–181.

Stein-Behrens, B., Elliott, E., Miller, C., Schilling, J., Newcombe, R. & Sapolsky, R. (1992) Glucocorticoids exacerbate kainic acid-induced extracellular accumulation of excitatory amino acids in the rat hippocampus. *Journal of Neurochemistry* **58**, 1730–1735.

Stein-Behrens, B., Lin, W. & Sapolsky, R. (1994a) Physiological elevations of glucocorticoids potentiate glutamate accumulation in the hippocampus. *Journal of Neurochemistry*, **63**, 596–602.

Stein-Behrens, B., Mattson, M., Chang, I., Yeh, M. & Sapolsky, R. (1994b) Stress exacerbates neuron loss and cytoskeletal pathology in the hippocampus. *Journal of Neuroscience*, **14**, 5373–5379.

Supko, D. & Johnston, M. (1994) Dexamethasone enhances NMDA receptor-mediated injury in the postnatal rat. *European Journal of Pharmacology*, **270**, 105–111.

Talmi, M., Carlier, E. & Soumireu-Mourat, B. (1993) Similar effects of aging and corticosterone treatment on mouse hippocampal function. *Neurobiology of Aging*, **14**, 239–245.

Theoret, Y., Caldwell-Kenkel, J. & Krigman, M. (1985) The role of neuronal metabolic insult in organometal neurotoxicity. *Toxicologist*, **6**, Abstract 491.

Tombaugh, G. & Sapolsky, R. (1992) Corticosterone accelerates hypoxia-induced ATP loss in cultured hippocampal astrocytes. *Brain Research*, **588**, 154–159.

Tombaugh, G. & Sapolsky, R. (1993) Endocrine features of glucocorticoid endangerment in hippocampal astrocytes. *Neuroendocrinology*, **57**, 7–13.

Tombaugh, G., Yang, S., Swanson, R. & Sapolsky, R. (1992) Glucocorticoids exacerbate hypoxic and hypoglycemic hippocampal injury *in vitro*: biochemical correlates and a role for astrocytes. *Journal of Neurochemistry*, **59**, 137–143.

Truelove, D., Shuaib, A., Ijaz, S., Richardson. S. & Kalra, J. (1994) Superoxide dismutase, catalase and U78517F attenuate neuronal damage in gerbils with repeated brief ischemic insults. *Neurochemical Research*, **19**, 665–673.

Tuor, U., Simone, C., Arellano, R., Tanswell, K. & Post, M. (1993) Glucocorticoids prevention of neonatal hypoxic-ischemic damage; role of hyperglycemia and antioxidant enzymes. *Brain Research*, **604**, 165–170.

Uhler, T., Frim, D., Pakzaban, P. & Isacson, O. (1994) The effects of mega-dose methylprednisolone and U-78517F on glutamate-receptormediated toxicity in the rat neostriatum. *Neurosurgery*, **34**, 122–128.

UNO, H., TARARA, R., ELSE, J., SULEMAN, M. & SAPOLSKY, R. (1989) Hippocampal damage associated with prolonged and fatal stress in primates. *Journal of Neuroscience*, **9**, 1705–1712.

UNO, H., FLUGGE, G., THIEME, C., JOHREN, O. & FUCHS, E. (1991) Degeneration of the hippocampal pyramidal neurons in the socially stressed tree shrew. *Society for Neuroscience Abstracts*, **17**, 52.20.

VERITY, M., BROWN, W. & CHEUNG, M. (1975) Organic mercurial encephalopathy: *in vivo* and *in vitro* effects of methyl mercury on synaptosomal respiration. *Journal of Neurochemistry*, **25**, 750–762.

VIRGIN, C., HA, T., PACKAN, D., TOMBAUGH, G., YANG, S., HORNER, H. & SAPOLSKY, R. (1991) Glucocorticoids inhibit glucose transport and glutamate uptake in hippocampal astrocytes: implications for glucocorticoid neurotoxicity. *Journal of Neurochemistry*, **57**, 1422–1428.

WASHKO, P. & LEVINE, M. (1992) Inhibition of ascorbic acid transport in human neutrophils by glucose. *Journal of Biological Chemistry*, **267**, 23568–23574.

WATANABE, Y., GOULD, E., CAMERON, H., DANIELS, D. & MCEWEN, B. (1992) Phenytoin prevents stress- and corticosterone-induced atrophy of CA3 pyramidal neurons. *Hippocampus*, **2**, 431–438.

WOOLLEY, C., GOULD, E. & MCEWEN, B. (1990) Exposure to excess glucocorticoids alters dendritic morphology of adult hippocampal pyramidal neurons. *Brain Research*, **531**, 225–232.

YAMAMOTO, M., SHIMA, T., UOZUMI, T., SOGABE, T., YAMADA, K. & KAWASAKI, T. (1983) A possible role of lipid peroxidation in cellular damage caused by cerebral ischemia and the protective effect of alpha-tocopherol administration. *Stroke*, **14**, 977–983.

YEE, S. & CHOI, B. H. (1994) Methylmercury poisoning induces oxidative stress in the mouse brain. *Experimental and Molecular Pathology*, **60**, 188–194.

YOSHIOKA, T., KAWAMURA, T., MEYRICK, B. O., BECKMAN, J. K., HOOVER, R. L., YOSHIDA, H. & ICHIKAWA, I. (1994) Induction of manganese superoxide dismutase by glucocorticoids in glomerular cells. *Kidney International*, **45**, 211–219.

ZAYED, J., GERIN, M., LORANGER, S., SIERRA, P., BEGIN, D. & KENNEDY, G. (1994) Occupational and environmental exposure of garage workers and taxi drivers to airborne manganese arising from the use of methylcyclopentadienyl manganese tricarbonyl in unleaded gasoline. *American Industrial Hygiene Association Journal*, **55**, 53–60.

ZEEVALK, G. & NICKLAS, W. (1991) Mechanisms underlying initiation of excitotoxicity associated with metabolic inhibition. *Journal of Pharmacology and Experimental Therapeutics*, **257**, 870–878.

9

Glucocorticosteroids, Stress and Developmental Toxicology

JOHN A. BALDWIN

Formerly of SmithKline Beecham Pharmaceuticals, Welwyn

9.1 Introduction

The major role of glucocorticoids in the foetus is to prepare it for transition from an intrauterine to an independent existence. At physiological doses, glucocorticoids are responsible for maturation of the foetal lung, thyroid, gastrointestinal tract, adrenal medulla and numerous other tissues. They are responsible for the induction of various enzyme systems that are necessary after birth but which have little or no function prenatally, and they also stimulate the deposition of glycogen in the foetal liver. In some species these functions are closely linked to the mechanisms involving parturition. At pharmacological doses and beyond, however, glucocorticoids (both natural and synthetic) are potentially harmful to the developing embryo/foetus and a review of such consequences is the primary aim of this chapter. Since corticosteroids play an essential and intrinsic role in the stress response in mammalian species, the effects of stress on *in utero* development are also reviewed.

9.2 Glucocorticoids and Developmental Toxicity in Laboratory Animals

9.2.1 *Cortisone*

Since its discovery as a potent cleft palate inducer in mice (Baxter & Fraser 1950) cortisone has become a major tool in research into the aetiology of facial clefts. Different strains of inbred mice exhibit differing susceptibilities to cortisone-induced cleft palate (Kalter 1954, 1962, Fraser & Fainstat 1951). In the A/J strain all surviving offspring of mothers treated with a standard dose of cortisone on Days 11 to 15 *post coitum* had cleft palate but, in the C57 BL/6J (C57) mouse treated under identical conditions, only 20–25% of offspring were affected (Kalter 1954, 1981, Biddle & Fraser 1976). The 50% effective dose (ED_{50}) of cortisone for cleft palate induction in the A/J strain was 115 mg/kg compared with 687 mg/kg in the C57 strain (Biddle & Fraser 1976). A difference in timing of palatal closure has been established – the A/J strain closes later – which may make A/J mice more susceptible than the C57 strain (Biddle 1977). It has also been established that the synthesis and metabolism of glucocorticoids differ among adult mice of varying strains (Badr & Spickett 1965). Another approach to explaining differences in strain sensitivity may lie in the demonstration that significantly different levels of cytoplasmic glucocorticoid receptors exist in the oro-facial region of these strains (Salomon & Pratt 1976, 1979). Significant differences in sensitivity to cortisone-induced cleft palate have also been noted with other strains (Vekemans *et al.* 1979).

Cortisone has been shown to produce cleft palate in rabbits (Fainstat 1954, Walker 1967) and multiple malformations in dogs (Nakayama *et al.* 1978). Despite initial failings to produce cleft palate in rats (Gunberg 1957, Curry & Beaton 1958, Czabo *et al.* 1967, Walker 1971), administration of very high doses has been found to elicit cleft palate, eye defects and CNS malformations (Buresh & Urban 1970, Wilson *et al.* 1970). The hamster appears quite refractory to cleft palate induced by cortisone (but not some other corticosteroids) (Shah & Kilistoff 1976) but, in this species, cortisone treatment has led to retardation of skeletal growth (Shah & Kilistoff 1976). Retarded growth has also been found to occur in rats (Walker 1971), mice (Kaduri & Ornoy 1974) and in chicks (Moscona & Karnofsky 1960). Ornoy & Horowitz (1972) found that administration of cortisone to rats from Days 8 to 20 *post coitum* led to changes in cartilage and bone ground substance and inhibited ossification, but postnatal studies revealed these effects to be reversible.

9.2.2 Corticosterone

Corticosterone, the primary glucocorticoid in the mouse, has also been found to produce cleft palate in rats (Buresh & Urban 1970), mice (Blaustein *et al.* 1971) and hamsters (Shah & Kilistoff 1976). Administration to rats during the second half of pregnancy resulted in significantly smaller foetal adrenals at term (Lemmen *et al.* 1977). There was also atrophy of the adrenal cortex. There appeared to be little, if any recovery by the 14th day postnatally.

9.2.3 Deoxycorticosterone

Deoxycorticosterone has been reported to induce cleft palate, CNS malformations and eye defects in rats (Buresh & Urban 1970) and cleft palate in hamsters (Tedford & Risley 1950). Walker (1965) was not able to induce malformations in mice with this corticosteroid. Grollman & Grollman (1962) reported persistent hypertension in offspring of rats treated with deoxycorticosterone during pregnancy. Although a mineralocorticoid, deoxycorticosterone produces similar effects to the principal glucocorticoids. It is, however, the only corticosteroid reviewed that has not been shown to produce cleft palate in mice.

9.2.4 Hydrocortisone

Like cortisone, hydrocortisone is teratogenic in mice (Kalter & Fraser 1952, Pinsky & DiGeorge 1965) and differing susceptibilities have been established between inbred mice strains. Hydrocortisone administered on Days 11–14 *post coitum* results in 20% cleft palate in the C57 mouse but 95% in the brachymorphic bm/bm strain (Pratt *et al.* 1980). The ED_{50} for cleft palate induction was ~40 mg/kg for A/J and bm/bm mice compared to 325 mg/kg for C57 mice. The findings also indicate that both A/J and bm/bm strains normally exhibit a delay in palatal shelf rotation and elevated levels of cyclic AMP which appear to be predisposing factors for cleft palate induction. Cytoplasmic glucocorticoid receptor densities in bm/bm mice were, however, more similar to C57 mice than to the A/J strain, suggesting that this is not a causative factor in strain difference. Cleft palates have been produced in hamsters (Shah & Travill 1976) and multiple malformations in rats (Gunberg 1957) and guinea pigs (Hoar 1962), and brain malformations in rabbits (Kasirsky & Lombardi 1970). Aoyama *et al.* (1974) found hydrocortisone-17-α-butyrate ineffective in producing malformations in rats and mice whereas Yamada *et al.* (1981) found omphalocele, cleft palate and oedema in rats and cleft palate, open neural tube defects and umbilical hernia in rabbits with hydrocortisone-17-butyrate 21-propionate. Hydrocortisone sodium phosphate produced rib and vertebral defects in mice, but no facial clefts (Fujii *et al.* 1973). When administered from Day 14 *post coitum* onwards to rats, hydrocortisone caused demasculinization of male offspring as evidenced by shortened anogenital distance and lowered testis weight (Dahlof *et al.* 1978).

9.2.5 *Betamethasone*

Betamethasone has been shown to cause cleft palate in rats (Walker 1971, Mosier *et al.* 1982), mice (Walker 1971) and rabbits (Ishimura *et al.* 1975). It has also been shown to cause omphalocele in rats in addition to retardation of skeletal growth and of calcification of bones, decreased foetal weight and impaired growth of foetal heart, liver, adrenals, kidneys and skeletal muscle (Mosier *et al.* 1981, 1982). Epstein *et al.* (1977) demonstrated increased liver growth and function in rhesus monkeys and also found evidence of neuronal injury in the brain. When given to this species during late gestation there were inhibitory effects on lung maturation and adrenal insufficiency (Johnson *et al.* 1981). The butyrate propionate form of betamethasone has produced growth retardation and foetal death in the absence of malformation in the rat (Takeshima *et al.* 1990) and a low incidence of cleft palate in rabbits (Saijo *et al.* 1990).

9.2.6 *Dexamethasone*

Dexamethasone has been found to produce cleft palate in mice (Pinsky & DiGeorge 1965), rats (Vannier *et al.* 1969), rabbits (Clavert *et al.* 1961), hamsters (Shah & Kilistoff 1976) and dogs (Robens 1974). Pinsky & DiGeorge (1965) considered this compound more potent at inducing cleft palate than hydrocortisone. Neural tube defects have been reported in the rabbit (Buck *et al.* 1962) although none were found in this species with the 17-valerate form (Esaki *et al.* 1981). As with betamethasone, dexamethasone caused omphalocele in rats in addition to retardation of skeletal growth and calcification and impaired growth of soft tissues (Mosier *et al.* 1981, 1982). Cranium bifidum and aplasia cutis congenita were produced in rhesus monkeys when dexamethasone was given during organogenesis (Jerome & Hendrickx 1988), and adrenal insufficiency was produced in this species when this compound was given late in gestation (Challis *et al.* 1974). There was atrophy of the foetal adrenal accompanied by marked reduction in secretion of C_{19} steroids for placental aromatization. Atrophy of the foetal adrenal has also been seen in rats following treatment throughout the second half of pregnancy (Lemmen *et al.* 1977). There were no signs of recovery by 14 days *post partum.*

9.2.7 Desoximethasone

Miyamoto *et al.* (1975) tested desoximethasone in pregnant rats and mice. Cleft palate and oedema were found in mice but only delayed ossification in rats.

9.2.8 Beclomethasone

Beclomethasone induced cleft palate and vertebral abnormalities in mice (Esaki *et al.* 1976) and cleft palate in rabbits (Oguro *et al.* 1970). There were no malformations in rhesus monkeys (Tanioka 1976) but abortion and reduced growth were observed. In rats, only reduced foetal growth was noted (Furuhashi *et al.* 1979).

9.2.9 Triamcinolone

Triamcinolone administration has resulted in cleft palate in mice (Walker 1965), rabbits (Walker 1967), rats (Rowland & Hendrickx 1983) and hamsters (Shah & Kilistoff 1976). In the rat, omphalocele, undescended testes and foetal growth retardation were frequent observations (Rowland & Hendrickx 1983). A range of craniofacial and brain abnormalities were found in three species of subhuman primates following triamcinolone treatment (Parker & Hendrickx 1983, Tarara *et al.* 1989). In mice, Walker (1965) found triamcinolone to be far more potent than cortisone at inducing cleft palate and this is evidently the case for all the above species. This appears to be due to pharmacokinetic factors, in the rat at least (Rowland *et al.* 1983), where plasma concentration and plasma and foetal half-life of triamcinolone were markedly higher than for cortisol. In hamsters, triamcinolone-induced palatal clefts resulted from delayed reorientation of the palatal shelves from a vertical to a horizontal plane in the foetus whereas hydrocortisone, administered under identical conditions, produced clefts by preventing fusion between the two horizontal shelves (Shah 1980).

9.2.10 Prednisolone

Prednisolone treatment has been found to result in cleft palate in rabbits (Walker 1967), hamsters (Shah & Kilistoff 1976) and mice (Pinsky & DiGeorge 1965). Pinsky & DiGeorge found prednisolone to be a more potent teratogen than hydrocortisone. Jaw, tongue and head defects, but no cleft palates, have been found in rats

(Kalter 1962). Using the 17-valerate 21-acetate form, Koga *et al.* (1980a,b) produced growth retardation and omphalocele in rats and cleft palate in rabbits.

9.2.11 *Methylprednisolone*

Treatment with methylprednisolone has resulted in cleft palate in mice (Walker 1971) but not rats (Walker 1971) nor rabbits (Walker 1967).

9.2.12 *Prednisone*

Growth of mice offspring exposed to prednisone during gestation is significantly retarded and results in diminished full-term birth weight (Reinisch *et al.* 1978). Cleft palates have been produced in mice (Zawoiski 1980). Zawoiski (1980) found that while L-glutamic acid attenuated the cleft-inducing activity of cortisone, it had no effect on the incidences of clefts following prednisone or prednisolone treatment, suggesting a different mechanism of action between these agents.

9.2.13 *Diflucortolone*

Diflucortolone administered subcutaneously has produced cleft palate in mice and rabbits (Ezumi *et al.* 1977a, 1978) but not rats (Ezumi *et al.* 1977b). Gunzel *et al.* (1976) found that dermal applications of diflucortolone produced cleft palate, omphalocele and vertebral defects in rabbits but absence of tail, omphalocele and growth retardation in rats.

Glucocorticoids may also play an indirect role in the teratogenicity of other chemicals. Sullivan-Jones *et al.* (1982) showed that three diverse agents, haloperidol, 2,4,5-trichlorophenoxyacetic acid (2,4,5-T) and phenytoin, could produce cleft palate and lip in A/J mice identical to those seen with cortisone. Corticosterone was measured in the treated dams and significantly increased plasma levels were found, there being a linear relationship between cleft lip/palate and maternal cortisone levels.

A summary of the teratogenic activity of glucocorticoids is presented in Table 9.1. The evidence shows, then, that glucocorticoids are structural teratogens in a wide variety of animal species; the mouse is the most susceptible, although there are marked strain differences within this species. The synthetic corticoids appear far more potent than do those compounds that occur naturally. For example, dexamethasone is 200 times as potent as cortisol at producing cleft palate in A/J mice but only 25 times as potent therapeutically in man. The relative cleft palate inducing potencies of triamcinolone and cortisol show differences of a similar order of magnitude (Rowland & Hendrickx 1983).

Corticosteroid administration results in embryotoxicity, embryolethality and teratogenicity. Whilst the most frequently reported malformation is cleft palate, it has not been as well recognized that a wide range of other abnormalities has been noted depending on the species, agent and time of administration. Cortisone inhibits the closure of the embryonic secondary palatal shelves but there are marked strain differences. In addition to delaying palatal shelf elevation, cortisone severely reduces the extent of contact between the palatal shelves in A/J embryos, with contact down

Table 9.1 Summary of Teratogenic Activity of Glucocorticoids[a]

Chemical	Species: Mouse	Rat	Rabbit	Hamster	Monkey	Key references
Cortisone	+					Baxter & Fraser (1950)
		+				Buresh & Urban (1970)
			+			Fainstat (1954)
				−		Shah & Kilistoff (1976)
Corticosterone	+					Blaustein *et al.* (1971)
		+				Buresh & Urban (1970)
				+		Shah & Kilistoff (1976)
Deoxycortico-sterone	−					Walker (1965)
		+				Buresh & Urban (1970)
				+		Tedford & Risley (1950)
Hydrocortisone	+					Kalter & Fraser (1952)
		+				Gunberg (1957)
			+			Kasirsky & Lombardi (1970)
				+		Shah & Travill (1976)
Betamethasone	+					Walker (1971)
		+				Walker (1971)
			+			Ishimura *et al.* (1975)
					+	Epstein (1977)
Dexamethasone	+					Pinsky & DiGeorge (1965)
		+				Vannier *et al.* (1969)
			+			Clavert *et al.* (1961)
				+		Shah & Kilistoff (1976)
					+	Jerome & Hendrickx (1988)
Desoximethasone	+					Miyamoto *et al.* (1975)
		−				Miyamoto *et al.* (1975)
Beclomethasone	+					Esaki *et al.* (1976)
		−				Furuhashi *et al.* (1979)
			+			Oguro *et al.* (1970)
					−	Tanioka (1976)
Triamcinolone	+					Walker (1965)
		+				Rowland & Hendrickx (1983)
			+			Walker (1967)
				+		Shah & Kilistoff (1976)
					+	Parker & Hendrickx (1983)
Prednisone	+					Zawoiski (1980)
Prednisolone	+					Pinsky & DiGeorge (1965)
		+				Kalter (1962)
			+			Walker (1967)
				+		Shah & Kilistoff (1976)

Table 9.1 (*Cont.*)

Methylpred-nisolone	+			Walker (1971)
		–		Walker (1971)
			–	Walker (1967)
Diflucortolone	+			Ezumi *et al.* (1977a)
		+		Gunzel *et al.* (1976)
			+	Ezumi *et al.* (1978)

Note: [a] Adapted from Schardein (1985).

to only 4% of that found in controls (Diewert & Pratt 1981). The palatal shelves in cortisone-treated animals are markedly smaller than those found in controls but cell density is much higher. This, associated with the reduced shelf size, suggests that formation and accumulation of extracellular matrix components such as hyaluronic acid is inhibited. Thus, glucocorticoids appear to act to cause cleft palate by interfering with palatal mesenchymal cell function (Salomon & Pratt 1978). However, mechanisms are not as simplistic as this – clefting occurs in a different way with, for example, triamcinolone (Shah 1980). It is advisable to refer to a specialist work on the mechanisms involved in cleft palate formation.

9.3 Stress and Developmental Toxicity in Laboratory Animals

Animal studies investigating the developmental toxicity of stress have produced conflicting results. By and large, the rat and mouse have been the only species subjected to close scrutiny with regard to effects evident at caesarean section or at birth, and stressors used have varied between noise, heat, food deprivation and restraint.

In mice, Nawrot *et al.* (1980) have produced embryolethality and retarded embryonic/foetal growth but no malformations using extremely high frequency (jet-engine-type) noise during organogenesis. Trains of 100 decibel white noise, administered half-hourly for 6 min at a time, during organogenesis of mice, led to reduced maternal weight and embryolethality but had no effect on embryonic growth or malformation rate (Kimmel *et al.* 1976). There have also been negative findings where high-frequency white noise during Days 6–15 of pregnancy in mice had no effect on the course and outcome of pregnancy (Nawrot *et al.* 1981). Terada (1974) imposed a daily routine of enforced physical activity on a treadmill for 30 min/day on Days 9–16 *post coitum*. Embryolethality, slight reduction in foetal weight and retarded skeletal development were found, but no malformations. Training before pregnancy was shown to abolish the embryolethality but skeletal retardation was still present.

There have been numerous publications reporting the occurrence of stress-induced malformations in mice. Beyer & Chernoff (1986) reported offspring with fused and/or supernumerary ribs and exencephaly following 12-h restraint. However, the skeletal changes may have been an indirect effect since they appeared to correlate well with loss of maternal weight. Cook *et al.* (1982) have reported exencephaly, open eye, cleft lip and fused sternebrae following high-frequency noise stress during pregnancy. Probably the most startling observation is that of

Rosenzweig & Blaustein (1970) who produced a 70% incidence of cleft palate in A/J mice using restraint and food deprivation: the control incidence was just 1%. As with the glucocorticoids, the A/J mouse seems to be susceptible to stress-induced facial clefting. Meskin & Shapiro (1971) have shown a significant increase in cleft lip and palate in this strain when they were transported by air from supplier to laboratory between Days 12–15 *post coitum*.

Cleft palate has also been seen at high incidence in mice by Barlow *et al.* (1975a, b) who demonstrated this to be a result of endogenous corticosterone release with findings similar to those resulting from exogenous corticosteroid administration. In mice, plasma corticosterone levels are elevated during pregnancy, reaching a peak of ~14 μg/ml on Day 16 which is ~60 times the basal rate in non-pregnant mice (Barlow *et al.* 1974). In mice subject to restraint stress for 24 h on Day 14 *post coitum* or to isolation plus food deprivation, cortisone levels were elevated to ~70 μg/ml of plasma compared with 12 μg/ml in non-stressed concurrent controls.

Increased resorption rate has been seen in rats subject to restraint stress, especially when applied late in gestation (Euker & Riegle 1973); conversely, Beyer & Chernoff (1986) were unable to produce any effect on rat embryos following 12-h immobilization coupled with food and water deprivation. Similarly conflicting results have been noted in rats subject to noise stress. Geber (1966) reported an increased resorption rate and decreased foetal weight following noise stress in combination with flashing lights for a 6-min period every half hour throughout pregnancy. A regime of 100 decibels white noise throughout pregnancy that was embryotoxic in mice had no effect in rats (Kimmel *et al.* 1976). Geber (1966) also demonstrated that his combination of audiogenic and visual stress could increase the incidence of malformations of varying morphology. However, no malformations have been detected in rats following restraint stress alone (Barlow *et al.* 1978) or noise stress alone (Kimmel *et al.* 1976).

Since stress will lead to a significant elevation in plasma corticosteroid levels, and since glucocorticoids are teratogenic and embryotoxic in laboratory animals, it is tempting to assume that the developmental toxicity described above following stress is in consequence of elevated glucocorticoid levels. Indeed, Barlow *et al.*'s (1974) data provide strong evidence for this argument. However, many stress procedures have resulted in reduced maternal weight gain and reduced food and water intakes or have used food and water deprivation, and these factors may have been indirectly or even directly involved aetiologically. With Gebers' experiments in rats, described above, the author concluded, but without empirical evidence, that malformations were consequent upon hypoxia in the uterine environment due, perhaps, to a catecholamine-mediated vasoconstriction. Cook *et al.* (1982) found that maternal catecholamine levels were measurably increased in mice following a regime of noise stress that induced a range of malformations.

The data are conflicting. Negative and positive studies are difficult to compare due to considerable interstudy difference in experimental variables. It does appear likely that, under the right conditions, stress is a developmental toxicant in both rats and mice. The implications when conducting developmental toxicity studies must be considered, especially for mice which show an exaggerated stress response. The procedure itself, if stressful, such as intravenous dosing, or maternal toxicity might lead indirectly to false positives. Most experienced investigators will have learnt to treat findings of cleft palate in mice with caution before concluding that their experimental drug is a teratogen.

Table 9.2 Examples of Pre-natal Stress and Post-natal Consequence

Species	Stress applied during pregnancy	Post-natal consequences	Key references
Rat	Nutritional deficiency	Increased aggression	Peters (1978)
		Increased error rate in maze	Baird *et al.* (1971)
		Slower extinction rates in conditioned avoidance test	Simonson & Chow (1970)
Rat	Conditioned avoidance stress	Decreased exploratory activity in open field	Thompson *et al.* (1962)
		Increased open field activity	Joffe (1965)
		Increased avoidance scores and decreased latencies in conditioned avoidance test	Joffe (1965)
		Increased exploratory behaviour in open field in 'grandpups'	Wehmer *et al.* (1970)
		Altered responses in open field, water maze and runway test	Thompson & Quinby (1964)
Rat	Restraint	Altered startle and cliff avoidance response	Barlow *et al.* (1978)
		Decreased copulatory and increased lordotic behaviour in males	Anderson *et al.* (1986)
Rat	Restraint under intense illumination	Failure of masculinization and of defeminisation of behaviour potentials	Ward & Weisz (1984)
Rat	Handling	Reduced emotionality in open field	Ader & Conklin (1963)
Mouse	Handling	Increased frequency of audiogenic seizure	Beck & Gavin (1976)
Mouse	Pre-weaning litter size	Altered open field behaviour	Akuta (1979)
Rat	Pre-weaning litter size/crowding	Altered maze behaviour	Robinson (1976)
Mouse	Crowding	Increased latency of emergence after prolonged food deprivation	Keeley (1962)
		Decreased aggression and sexual behaviour in males and delayed puberty and disrupted oestrous cycles in females	Harvey & Chevins (1984, 1985, 1987)
Rat	Injection of water/saline	Altered response on inclined plane	Havlena & Werboff (1963)
		Increased emotionality in open field	
		Increased error rate in maze	
		Increased susceptibility to audiogenic seizure	
Mouse	Swimming, tilting and noise stress	Reduced open field activity in BALB mouse; but increased in C57	DeFries *et al.* (1967)
Rat	REM sleep deprivation	Altered susceptibility to seizures	Suchecki & Neto (1991)
		Reduced activity in open field	

So far, this section has concentrated on those manifestations of stress-induced developmental toxicity that are readily observable at caesarian section or at birth. Prenatal stress has also been implicated as a 'behavioural teratogen', an all-embracing and somewhat misleading term which encompasses delayed or aberrant behavioural maturation, impaired activity, impaired rate or extent of learning, adaptation or problem-solving or any other measure of compromised behavioural competence. Behavioural teratogenicity follows the same rules as apply to classical structural teratogenicity, in particular the type and magnitude of the response being a function of the stage of development at which the agent acts. Indeed, there is a complex interaction of factors such as prior experience of the mother, strain of animal tested, the type of stressor, whether cross-fostering techniques are employed and the types of measurements used. There is an extensive literature on the post-natal effects of stress administered prenatally: many stressors have been employed at varying stages of gestation and a wide range of possible behavioural targets have been investigated. For this reason it is impracticable to provide a detailed review of the literature here, however examples of various prenatally administered stressors and their postnatal outcomes are presented in Table 9.2 in an attempt to demonstrate the scope of the problem. This list is by no means exhaustive, and the reader should also consider that the numerous negative reports have not been listed. The positive findings are sufficient to conclude that, in laboratory species at least, stress is a development toxicant both in terms of 'defects of the body as well as of the mind'. One important observation is that in both rats (Ader & Conklin 1963) and mice (Beck & Gavin 1976), the most commonly used species in experimental teratology, maternal handling during pregnancy has been shown to elicit postnatal behavioural deficits. Thus, when conducting developmental toxicity studies particular attention must be given to standardizing handling and other stressful experimental procedures to avoid false-positive findings.

9.4 Glucocorticoids and Developmental Toxicity in Man

The demonstration in the 1950s that corticosteroids could induce developmental toxicity in a wide range of laboratory animal species, coupled with the awareness of drug-induced teratogenicity in man elicited by the thalidomide tragedy, prompted exhaustive studies in man, and numerous reports have consequently appeared. Some, perhaps understandably, have concentrated on an association between glucocorticoid therapy and cleft lip and/or palate, whereas others have extended their remit to an association with any developmental defect.

As a class, glucocorticoids have not been implicated as causing developmental defects in man. Bongiovanni & McFadden (1960) found only two instances of cleft palate in their group of 260 women treated with corticosteroids during pregnancy. Only three cases of facial cleft out of a cohort of 428 mothers were reported by Serment & Ruf (1968): there was a small and insignificant number of defects of other systems. The Collaborative Perinatal Projects retrospective survey of 50 000 pregnancies in 12 US medical centres from 1951 to 1965 contained a cohort of 145 mothers treated with glucocorticoids and corticotropins during the first trimester. There was no evidence of association with malformations, although it must be cautioned that numbers were limited (Heinonen *et al.* 1977). Other major surveys providing a negative association between glucocorticoid therapy and developmental defects have included those of Kullander & Kallen (1976), Richards (1972), Walsh & Clark (1967) and Yackel *et al.* (1966). Where positive associations between corticosteroid therapy and developmental toxicity have been suggested, this has invariably been attributed to cortisone. Increased pregnancy wastage (abortion and still birth) and prematurity has been noted (Reilly 1958), and some authors (Harris & Ross 1956, Popert 1962) have reported an increased incidence of facial cleft but the data are far from convincing. Malformations other than facial cleft have also been reported to follow cortisone treatment during pregnancy. Hydrocephalus and gastroschisis have been noted in two separate cases (Malpas 1965), and other diverse defects have been reported elsewhere (Guilbeau 1953, Warrell & Taylor 1968). Individual case reports of malformation associated with cortisone therapy have appeared in the literature. In addition to facial cleft, isolated cases of limb and heart defects, anencephaly, cyclopia and Aicardi's syndrome have been presented (Doig & Coltman 1956, Popert 1962, Noda *et al.* 1963, Chhabria 1981). There was no morphological consistency in the defects noted. A more detailed collection of references to surveys and case reports on glucocorticoid therapy in pregnancy may be found in Schardein (1985).

With regard to other corticosteroids there have been few convincing case reports. A case of hydrocephalus has been noted following maternal treatment with prednisone throughout pregnancy (Editorial 1974). Reinisch *et al.* (1978) studied a large sample of children of women treated with low dosages of glucocorticoids for infertility and maintenance of pregnancy such that the offspring were exposed throughout gestation. The results indicate that growth of offspring exposed to prednisone during gestation is significantly retarded leading to diminished full-term birth weight. Conversely, the Collaborative Perinatal Project (Heinonen 1977) reported only one malformed child among 43 pregnancy exposures to prednisone. There have been numerous references to normal pregnancy outcome following prednisolone treatment during the susceptible period of pregnancy (Board *et al.* 1967, Schardein 1978, Snyder & Snyder 1978).

No malformations were found in a cohort of pregnant women given dexamethasone at a sensitive stage (Serment & Ruf 1968). The results of the Collaborative Perinatal Project (Heinonen *et al.* 1977) for 'other corticosteroids and corticotropins' (ie excluding cortisone, hydrocortisone and prednisone but including methyl prednisolone (16 cases), prednisolone (15), triamcinolone (8), dexamethasone (8), corticotropin (6), paramethasone (2), desoxycorticosterone (2), aldosterone (1) and betamethasone (1)] showed a normal incidence of two malformed children among 56 pregnancies.

Glucocorticoids have been used extensively during pregnancy for many years and there is a wealth of literature concerning pregnancy outcome. It would seem most likely that these agents, administered under normal therapeutic circumstances, do not significantly increase the rate of spontaneously occurring developmental defects in man.

9.5 Stress During Pregnancy and Developmental Toxicity in Man

The belief that a profound and sudden impression made on the mothers mind, be it by sight, sound or touch, can in some way alter the development of the child with which she is pregnant has been established since ancient times and has become known as the theory of Maternal Impression (Ballantyne 1904, Warkany 1977). It was by maternal impression that ancient medicine, legend and folklore accounted for the occurrence of all kinds of congenital abnormalities, from birthmarks to monsters. In its original concept, impressions during pregnancy, or even at the time of conception, will influence the development of the child by producing a kind of photographic image on the offspring. There is thus a marked similarity between the thing producing the impression and the resulting defect in the offspring, so that the sight of a hare during pregnancy can be responsible for a hare-lip in the child or the sight of a man with an amputated leg can be responsible for a congenital loss of limb. The theory is not restricted to the past nor to primitive minds. Until relatively recently, there have been frequent accounts in the medical press. Two such stories, published in 1960, concern children with phocomelia. In the first case (Lussier 1960), the mother was convinced that her son's deformities had been caused by her mother-in-law who had a habit of rowing with her. In one argument 3 months into the pregnancy, the mother-in-law seized her by the shoulders to stop her walking past. She was not injured but was emotionally disturbed by the encounter. In the second account, Turner (1960) describes 'a very specific and severe emotional stress, applied early in an otherwise normal pregnancy, which may have caused profound teratogenic effects of a predicted type in the offspring'. The mother, a 16-year-old unmarried Australian, told her own mother that she was expecting when she was ~5–6 weeks pregnant. In the hearing of several witnesses the mother then cursed the girl, saying that if she continued with the pregnancy the baby would be born 'without arms and legs and blind'. A strikingly similar case has been reported more recently (Stevenson 1985). That a belief in maternal impressions still exists in modern civilized society has been amply demonstrated by two surveys conducted in Australian and American antenatal clinics (Pearn & Pavlin 1971, Snow *et al.* 1978). The results of the survey were broadly in agreement: in the USA 77% of the patients interviewed believed that external factors such as strong emotional states on the part of the mother could permanently disfigure or even kill her unborn child.

The ancient belief in maternal impression has evolved into a modern and perhaps more acceptable counterpart with the tenet that maternal frights, worries, shocks or stress can produce malformation or some other unwanted pregnancy outcome in a non-specific way so that the impression and the defect are in no way similar. Sensory perception activates the hypothalamic-pituitary-adrenal (HPA) axis, among other centres, which results in increased maternal adrenal hormone release (Moore 1963). Steroids in the serum of pregnant women may reach four-fold the levels recorded in non-pregnant females (Robinson *et al.* 1955), so additional stress-induced quantities may exceed theoretical levels for the production of abnormalities.

In a retrospective analysis conducted by Stott (1957), the data indicated that early childhood, mental retardation and congenital malformation were significantly more common for pregnancies associated with stress such as illness, harassment and/or distress of the expectant mother, shocks and accidents. In another retrospective survey the same worker also reported that emotional shocks during early pregnancy occurred more frequently in the mothers of Down syndrome children than in mothers of children suffering from other types of mental retardation (Stott 1961). Strean & Peer (1956) concluded from their retrospective study of children with cleft palate or hare-lip that the mothers suffered from a high incidence of physiological, emotional or traumatic stress whilst pregnant. Other retrospective surveys of this type (Gunder 1963, Newton *et al.* 1979, Berkowitz 1981) have indicated that mothers of pre-term babies had a higher rate of stressful events or, at least, reported greater stress levels during pregnancy than did controls. A similar survey (Williams *et al.* 1975) failed to show any such correlation. Several surveys have concentrated on pregnancies among inmates in mental hospitals. Turner (1956) found that among 100 psychotic patients suffering severe emotional stress during pregnancy, 13 infants with 'difficult behaviour' were born. Blomberg (1980) reported a higher incidence of malformations among patients suffering emotional stress due to unwanted pregnancies. The problem with the majority of these retrospective surveys is that other, possibly confounding, variables were uncontrolled. In the studies on mental patients, for example, there remains the distinct possibility that malformation and behavioural disturbance in offspring was due to antipsychotic therapy to control stress rather than the stress itself. Furthermore, retrospective studies run the risk that mothers will be quick to come up with some cause or other once they knew their baby was malformed, but are less likely to suggest any pregnancy complication if they thought that their baby was normal. One retrospective analysis that did not rely on the vagaries of the mind, comparing only fact with fact, analysed the birth records in Los Angeles County for the years 1970–1972 taking into account the area where the mother lived in relation to Los Angeles International Airport (Jones & Tauscher 1977). The data suggested that residence under an airport landing pattern where noise exceeds 90 decibels may be a factor causing birth defects.

Prospective surveys have also implicated stress as a developmental toxicant. Davids (1961, 1962) indicated that women who were later to experience complications in the delivery room or were to give birth to children with abnormalities were markedly more anxious during pregnancy. Crandon (1979) used an anxiety test administered in late pregnancy in conjunction with Apgar scores in the offspring. Apgar scores in a group of severely anxious patients were significantly lower than in normal patients. Another study (Grimm 1961) found that in 11 patients exhibiting marked anxiety, 10 went on to have adverse pregnancy outcomes including one case of congenital defect, four miscarriages and five neonatal deaths. Other prospective clinical surveys have demonstrated a relationship between stress and adverse pregnancy outcome (Gorsuch & Key 1974, Standley *et al.* 1979, Omer *et al.* 1986). Grimm & Venet (1966) and Newton & Hunt (1984) in their prospective surveys were unable to establish a clear relationship between maternal stress and adverse pregnancy outcome, although the latter report did show that major events such as the death of a family member or friend were associated with a greater risk of low birth weight or prematurity.

The reports described above by no means represent an exhaustive account of the literature but, in general, there are numerous accounts associating stress, in the broadest sense, with developmental toxicity. Data where no association has been found are considerably less common, but perhaps only since negative results tend to appear of little interest and are frequently not published.

References

ADER, R. & CONKLIN, P. M. (1963) Handling of pregnant rats: effects on emotionality of their offspring. *Science*, **142**, 411–412.

AKUTA, T. (1979) Effects of rearing conditions on the behaviour of mothers and on the offspring in the mouse. *Japanese Journal of Psychology*, **50**, 73–81.

ANDERSON, R. H., FLEMING, D. E., RHEES, R. W. & KINGHORN, E. (1986) Relationships between sexual activity, plasma testosterone and the volume of the sexually dimorphic nucleus of the pre-optic area in prenatally stressed and nonstressed rats. *Brain Research*, **370**, 1–10.

AOYAMA, T., FURUOKA, R., HASEGAWA, N. & TERABAYASHI, M. (1974) Teratological studies on hydrocortisone-17-α-butyrate (H-17B) in mice and rats. *Oyo Yakuri*, **8**, 1035–1047.

BADR, F. M. & SPICKETT, S. G. (1965) Genetic variation in the biosynthesis of corticosteroids in *Mus musculus*. *Nature*, **205**, 1088–1090.

BAIRD, A., WIDDOWSON, E. M. & COWLEY, J. J. (1971) Effects of calorie and protein deficiencies early in life on the subsequent learning ability of rats. *British Journal of Nutrition*, **25**, 391–403.

BALLANTYNE, J. W. (1904) *Manual of Antenatal Pathology and Hygiene*, Edingburgh: W. Green and Sons.

BARLOW, S., MCELHATTON, P., MORRISON, P. & SULLIVAN, F. M. (1974) Effects of stress during pregnancy on plasma corticosterone levels and foetal development in the mouse. *Journal of Physiology*, **229**, 55P–56P.

BARLOW, S. M., MCELHATTON, P. R. & SULLIVAN, F. M. (1975a) The relationship between maternal restraint and food deprivation, plasma corticosterone and induction of cleft palate in the offspring of mice. *Teratology*, **12**, 97–104.

BARLOW, S. M., MORRISON, P. J. & SULLIVAN, F. M. (1975b) Effects of acute and chronic stress on plasma corticosterone levels in the pregnant and non-pregnant mouse. *Journal of Endocrinology*, **66**, 93–99.

BARLOW, S. M., KNIGHT, A. F. & SULLIVAN, F. M. (1978) Delay in postnatal growth and development of offspring produced by maternal restraint stress during pregnancy in the rat. *Teratology*, **18**, 211–218.

BAXTER, H. & FRASER, F. C. (1950) Production of congenital defects in offspring of female mice treated with cortisone. *McGill Medical Journal*, **19**, 245–249.

BECK, S. L. & GAVIN, D. L. (1976) Susceptibility of mice to audiogenic seizures is increased by handling their dams during gestation. *Science*, **193**, 427–428.

BERKOWITZ, G. S. (1981) An epidemiologic study of preterm delivery. *American Journal of Epidemiology*, **113**, 81.

BEYER, P. E. & CHERNOFF, N. (1986) The induction of supernumerary ribs in rodents: role of maternal stress. *Teratogenesis, Carcinogenesis, Mutagenesis*, **6**, 419.

BIDDLE, F. G. (1977) 6-Aminonicotinamide-induced cleft palate in the mouse: the nature of the difference between the A/J and C57 BL/6J strains in frequency of response and its genetic basis. *Teratology*, **16**, 301–312.

BIDDLE, F. G. & FRASER, F. C. (1976) Genetics of cortisone-induced cleft palate in the mouse; embryonic and maternal effects. *Genetics*, **84**, 743–754.

BLAUSTEIN, F. M., FELLER, R. & ROSENZWEIG, S. (1971) Effect of ACTH and adrenal hormones on cleft palate frequency in CD1 mice. *Journal of Dental Research*, **50**, 609–612.

BLOMBERG, S. (1980) Influence of maternal distress during pregnancy on fetal malformations. *Acta Psychiatrica Scandinavica*, **62**, 315–330.

BOARD, J. A., LEE, H. M., DRAPER, D. A. & HUME, D. M. (1967) Pregnancy following kidney homotransplantation from a non-twin. Report of a case with concurrent administration of azathioprine and prednisone. *Obstetrics and Gynecology*, **29**, 318–323.

BONGIOVANNI, A. M. & MCFADDEN, A. J. (1960) Steroids during pregnancy and possible fetal consequences. *Fertility and Sterility*, **11**, 181–186.

BUCK, P., CLAVERT, J. & RUMPLER, Y. (1962) Action teratogenique des corticoides chez la lapine. *Annales de Chirurgerie de l'Enfant*, **3**, 73–87.

BURESH, J. J. & URBAN, T. T. (1970) Palatal abnormalities induced by cortisone and corticosterone in the rat. *Journal of the Academy of General Dentistry*, **18**, 34–37.

CHALLIS, J. R. G., DAVIES, J., BENIRSCHKE, K., HENDRICKX, A. G. & RYAN, K. J. (1974) The effects of dexamethasone on plasma steroid levels and fetal adrenal histology in the pregnant rhesus monkey. *Endocrinology*, **95**, 1300–1305.

CHHABRIA, S. (1981) Aicardi's syndrome: are corticosteroids teratogens? *Archives of Neurology*, **38**, 70.

CLAVERT, J., BUCK, P. & RUMPLER, C. (1961) Action teratogenique de soludecadron chez la lapine. *Comptes Rendus des Séances de la Societé de Biologie et de Ses Filiales*, **155**, 1569–1571.

COOK, R. O., NAWROT, P. S. & HAMM, C. W. (1982) Effects of high frequency noise on prenatal development and maternal plasma and uterine catecholamine concentrations in the CD-1 mouse. *Toxicology and Applied Pharmacology*, **66**, 338.

CRANDON, A. J. (1979) Maternal anxiety and obstetric complications. *Journal of Psychosomatic Research*, **23**, 109.

CURRY, D. M. & BEATON, G. H. (1958) Cortisone resistance in pregnant rats. *Endocrinology*, **63**, 155–161.

CZABO, G., TORO, I. & FISCHER, J. (1967) Effect of cortisone on the foetus of pregnant rats. An attempt at widening the concept of teratogenesis. *Acta Paediatrica Academica Scientiae Hungarica*, **8**, 217–223.

DAHLOF, L.-G., HARD, E. & LARSSON, K. (1978) Sexual differentiation of offspring of mothers treated with cortisone during pregnancy. *Physiology and Behavior*, **21**, 673–674.

DAVIDS, A. & DE VAULT, S. (1962) Maternal anxiety during pregnancy and childbirth abnormalities. *Psychosomatic Medicine*, **24**, 464–470.

DAVIDS, A., DE VAULT, S. & TALMADGE, M. (1961) Anxiety, pregnancy, and childbirth abnormalities. *Journal of Consulting Psychology*, **25**, 74–77.

DEFRIES, J. C., WEIR, J. M. & HEGMANN, J. P. (1967) Differential effects of prenatal maternal stress on offspring behaviour in mice as a function of genotype and stress. *Journal of Comparative Physiology and Psychology*, **63**, 332–334.

DIEWERT, V. M. & PRATT, R. M. (1981) Cortisone-induced cleft palate in A/J mice: failure of palatal shelf contact. *Teratology*, **24**, 149–162.

DOIG, R. K. & COLTMAN, O. M. (1956) Cleft palate following cortisone therapy in early pregnancy. *Lancet*, **2**, 730.

EDITORIAL (1974) Are sex hormones teratogenic? *Lancet*, **2**, 1489–1490.

EPSTEIN, M. F., FARRELL, P. M., SPARKS, J. W., PEPE, G., DRISCOLL, S. G. & CHEZ, R. A. (1977) Maternal betamethasone and fetal growth and development in the monkey. *American Journal of Obstetrics and Gynecology*, **127**, 261–263.

ESAKI, K., IZUMIYAMA, K. & YASUDA, Y. (1976) Effects of inhalant administration of beclomethasone diproprionate on the reproduction in mice. *Preclinical Report of the Central Institute of Experiments on Animals*, **2**, 213–222.

ESAKI, K., SHIKATA, Y. & YANAGITA, T. (1981) Effects of dermal administration of dexamethasone-17-valerate in rabbit fetuses. *Preclinical Report of the Central Institute of Experiments on Animals*, **7**, 245–256.

EUKER, J. S. & RIEGLE, G. D. (1973) Effects of stress on pregnancy in the rat. *Journal of Reproduction and Fertility*, **34**, 343–346.

EZUMI, Y., TOMOYAMA, J. & KODAMA, N. (1977a) Teratogenicity, especially on the formation of cleft palate in mouse embryo of diflucortolone valerate by a single administration. *Yakabutsu Ryoho*, **10**, 1585–1594.

EZUMI, Y., TOMOYAMA, J. & KODAMA, N. (1977b) Effects of diflucortolone valerate subcutaneously injected into rats in mid gestation (period of organogenesis) or late gestation on the pre- and postnatal development of their offspring. *Yakabutsu Ryoho*, **10**, 1357–1365.

EZUMI, Y., TOMOYAMA, J. & KODAMA, N. (1978) Effects of diflucortolone valerate subcutaneously injected to rabbits in mid gestation on the prenatal development of their offspring. *Yakabutsu Ryoho*, **11**, 229–326.

FAINSTAT, T. (1954) Cortisone-induced cleft palate in rabbits. *Endocrinology*, **55**, 502–508.

FRASER, F. C. & FAINSTAT, T. D. (1951) Production of congenital defects in the offspring of pregnant mice treated with cortisone. *Pediatrics*, **8**, 527–533.

FUJII, T., KITAGAWA, M. & YOKOYAMA, Y. (1973) Comparative teratogenic effects of water-soluble and insoluble hydrocortisone in the mouse embryo. *Teratology*, **8**, 92.

FURUHASHI, T., NOMURA, A., MIYOSHI, K., IKEYA, E. & NAKAYOSHI, H. (1979) Teratologic and fertility studies on beclomethasone diproprionate: 2. Teratological studies by oral administration. *Oyo Yakuri*, **18**, 1021–1038.

GEBER, W. F. (1966) Developmental effects of chronic maternal audiovisual stress on the rat fetus. *Journal of Embryology and Experimental Morphology*, **16**, 1–16.

GORSUCH, R. L. & KEY, M. K. (1974) Abnormalities of pregnancy as a function of anxiety and life stress. *Psychosomatic Medicine*, **36**, 352–362.

GRIMM, E. R. (1961) Psychological tension in pregnancy. *Psychosomatic Medicine*, **23**, 520–527.

GRIMM, E. R. & VENET, W. R. (1966) The relationship of emotional adjustment and attitudes to the course and outcome of pregnancy. *Psychosomatic Medicine*, **28**, 34.

GROLLMAN, A. & GROLLMAN, E. F. (1962) Teratogenic induction of hypertension. *Journal of Clinical Investigation*, **41**, 710–714.

GUILBEAU, J. A. (1953) Effects of cortisone on the fetus. *American Journal of Obstetrics and Gynecology*, **65**, 227.

GUNBERG, D. L. (1957) Some effects of exogenous hydrocortisone on pregnancy in the rat. *Anatomical Records*, **129**, 133–153.

GUNDER, L. M. (1963) Psychopathology and stress in the life experience of mothers of premature infants. A comparative study. *American Journal of Obstetrics and Gynecology*, **86**, 333.

GUNZEL, P., EL ETREBY, M. F., BHARGAVA, A. S., POGGEL, H. A., SCHOBEL, C., SCHUPPLER, J., SIEGMUND, F. & STABEN, P. (1976) Tier experimentelle vertraglichkertsprufung von Diflucortolonvalerianat als reiner Wirkstoff und als Salbe, Fettsalbe und Creme. *Arzneimittel Forschung*, **26**, 1476–1479.

HARRIS, J. W. S. & ROSS, I. P. (1956) Cortisone therapy in early pregnancy: relation to cleft palate. *Lancet*, **1**, 1045–1047.

HARVEY, P. W. & CHEVINS, P. F. D. (1984) Crowding or ACTH treatment of pregnant mice affects adult copulatory behaviour of male offspring. *Hormones & Behaviour*, **18**, 101–110.

HARVEY, P. W. & CHEVINS, P. F. D. (1985) Crowding pregnant mice affects attack and threat behaviour of male offspring. *Hormones & Behaviour*, **19**, 86–97.

HARVEY, P. W. & CHEVINS, P. F. D. (1987) Crowding during pregnancy delays puberty and alters estrous cycles of female offspring in mice. *Experientia*, **43**, 306–308.

HAVLENA, J. & WERBOFF, J. (1963) Postnatal effects of control fluids administered to gravid rats. *Psychological Reports*, **12**, 127–131.

HEINONEN, O. P., SLONE, D. & SHAPIRO, S. (1977) *Birth Defects and Drugs in Pregnancy*, Littleton, MA: Publishing Sciences Group.

HOAR, R. M. (1962) Similarity of congenital malformations produced by hydrocortisone to those produced by adrenalectomy in guinea pigs. *Anatomical Record*, **144**, 155–164.

ISHIMURA, K., HONDA, Y., NEDA, K., ISHIKAWA, I., OTAWA, T., KAWAGUCHI, Y., SATO, H. & HENMI, Z. (1975) Teratological studies on betamethasone 17-benzoate (MS 1112). II. Teratogenicity test in rabbits. *Oyo Yakuri*, **10**, 685–694.

JEROME, C. P. & HENDRICKX, A. G. (1988) Comparative teratogenicity of triamcinolone acetonide and dexamethasone in the rhesus monkey (*Macaca mulatta*). *Journal of Medical Primatology*, **17**, 195–203.

JOFFE, J. M. (1965) Genotype and prenatal and premating stress interact to affect adult behaviour in rats. *Science*, **150**, 1844–1845.

JOHNSON, J. W. C., MITZNER, W., BECK, J. C., LONDON, W. T., SLY, D. L., LEE,

P. A., KHOUZAMI, V. A. & CAVALIERI, R. L. (1981) Long-term effects of betamethasone on fetal development. *American Journal of Obstetrics and Gynecology*, **141**, 1053–1064.

JONES, F. N. & TAUSCHER, A. (1977) Residence under an airport landing pattern as a factor in teratism. *Archives of Environmental Health*, **33**, 10–12.

KADURI, A. J. & ORNOY, A. (1974) Impaired osteogenesis in the fetus induced by administration of cortisone to pregnant mice. *Israel Journal of Medical Science*, **10**, 476–481.

KALTER, H. (1954) Inheritance of susceptibility to the teratogenic action of cortisone in mice. *Genetics*, **39**, 185–196.

KALTER, H. (1962) No cleft palate with prednisolone in the rat. *Anatomical Record*, **142**, 311.

KALTER, H. (1981) Dose–response studies with genetically homogeneous lines of mice as a teratology testing and risk-assessment procedure. *Teratology*, **24**, 743–754.

KALTER, H. & FRASER, F. C. (1952) Production of congenital defects in offspring of pregnant mice treated with compound F. *Nature*, **169**, 665.

KASIRSKY, G. & LOMBARDI, L. (1970) Comparative teratogenic study of various corticoid ophthalmics. *Toxicology and Applied Pharmacology*, **16**, 773–778.

KEELY, K. (1962) Prenatal influence on behaviour of offspring of crowded mice. *Science*, **135**, 44–45.

KIMMEL, C. A., COOK, R. O. & STAPLES, R. E. (1976) Teratogenic potential of noise in mice and rats. *Toxicology and Applied Pharmacology*, **36**, 239–245.

KOGA, T., OTA, T., AOKI, Y., NISHIGAKI, K. & SUGANUMA, Y. (1980a) Reproductive studies of prednisolone 17-valerate-21-acetate: teratologic studies in the rat. *Oyo Yakuri*, **20**, 67–86.

KOGA, T., OTA, T., AOKI, Y. & SUGANUMA, Y. (1980b) Reproductive studies of prednisolone 17-valerate-21-acetate: teratologic studies in rabbits. *Oyo Yakuri*, **20**, 87–98.

KULLANDER, S. & KALLEN, B. (1976) A prospective study of drugs and pregnancy. 3. Hormones. *Acta Obstetrica et Gynecologica Scandinavica*, **55**, 221–224.

LEMMEN, K., MAURER, W., TRIEB, H., UEBERBERG, H. & SEELIGER, H. (1977) Morphologic changes in the adrenal glands of fetal and newborn rats following administration of glucocorticoids to the mother during pregnancy. *Bertrage Pathologie*, **160**, 361–380.

LUSSIER, A. (1960) The analysis of a boy with a congenital deformity. *The Psychosomatic Study of the Child*, **25**, 430.

MALPAS, P. (1965) Foetal malformation and cortisone therapy. *British Medical Journal*, **1**, 795.

MESKIN, L. H. & SHAPIRO, B. L. (1971) Teratogenic effect of air shipment on A/Jax mice. *Journal of Dental Research*, **50**, 169.

MIYAMOTO, M., OHTSU, M., SUGISAKI, T. & SAKAGUCHI, T. (1975) Teratogenic effect of 9-fluoro-11-β,21-dihydroxy-16-α-methylpregna-1,4-diene,3,20-dione (A41304) a new anti-inflammatory agent, and of dexamethasone in rats and mice. *Folia Pharmacologia Japonica*, **71**, 367–378.

MOORE, W. W. (1963) In SELKURT, E. E. (Ed.), *Physiology*, pp. 727–741, Boston: Little, Brown.

MOSCONA, M. H. & KARNOFSKY, D. A. (1960) Cortisone induced modifications in the development of the chick embryo. *Endocrinology*, **66**, 533–549.

MOSIER, H. D., DEARDEN, L. C., ROBERTS, R. C., JANSONS, R. A. & BIGGS, C. S. (1981) Regional differences in the effects of glucocorticoids on maturation of the fetal skeleton of the rat. *Teratology*, **23**, 13–24.

MOSIER, H. D., DEARDEN, L. C., JANSONS, R. A., ROBERTS, R. C. & BIGGS, C. S. (1982) Disproportionate growth of organs and body weight following glucocorticoid treatment of the rat fetus. *Developmental Pharamcology and Therapeutics*, **4**, 89–105.

NAKAYAMA, T., HIRAYAMA, M. & ESAKI, K. (1978) Effects of cortisone acetate in the beagle fetus. *Teratology*, **18**, 149.

NAWROT, P. S., COOK, R. O. & STAPLES, R. E. (1980) Embryotoxicity of various noise stimuli in the mouse. *Teratology*, **22**, 279–289.

NAWROT, P. S., COOK, R. O. & HAMM, C. W. (1981) Embryotoxocity of broadband high-frequency noise in the CD-1 mouse. *Journal of Toxicology and Environmental Health*, **8**, 151.

NEWTON, R. W. & HUNT, L. P. (1984) Psychological stress in pregnancy and its relations to low birth weight. *British Medical Journal*, **288**, 1191.

NEWTON, R. W., WEBSTER, P. A., BINU, P. S., MASKREY, N. & PHILLIPS, A. B. (1979) Psychological stress in pregnancy and its relation to the onset of premature labour. *British Medical Journal*, **2**, 411.

NODA, T., UEDA, K. & SATOYAMA, M. (1963) A case of malformed infant born to a mother treated with adrenocorticoids during pregnancy. *Sanfujinka No Shimpo*, **15**, 189.

OGURU, Y., KIYOHARA, A., MIYAGAWA, A., IMAMURA, S., KOYAMA, K. & HARA, T. (1970) Pharmacological and toxicological studies on beclomethasone dipropionate. *Yamaguchi Igaku*, **19**, 65–86.

OMER, H., ELIZUR, Y., BARNEA, T., FRIEDLANDER, D. & PALTI, Z. (1986) *Journal of Psychosomatic Research*, **30**, 559.

ORNOY, A. & HOROWITZ, A. (1972) Post-natal effects of maternal hypercortisonism on skeletal development in new born rats. *Teratology*, **6**, 153–158.

PARKER, R. M. & HENDRICKX, A. G. (1983) Craniofacial and central nervous system malformations induced by triamcinolone acetonide in non-human primates: 2. Craniofacial pathogensis. *Teratology*, **28**, 35–44.

PEARN, J. H. & PAVLIN, H. (1971) 'Maternal Impression' in a modern Australian community. *Medical Journal of Australia*, **2**, 1123–1126.

PETERS, D. P. (1978) Effects of prenatal nutritional deficiency on affiliation and aggression in rats. *Physiology and Behavior*, **20**, 359–362.

PINSKY, L. & DIGEORGE, A. M. (1965) Cleft palate in the mouse: a teratogenic index of glucocorticoid potency. *Science*, **147**, 402–403.

POPERT, A. J. (1962) Pregnancy and adrenocortical hormones. Some apsects of their interaction in rheumatic diseases. *British Medical Journal*, **1**, 967–972.

PRATT, R. M., SALOMON, D. S., DIEWERT, V. M., ERICKSON, R. P., BURNS, R. & BROWN, K. S. (1980) Cortisone-induced cleft palate in the brachymorphic mouse. *Teratogenesis, Carcinogenesis and Mutagenesis*, **1**, 15–23.

REILLY, W. A. (1958) Hormone therapy during pregnancy: effects on the fetus and newborn. *Quarterly Review of Pediatrics*, **13**, 198–202.

REINISCH, J. M., SIMON, N. G., KAROW, W. G. & GANDELMAN, R. (1978) Prenatal exposure to prednisone in humans and animals retards intrauterine growth. *Science*, **202**, 436–438.

RICHARDS, I. D. G. (1972) A retrospective enquiry into possible teratogenic effects of drugs in pregnancy. In KLINBERG, M. A., ABRAMOVICI, A. & CHEMKE, J. (Eds), *Drugs and Fetal Development*, pp. 441–445, New York: Plenum.

ROBENS, J. F. (1974) Teratogenesis. In KIRK, R. W. (Ed.), *Current Veterinary Therapy*, pp. 152–154, Philadelphia: W. B. Saunders.

ROBINSON, E. (1976) The effects of litter size and crowding on position learning by male and female albino rats. *Psychology Record*, **26**, 61–66.

ROBINSON, H. J., BERNHARD, W. G., GRUBIN, H., SERNIKOW, G. W. & SILBER, R. H. (1955) 17,21-Dihydroxy-20-ketosteroids in plasma during and after pregnancy. *Journal of Clinical Endocrinology*, **15**, 317–323.

ROSENZWEIG, S. & BLAUSTEIN, F. M. (1970) Cleft palate in A/J mice resulting from restraint and deprivation of food and water. *Teratology*, **3**, 47–52.

ROWLAND, J. M. & HENDRICKX, A. G. (1983) Teratogenicity of triamcinolone acetonide in rats. *Teratology*, **27**, 13–18.

ROWLAND, J. M., ALTHAUS, Z. R., SLIKKER, W. & HENDRICKX, A. G. (1983) Comparative distribution and metabolism of triamcinolone acetonide and cortisol in the rat embryomaternal unit. *Teratology*, **27**, 333–341.

SAIJO, T., FUJITA, T., SADANAGA, O. & DEGUCHI, T. (1990) Reproduction study of betamethasone butyrate propionate (bbp). 4-Teratogenicity study in rabbits by subcutaneous administration. *Kiso to Rinsho*, **24**, 5779–5786.

SALOMON, D. S. & PRATT, R. M. (1976) Glucocorticoid receptors in murine embryonic facial mesenchyme cells. *Nature*, **264**, 174–177.

SALOMON, D. S. & PRATT, R. M. (1978) Inhibition of growth *in vitro* by glucocorticoids in mouse embryonic and facial mesenchyme cells. *Journal of Cell Physiology*, **97**, 315–328.

SALOMON, D. S. & PRATT, R. M. (1979) Involvement of glucocorticoids in the development of the secondary palate. *Differentiation*, **13**, 141–154.

SCHARDEIN, J. L. (1985) *Chemically Induced Birth Defects*, New York and Basel: Marcel Dekker.

SERMENT, H. & RUF, H. (1968) Corticotherapie et grossesse. *Bulletin Federation des Societés Gynecologique et Obstetrique de Langues Françaises*, **20**, 77–85.

SHAH, R. M. (1980) Ultrastructural observations on the development of triamcinolone-induced cleft palate in hamsters. *Investigative Cell Pathology*, **3**, 281–294.

SHAH, R. M. & KILISTOFF, A. (1976) Cleft palate induction in hamster foetuses by glucocorticoid hormones and their synthetic analogues. *Journal of Embryology and Experimental Morphology*, **36**, 101–108.

SHAH, R. M. & TRAVILL, A. A. (1976) Morphogenesis of the secondary palate in normal and hydrocortisone treated hamsters. *Teratology*, **13**, 71–84.

SIMONSON, M. & CHOW, B. F. (1970) Maze studies on progeny of underfed mother rats. *Journal of Nutrition*, **100**, 685–690.

SNOW, L. F., JOHNSON, S. M. & MAYHEW, H. E. (1978) The behavioural implications of some old wives' tales. *Obstetrics and Gynecology*, **51**, 727–732.

SNYDER, R. D. & SNYDER, D. (1978) Corticosteroids for asthma during pregnancy. *Annals of Allergy*, **41**, 340–341.

STANDLEY, K., SOULE, B. & COPANS, S. A. (1979) Dimensions of prenatal anxiety and their influence on pregnancy outcome. *American Journal of Obstetrics and Gynecology*, **135**, 22.

STEVENSON, I. (1985) Clinical curio: birth defects from cursing? A case report. *British Medical Journal*, **290**, 1813.

STOTT, D. H. (1957) Physical and mental handicaps following a disturbed pregnancy. *Lancet*, **1**, 1006–1012.

STOTT, D. H. (1961) Mongolism related to emotional shock in early pregnancy. *Vita Humana*, **4**, 57.

STREAN, L. P. & PEER, L. A. (1956) Stress as an etiologic factor in the development of cleft palate. *Plastic Reconstruction Surgery*, **18**, 1–8.

SUCHECKI, D. & NETO, J. P. (1991) Prenatal stress and emotional response of adult offspring. *Physiology and Behavior*, **49**, 423–426.

SULLIVAN-JONES, P., HANSEN, D. K., SHEEHAN, D. M. & HOLSON, R. R. (1982) The effect of teratogens on maternal corticosterone levels and cleft incidence in A/J mice. *Journal of Craniofacial Genetics and Developmental Biology*, **12**, 183–189.

TAKESHIMA, T., TAUCHI, K. & IMAI, S. (1990) Reproduction study of betamethasone butyrate propionate (bbp). 2-Teratogenicity study in rats by subcutaneous administration. *Kiso to Rinsho*, **24**, 5747–5763.

TANIOKA, Y. (1976) Teratogenicity test on beclomethasone diproprionate by inhalation in rhesus monkeys. *Preclinical Report of the Central Institute of Experiments on Animals*, **2**, 155–164.

TARARA, R. P., CORDY, D. R. & HENDRICKX, A. G. (1989) Central nervous system malformations induced by triamcinolone acetonide in non-human primates: pathology. *Teratology*, **39**, 75–84.

TEDFORD, M. D. & RISLEY, P. L. (1950) Deoxycorticosterone and pregnancy in ovariectomised hamsters. *Anatomical Record*, **108**, 596.

TERADA, M. (1974) Effect of physical activity before pregnancy on fetuses of mice exercised forcibly during pregnancy. *Teratology*, **10**, 141–144.

THOMPSON, W. R. & QUINBY, S. (1964) Prenatal maternal anxiety and offspring behavior: parental activity and level of anxiety. *Journal of Genetic Psychology*, **106**, 359–371.

THOMPSON, W. R., WATSON, J. & CHARLESWORTH, W. R. (1962) The effects of prenatal maternal stress on offspring behavior in rats. *Psychology Monographs*, **76**, 1–26.

TURNER, E. K. (1956) The syndrome in the infant resulting from maternal emotional stress during pregnancy. *Medical Journal of Australia*, **1**, 221–222.

TURNER, E. K. (1960) Teratogenic effects on the human foetus through maternal emotional stress: report of a case. *Medical Journal of Australia*, **2**, 502–503.

VANNIER, B., JEQUIER, R. & JUDE, A. (1969) [Sensibility of wistar rat to teratogenic action of dexamethasone.] *Comptes Rendus des Séances de la Societé de Biologie et de Ses Filiales*, **163**, 1269–1272.

VEKEMANS, M., TAYLOR, B. A. & FRASER, F. C. (1979) Evidence against a simple genetic model in cortisone-induced cleft palate in the mouse. *Teratology*, **19**, 51A–52A.

WALKER, B. E. (1965) Cleft palate produced in mice by human-equivalent dosage with triamcinolone. *Science*, **149**, 682–683.

WALKER, B. E. (1967) Induction of cleft palate in rabbits by several glucocorticoids. *Proceedings of the Society of Experimental Biology and Medicine*, **125**, 1281–1284.

WALKER, B. E. (1971) Induction of cleft palate in rats with anti-inflammatory drugs. *Teratology*, **4**, 39–42.

WALSH, S. D. & CLARK, F. R. (1967) Pregnancy in patients on long-term corticosteroid therapy. *Scottish Medical Journal*, **12**, 302–306.

WARD, I. L. & WEISZ, J. (1984) Differential effects of maternal stress on circulating levels of corticosterone, progesterone and testosterone in male and female rat fetuses and their mothers. *Endocrinology*, **114**, 1635–1644.

WARKANY, J. (1977) History of teratology. In WILSON, J. G. & FRASER, F. C. (Eds), *Handbook of Teratology*, Vol. 1, pp. 3–45, New York and London: Academic Press.

WARRELL, D. W. & TAYLOR, R. (1968) Outcome for the foetus of mothers receiving prednisolone during pregnancy. *Lancet*, **1**, 117–118.

WEHMER, F., PORTER, R. H. & SCALES, B. (1970) Pre-mating and pregnancy stress in rats affects behaviour of grandpups. *Nature*, **227**, 622.

WILLIAMS, C. C., WILLIAMS, R. A., GRISWOLD, M. J. & HOLMES. T. H. (1975) Pregnancy and life change. *Journal of Psychosomatic Research*, **19**, 123.

WILSON, J. G., FRADKIN, R. & SCHUMACHER, H. J. (1970) Influence of drug pretreatment on the effectiveness of known teratogenic agents. *Teratology*, **3**, 210–211.

YACKEL, D. B., KEMPERS, R. D. & MCCONAHEY, W. M. (1966) Adrenocorticosteroid therapy in pregnancy. *American Journal of Obstetrics and Gynecology*, **96**, 985–989.

YAMADA, T., SUZUKI, H., MATSUMOTO, S., NAKANE, S., SASAJIMA, M. & OHZEKI, M. (1981) Reproductive studies of hydrocortisone 17-butyrate 21-propionate in rats and rabbits. *Oyo Yakuri*, **21**, 427–482.

ZAWOISKI, E. J. (1980) Effect of L-glutamic acid on glucocorticoid-induced cleft palate in gestating albino mice. *Toxicology and Applied Pharmacology*, **56**, 23–27.

10

Glucocorticosteroids and Immunotoxicity

IAN KIMBER

Zeneca Central Toxicology Laboratory, Alderley Park, Macclesfield

10.1 Introduction

Activation of the hypothalamus-pituitary-adrenal (HPA) axis by a variety of stressors results in the secretion by the adrenal cortex of glucocorticoids. The primary physiological function of these hormones is in the preparation for physical activity and the stimulation of blood glucose. It is well established that glucocorticoids influence virtually all aspects of immune and inflammatory responses, the consensus being that in most instances the effects are immunosuppressive or anti-inflammatory in nature. Indeed, synthetic corticosteroids have found widespread application in the clinical manipulation of immune reactivity and remain one of the most effective treatments available for allergic disease. The purpose of this chapter is to consider the nature, relevance and mechanisms of action of the immunotoxic and immune-modifying properties of glucocorticoids.

10.2 The Immune System

Some form of host recognition of, and host response to, foreign material is phylogenetically ancient. Even the most primitive members of the animal kingdom display rudimentary natural immunity. Such innate or natural immune function is preserved in the vertebrates, but here there has evolved also a sophisticated adaptive immune system capable of mounting specific responses that confer protection against potentially harmful antigens. The cellular vector of adaptive immune responses is the lymphocyte and it is this cell that is able to recognize and respond to antigen in a specific manner and that provides the immunological memory necessary for the subsequent provocation of secondary immune responses. While lymphocytes play a pivotal role in the induction and expression of immune reactivity they do not operate in isolation, and effective immunity is dependent upon complex and concerted interactions between various tissues, cells and molecules that are regulated in time and space. The primary purpose of the immune system is protection of the host from infection and malignant disease. To achieve this two major arms of the immune system have developed and these are reflected in the main populations of lymphoid cells. B lymphocytes are required for humoral (serum-borne) immunity that provides protection against extracellular bacterial infection. In response to antigen B lymphocytes divide and differentiate. The end-cell of B lymphocyte differentiation is the plasma cell that has the synthetic and secretory machinery to manufacture and export large amounts of antibody reactive with the inducing antigen. With the participation of other molecules (complement) and host cells (phagocytes), antibody serves to eliminate bacteria and other pathogenic micro-organisms. Other infectious agents reside within cells (facultative intracellular bacteria) or are obligate intracellular parasites (viruses). The cell membrane protects such pathogens from the attention of antibody and here the mechanism of host resistance is dependent upon cell-mediated immunity effected by T lymphocytes. These cells have two primary responsibilities, the first of which is recognition, and subsequent elimination, of host cells that display novel antigens as a consequence of either infection or malignant transformation. The second major function of T lymphocytes is the qualitative and quantitative regulation of immune responses. Many of the functions of T lymphocytes are effected via, and are regulated by, cytokines: a family of proteins and glycoproteins produced by lymphoid and non-lymphoid cells that collectively play a pivotal role in the induction and elicitation of controlled and directed immune responses.

If the immune system is impaired then susceptibility to infectious and/or malignant disease is increased significantly. There exist many examples of congenital and acquired immunodeficiency disorders where the absence or functional inadequacy of one or more components of the immune system is associated with reduced host resistance. It is now apparent that the integrity of immune system can be influenced markedly by a number of factors and that chemicals and drugs may have profound effects on immune responses.

10.3 Immunotoxicology

Immunotoxicology may be defined as the study of adverse health effects resulting from the interaction of chemicals and drugs with the immune system. As such immunotoxicology embraces two main forms of insult. Immunotoxicity denotes the potential of drugs or chemicals to cause adverse effects secondary to immunosuppression resulting from damage to, or operational impairment of, immune function. Here the concern is that the induced immune deficit will translate into a reduced capacity to resist infectious disease or tumour development. There are available myriad methods for experimental evaluation of the immunotoxic properties of chemicals. These include examination of the ability of lymphocytes to respond to antigenic stimuli by proliferation, differentiation or the production of cytokines or antibody. Other methods can be used to examine the integrity of phagocyte function, the complement system and other aspects of adaptive or natural immune activity.

The other way in which chemicals and drugs may cause adverse effects through interaction with the immune system is by stimulation of specific immunological responses. The adverse health consequences that result from the induction of specific immune responses are known collectively as allergy. Xenobiotics may provoke various forms of allergic disease including contact hypersensitivity (allergic contact dermatitis), hypersensitivity of the respiratory tract (allergic asthma and rhinitis) and a variety of systemic allergic reactions that not infrequently resemble autoimmune disease.

The immunotoxicity of glucocorticoids will be considered here in the context of both induced immunomodulation and allergic disease. Attention will focus on the nature and significance of immunosuppressive actions, the induction of lymphoid cell apoptosis, and the influence of glucocorticoids on immediate-type (IgE antibody-mediated) hypersensitivity reactions.

10.4 Immunosuppressive and Immunoregulatory Properties of Glucocorticoids

The influence of glucocorticoids on immune function has a long history and some examples of the changes reported are summarized in Table 10.1. A number of general points can be made regarding the effects observed. First, different immune cell populations and different immune processes vary with respect to their sensitivity to the action of glucocorticoids. Thus, for instance, T lymphocyte function and cell-mediated immunity appear in general terms to be more sensitive than do B lymphocytes and humoral immunity. It is the case also that the influence of glucocorticoids on the induction of primary immune responses is greater than on established or secondary responses. The corollary is that in experimental studies the magnitude of induced alterations is to an important extent dependent upon the time of administration of glucocorticoids. There are, in addition, significant species differences in susceptibility to the action of these hormones. For instance, species differ with respect to the magnitude, duration and reversibility of lymphopaenia associated

Table 10.1 Changes in immune function induced by corticosteroids

Cell-mediated immunity and T lymphocyte function
Inhibition of mitogen (phytohaemagglutinin, concanavalin A) induced proliferation *in vitro*
Impaired generation of cytotoxic T lymphocytes *in vitro*
Suppression of allogeneic and autologous mixed lymphocyte reactions *in vitro*
Reduced cellularity of thymic cortex; loss and/or redistribution of peripheral lymphocytes
Reduced cytokine production
Humoral immunity
Altered (reduced or increased) antibody production
Decreased serum immunoglobulin concentration
Natural immune function
Reduced natural killer (NK) cell function *in vivo* and *in vitro*
Monocytopaenia
Altered macrophage function (reduced phagocytic and microbicidal activity)

Note: Summarized from Fauci 1978/9, Gillis *et al.* 1979, Cupps & Fauci 1982, Hall & Goldstein 1984, Kelso & Munck 1984, Munck *et al.* 1984, Spreafico *et al.* 1985, Ader *et al.* 1990, Descotes 1990, Stam *et al.* 1993.

with glucocorticoid treatment (Spreafico *et al.* 1985). Finally, accurate interpretation of the direct effects of corticosteroids on aspects of immune function is complicated frequently by their potent anti-inflammatory activity.

The therapeutic benefits that derive from glucocorticoids, such as prevention of transplant rejection and treatment of allergic disease, are a function of their immunosuppressive and anti-inflammatory properties. These same properties can be modelled *in vivo* and *in vitro* using pharmacological concentrations of the drugs. There is evidence, however, that at physiological concentrations glucocorticoids exert more subtle regulatory influences on the immune system and under such conditions may enhance some immunological responses (Hall & Goldstein 1984). Indeed, the complete absence of adrenal and circulating corticosterone may be associated with depressed immune reactivity (Wiegers *et al.* 1993, Kusnecov & Rabin 1994). Under normal circumstances the immunosuppressive properties of glucocorticoids possibly serve to maintain homeostatic control and prevent an unwanted over-reaction of the immune system that could result in autoimmunity (Munck *et al.* 1984, Besedovsky & del Rey 1991, Ader *et al.* 1995). Such a role is reflected in experimental models of autoimmunity. The Lewis strain of rat is susceptible to the induction of experimental allergic encephalomyelitis (EAE). This is a T lymphocyte-mediated autoimmune condition, characterized by progressive neurological dysfunction and an inflammatory cell infiltration of the central nervous tissue, that is induced by injection of encephalitogen (heterologous spinal cord tissue) emulsified in adjuvant. The susceptibility of Lewis strain rats is due, at least in part, to a defective HPA axis resulting in reduced levels of glucocorticoids (Sternberg *et al.* 1989). Spontaneous recovery of Lewis rats from clinical symptoms is dependent upon endogenous corticosteroids. Adrenalectomy or treatment of rats with a glucocorticoid receptor antagonist results in exacerbated disease and death (Bolton & Flower 1989, MacPhee *et al.* 1989). In contrast, other strains of rat such as Brown Norway and PVG are resistant to the development of EAE and this is the result of relatively high concentrations of circulating corticosteroids that serve to protect completely against the development of symptoms (Mason *et al.* 1990). In this case the critical role of endogenous steroids is presumed to be control of T lymphocyte responses (Peers *et al.* 1995).

There is increasing evidence that signalling pathways between the endocrine and immune systems are bidirectional and that products of the immune response may regulate the HPA axis by triggering the secretion of hypothalamic corticotrophin releasing factor (CRF) and/or pituitary hormones. It was observed some time ago that maximal antibody responses in mice and rats were associated with a transient increase in the concentration of serum cortisone, and that an elevation of serum cortisone could be stimulated in animals by the administration of soluble products of activated T lymphocytes (Besedovsky *et al.* 1975, 1981, Besedovsky & Sorkin 1977). The thesis is that immune activation triggers an increased secretion by the adrenal cortex of glucocorticoids that in turn serve to constrain the vigour of induced immune responses (Besedovsky *et al.* 1979).

The molecular basis for stimulation during immune responses of increased glucocorticoid levels is almost certainly cytokines produced by activated cells. Interleukin 1 (IL-1) stimulates secretion of CRF and increases the plasma concentration of adrenocorticotrophic hormone (ACTH) (Besedovsky *et al.* 1986, Berkenbosch *et al.* 1987, Sapolsky *et al.* 1987). Balance is provided by the transcriptional and translational regulation of IL-1 production by glucocorticoids (Knudsen *et al.* 1987, Kern *et al.* 1988, Lew *et al.* 1988). The negative regulation by corticosteroids of cytokine synthesis may represent the pivotal event in many of their immunosuppressive properties. In addition to blocking IL-1 production, glucocorticoids have been reported to inhibit the production of, among others, interleukins 2, 3 and 4 (IL-2, IL-3 and IL-4), interferon γ (IFN-γ) and granulocyte/macrophage colony-stimulating factor (GM-CSF) (Gillis *et al.* 1979, Arya 1984, Munck *et al.* 1984, Culpepper & Lee 1985, Grabstein *et al.* 1986, Gessani *et al.* 1988, Ader *et al.* 1990, Stam *et al.* 1993). The regulation by glucocorticoids of the T cell growth factor IL-2 has attracted particular attention. The transcription of the human IL-2 gene induced by treatment with phorbol ester and calcium ionophore is inhibited by glucocorticoid hormones (Vacca *et al.* 1990). Recently, a more detailed understanding of the mechanism has been derived. Glucocorticoid hormones are known to control gene expression by activating intracellular receptors (glucocorticoid receptors; GR) belonging to the steroid/thyroid hormone/retinoic acid receptor superfamily of nuclear *trans*-acting factors that bind to specific consensus sequences (Wright *et al.* 1993). It has been shown that the dexamethasone-activated GR selectively impairs the synergistic co-operation between nuclear factor of activated T cells (NAFT) and AP-1 enhancer elements to inhibit IL-2 gene transcription (Vacca *et al.* 1992). More recently still, it has been proposed that a critical event in the induction of immunosuppression by glucocorticoids is regulation of the transcription factor NF-κB that is required for the synthesis of many cytokines. It has been shown that dexamethasone induces the synthesis of IκBα, an inhibitory protein that associates with NF-κB and prevents its translocation to the nucleus (Scheinman *et al.* 1995; Auphan *et al.* 1995).

One interesting observation is worthy of consideration in the context of glucocorticoid-mediated regulation of cytokine synthesis. Daynes & Araneo (1989) demonstrated in immunized mice that treatment with glucocorticoids has a differential influence on cytokine growth factors; production of IL-4 was increased, but IL-2 was significantly depressed. These data are of potential importance as an increased synthesis of IL-4 will favour the development of Th_2-type T helper (Th) cell responses and the stimulation of humoral immunity. A cautionary note is necessary, however, since in other investigations it has been found that the production *in vitro* by activated human peripheral blood T lymphocytes of IL-4 is inhibited in a dose-dependent manner by glucocorticoids (Stam *et al.* 1993). These apparently conflicting data may reflect species differences in susceptibility to glucocorticoids or more subtle variations in experimental design. Notwithstanding these considerations, a variable susceptibility of immunoregulatory cytokines to the action of glucocorticoids may have important implications for the development of allergic disease, and this is discussed later. In summary, the immunosuppressive properties of glucocorticoids at pharmacological levels that have been of proven value in the clinic reflect a more subtle immunoregulatory influence at physiological concentrations. Complex interactions between the HPA axis and the immune system exist and are characterized by hormonal cross-regulation. A key element of the immunoregulatory properties of glucocorticoids resides in their ability to influence negatively the transcription and translation of cytokines. This property, combined with a potent capacity to induce the programmed cell death of certain lymphoid populations, forms the basis of glucocorticoid-mediated immunotoxicity.

10.5 Glucocorticoids and Lymphoid Cell Apoptosis

It is increasingly apparent that programmed cell death of thymocytes and of T lymphocytes plays a vital role in the development and regulation of the immune system and in the pathogenesis of chemical-driven immunotoxicity (Cohen *et al.* 1992, Schwartz & Osborne 1993, Ashwell *et al.* 1994, McConkey *et al.* 1994, Kroemer 1995). It is well established that immature cortical thymocytes are sensitive to glucocorticoid-induced apoptosis (Wyllie 1980, Nieto *et al.* 1992, Schwartzman & Cidlowski 1994). Bilateral adrenalectomy or treatment with a glucocorticoid receptor antagonist has been found to cause a rapid increase in the size of the thymus followed by a depletion of thymocytes through programmed cell death (Gonzalo *et al.* 1993, Kroemer 1995). There is increasing evidence also that certain populations of mature T lymphocytes, immature B cells and natural killer (NK) cells may be sensitive to glucocorticoid-induced apoptosis (Garvy *et al.* 1993, Migliorati *et al.* 1994). There is no doubt that apoptosis caused by glucocorticoids is stimulated via the glucocorticoid receptor and that the vigour of the response correlates closely with affinity of the ligand for the GR (Cohen & Duke 1984, Compton & Cidlowski 1986, Schwartzman & Cidlowski 1994). Apoptosis is dependent upon downstream events resulting from receptor activation for which the DNA binding domain of the GR is required (Nazareth *et al.* 1991). This observation, combined with the fact that *de novo* protein and RNA synthesis are necessary, indicates that apoptosis is dependent upon altered gene expression. In thymocytes treated with glucocorticoids, cell death is associated with the elevated expression of a comparatively small number of genes (Baughman *et al.* 1991, Owens *et al.* 1991). However, the gene products stimulated by steroids and their relevance for apoptosis have still to be elucidated. Clearly the mechanisms involved in triggering cell death are complicated and subject to regulatory events which prevent glucocorticoids inducing apoptosis in all cells that express functional receptors. A central event appears to be elevated concentrations of calcium in target cells. It is proposed that a cycloheximide-sensitive increase in cytosolic calcium induces, via a calmodulin-dependent mechanism, the activation of endonucleases and the interneucleosomal cleavage of chromatin (Kizaki *et al.* 1989, McConkey *et al.* 1989, 1994).

In many cell types it has been found that oncogene expression may be decisive in the regulation of programmed cell death. Much attention has focused on *bcl*-2, the expression of which blocks apoptosis, extending the lifespan of cells that would otherwise have been triggered for programmed cell death. Resistance to glucocorticoid-induced DNA fragmentation and cell death is conferred by the *bcl*-2 gene (Miyashita & Reed 1992). The differential susceptibility of lymphoid cell populations to programmed cell death triggered by glucocorticoids may at least in part be a function of *bcl*-2 expression (Schwartzman & Cidlowski 1994). An additional influence is provided by growth factors that protect against apoptosis. In the case of immune cells it has been shown that the T cell growth factors IL-2 and IL-4 can rescue thymocytes and more mature T lymphocytes, including cytotoxic T cells, from glucocorticoid-induced programmed cell death (Nieto & Lopez-Rivas 1989, Nieto *et al.* 1990, Migliorati *et al.* 1992, 1994), probably via a protein kinase C (PKC)-dependent mechanism. It is of considerable interest that the survival factors for functional subpopulations of Th cells differ. Th_2 cells are rescued from glucocorticoid-induced apoptosis by IL-4, but not by either IL-1 or IL-2. In contrast, the equivalent rescue factor for Th_1 cell populations is IL-2 (Zubiaga *et al.*

1992). The ability of their respective cytokine products (IL-2 and IL-4, respectively) to protect Th_1 and Th_2 cells from induced DNA damage and cell death may have important implications for the development of selective immune responses.

Much of the information regarding induced lymphoid cell death derives from investigations employing supraphysiological concentrations of glucocorticoids. There is evidence, however, that increased levels of endogenous glucocorticoids, stimulated for instance by the administration of mitogen-stimulated lymphocyte supernatants, are also able to cause thymocyte apoptosis *in situ* (Gruber *et al.* 1994).

Apoptosis of thymocytes represents an important component of the developmental process during the maturation of immune function and provides a mechanism whereby those cells that express a receptor with too high an affinity for 'self' (and consequently represent a threat of autoimmunity) can be eliminated. Although, as described above, mature T lymphocytes are less susceptible, it is likely that here as well programmed cell death may be important for maintenance of the integrity of immune function (Cohen *et al.* 1992). The potential role of endogenous glucocorticoids in these processes is presently unclear, although they may serve to accentuate or modify the susceptibility of thymocytes and mature lymphocytes to programmed cell death induced by other stimuli (Kroemer 1995).

A variety of chemicals may exert their immunotoxic effects by the stimulation of thymocyte apoptosis. One such chemical is 2,3,7,8-tetrachlorodibenzo-*p*-dioxin (TCDD), a very potent rodent immunotoxicant that causes thymic atrophy secondary to depletion of cortical thymocytes. This depletion of immature thymocytes is mediated via the induction of apoptosis (McConkey *et al.* 1988). There exist many similarities between glucocorticoid-induced thymocyte cell death and that stimulated by TCDD (McConkey *et al.* 1988, 1994, McConkey & Orrenius 1989). It appears that TCDD may mimic the responses induced by glucocorticoids and McConkey *et al.* (1994) have suggested that potential synergism between TCDD and glucocorticoids, rather than an independent effect of the former alone, may be responsible for the thymic atrophy associated with exposure to this chemical.

10.6 Glucocorticoids and IgE-Mediated Hypersensitivity Reactions

As defined above, allergy is a collective term used to describe the adverse health effects that may result from stimulation of a specific immune response. Allergic disease characteristically develops in two phases. Following first exposure of the susceptible individual to an inducing protein or chemical allergen a specific immune response is initiated which, if of sufficient magnitude, renders the subject sensitized. Following subsequent exposure of the now sensitized individual to the same inducing agent an accelerated and more vigourous secondary immune response is provoked that results in elicitation of a hypersensitivity (allergic) reaction. Allergic diseases are classified conventionally on the basis of the speed with which symptoms appear following challenge of the sensitized subject. Those that develop most rapidly after exposure (within minutes or hours) are described as immediate-type hypersensitivity reactions. Examples include atopic dermatitis, allergic conjunctivitis and respiratory hypersensitivity reactions such as asthma and rhinitis. The most important effector molecule of immediate-type hypersensitivity is IgE antibody. After exposure to the inducing allergen the susceptible individual will mount a specific IgE antibody response. This antibody distributes systemically and associates via membrane receptors with mast cells that are found in virtually all vascularized tissue, including the skin and respiratory tract. During the elicitation phase, mast cell-bound IgE is cross-linked by the allergen. This in turn triggers mast cell degranulation and the release of preformed and newly synthesized inflammatory mediators such as histamine and leukotrienes. These factors initiate a local inflammatory response at sites of allergen exposure that is recognized clinically as allergic hypersensitivity.

There has been considerable interest in the influence of glucocorticoids on IgE antibody production and immediate-type hypersensitivity reactions. As mentioned previously, glucocorticoids remain one of the most effective treatments for allergy. Herein lies an apparent paradox as there is reason to believe that glucocorticoids are able to augment IgE antibody responses.

The production of IgE antibody is regulated actively during immune responses. It has been shown in the mouse that the stimulation and maintenance of IgE responses is dependent upon the availability of IL-4, a cytokine produced by Th_2 lymphocytes and by other cells (Finkelman *et al.* 1988b, Mosmann & Coffman 1989, Kuhn *et al.* 1991, Mosmann *et al.* 1991). Conversely, IFN-γ, a product of Th_1 cells, inhibits IgE antibody formation (Finkelman *et al.* 1988a). These same cytokines also serve to regulate reciprocally IgE responses in humans (Del Prete *et al.* 1988, Pene *et al.* 1988). The vigour of induced IgE antibody production in response to antigenic stimulation will as a consequence reflect the relative availability of IL-4 and IFN-γ in the micro-environment (Kimber & Dearman 1994).

It has been found that glucocorticoids are able to potentiate *in vitro* the IL-4-dependent production of IgE by human peripheral blood mononuclear cells (Fischer & Konig 1991, Nusslein *et al.* 1992, 1994, Klebl *et al.* 1994). The influence of glucocorticoids on IgE synthesis *in vivo* is more controversial. Consistent with the results of some previous studies (Posey *et al.* 1978, Settipane *et al.* 1978), it has been reported recently that treatment of asthma patients with a 7-day course of prednisone caused an increase in serum IgE concentrations (Zieg *et al.* 1994). In other investigations, however, similar treatment was found not to influence significantly IgE levels (Johansson & Juhlin 1970, Henderson *et al.* 1973, Klebl *et al.* 1994). The reasons for such variations are unclear but possibly reflect differences in experimental design and/or patient selection. The studies of Zieg *et al.* (1994) are of particular interest as here a significant elevation of serum IgE was associated with alterations in cytokine secretion. Comparisons were made of the mitogen-stimulated production of IL-4 and IFN-γ by lymphocytes isolated from patients before and after treatment with prednisone. While treatment resulted in a very substantial impairment of IFN-γ production, the synthesis of IL-4 was unaffected. Changes in cytokine production were paralleled by a reduction in the frequency of peripheral lymphocytes expressing IFN-γ as judged by immunocytochemical staining (Zieg *et al.* 1994). These data suggest that an important component of the influence of glucocorticoids on IgE synthesis may be alterations in the balance between immunoregulatory cytokines. Although some investigators have found glucocorticoids to suppress IL-4 production *in vitro* (Wu *et al.* 1991, Byron *et al.* 1992), comparatively high (supraphysiological) concentrations are required. Importantly, glucocorticoids even at nanomolar concentrations inhibit mitogen-driven IFN-γ transcription and synthesis (Gessani *et al.* 1988). It can be proposed that cytokines differ with respect to glucocorticoid-induced inhibition of production, with the synthesis of IFN-γ being more sensitive than that of IL-4. This may be one reason why at some concentrations glucocorticoids promote humoral immune responses and the synthesis of IgE. The apparent inconsistency arising from the value of corticosteroids in the treatment of allergy, despite the fact that they may serve to potentiate IgE antibody production, can be explained on the basis that the main therapeutic benefits derive from depression of local inflammatory reactions. The development of IgE-dependent immediate-type hypersensitivity reactions is associated with an accumulation at the inflammatory site of eosinophils and T lymphocytes. In respiratory hypersensitivity, eosinophils acting together with infiltrating T lymphocytes are required for chronic bronchial inflammation and injury (Gleich 1990, Bentley *et al.* 1992, Corrigan & Kay 1992). The toxic granule proteins of eosinophils, such as major basic protein, and eosinophil-derived leukotrienes have been implicated in many of the pathologic features of asthma, including airway hyperreactivity, vascular leakage and the damage and sloughing of epithelium (Flavahan *et al.* 1988, Gundel *et al.* 1991). For their participation in inflammatory reactions eosinophils must be recruited from the blood. Eosinophils express the GR and glucocorticoids are known to inhibit this process (Schleimer & Bochner 1994). Several investigations have demonstrated that glucocorticoid treatment reduces the number of eosinophils present in the airways of patients with asthma (Adelroth *et al.* 1990, Laitinen *et al.* 1992). Glucocorticoids may affect directly the ability of eosinophils to adhere to endothelial cells, to migrate along chemotactic gradients, or to survive within inflammatory sites. Additionally, these processes could be influenced indirectly by the glucocorticoid-induced inhibition of cytokines known to regulate the development, migration and function of eosinophils (Schleimer & Bochner 1994). For instance, the survival of eosinophils is maintained by relevant growth factors such as GM-CSF and IL-5: cytokines that are down-regulated by corticosteroids.

In summary, glucocorticoids have complex effects on the induction and elicitation of immediate-type hypersensitivity responses. Under certain conditions glucocorticoids may potentiate IgE antibody responses, secondary possibly to inhibition of IFN-γ production. Depression of IFN-γ synthesis by glucocorticoids, while IL-4 production is spared, may have an important influence on Th cell function and the quality of immune responses. Certainly such an observation would be consistent with the view that cell-mediated immune responses exhibit a greater susceptibility to glucocorticoids than does humoral immunity. Despite the potential to augment IgE production, glucocorticoids have proven value in the treatment of asthma and other allergic diseases. This therapeutic benefit derives from anti-inflammatory properties and, importantly, the ability of glucocorticoids to modify eosinophil function.

Given the anti-inflammatory effects of glucocorticoids it is of interest that they are sometimes implicated as the cause of human allergic disease. A number of synthetic corticosteroids have been associated with cutaneous allergic reactions, including contact hypersensitivity and contact urticaria (Whitmore 1995).

10.7 Concluding Comments

There is no doubt that glucocorticoids can interact with and influence the immune system in many ways and that the results of such interactions may be manifested as immunosuppression or immunotoxicity. Such adverse effects represent a pharmacological extension of the more subtle immunoregulatory properties of glucocorticoids that are displayed under physiological conditions and that are required for the normal homeostatic regulation of immune function. It is probably wisest to view glucocorticoids as an important component of the immune–endocrine network that have the potential to stimulate significant changes in the immune system.

References

ADELROTH, E., ROSENHALL, L., JOHANSSON, S.-A., LINDEN, M. & VENGE, P. (1990) Inflammatory cells and eosinophilic activity in asthmatics investigated by bronochoalveolar lavage. The effects of antiasthmatic treatment with budesonide or terbutaline. *American Review of Respiratory Disease*, **142**, 91–99.

ADER, R., FELTEN, D. & COHEN, N. (1990) Interactions between the brain and immune system. *Annual Review of Pharmacology and Toxicology*, **30**, 561–602.

ADER, R., COHEN, N. & FELTEN, D. (1995) Psychoneuroimmunology: interactions between the nervous system and the immune system. *The Lancet*, **345**, 99–103.

ARYA, S. K., WONG-STAAL, F. & GALLO, R. C. (1984) Dexamethasone-mediated inhibition of human T cell growth factor and gamma-interferon messenger mRNA. *Journal of Immunology*, **133**, 273–276.

ASHWELL, J. D., BERGER, N. A., CIDLOWSKI, J. A., LANE, D. P. & KORSMEYER, S. J. (1994) Coming to terms with death: apoptosis in cancer and immune development. *Immunology Today*, **15**, 147–151.

AUPHAN, N., DIDONATO, J. A., ROSETTE, C., HELMBERG, A. & KARIN, M. (1995) Immunosuppression by glucocorticoids: inhibition of NF-κB activity through induction of IκBα synthesis, *Science*, **270**, 286–290.

BAUGHMAN, G., HARRIGAN, M. T., CAMPBELL, N. F., NURRISH, S. J. & BOURGEOIS, S. (1991) Genes newly identified as regulated by glucocorticoids in murine thymocytes. *Molecular Endocrinology*, **5**, 637–644.

BENTLEY, A. M., MAESTRELLI, P., SAETTA, M., FABBRI, L. M., ROBINSON, D. S., BRADLEY, B. L., JEFFREY, P. K., DURHAM, S. R. & KAY, A. B. (1992) Activated T-lymphocytes and eosinophils in the bronchial mucosa in isocyanate-induced asthma. *Journal of Allergy and Clinical Immunology*, **89**, 821–829.

BERKENBOSCH, J., VAN OERS, J., DEL RAY, A., TILDERS, F. & BESEDOVSKY, H. (1987) Corticotropin-releasing factor-producing neurones in the rat activated by interleukin-1. *Science*, **238**, 524–526.

BESEDOVSKY, H. O. & DEL REY, A. (1991) Physiologic implications of the immuno-neuroendocrine network. In ADER, R., COHEN, N. & FELTEN, D. L. (Eds), *Psychoneuroimmunology*, 2nd Edn, pp. 589–608, New York: Academic Press.

BESEDOVSKY, H. O. & SORKIN, E. (1977) Network of immune neuroendocrine interactions. *Clinical and Experimental Immunology*, **27**, 1–12.

BESEDOVSKY, H. O., SORKIN, E., KELLER, M. & MULLER, J. (1975) Changes in blood hormone levels during the immune response. *Proceedings of the Society of Experimental Biology and Medicine*, **150**, 466–470.

BESEDOVSKY, H. O., DEL REY, A. & SORKIN, E. (1979) Antigenic competition between horse and sheep red blood cells as hormone-dependent phenomenon. *Clinical and Experimental Immunology*, **37**, 106–113.

BESEDOVSKY, H. O., DEL REY, A. & SORKIN, E. (1981) Lymphokine containing supernatants from con A-stimulated cells increase corticosterone blood levels. *Journal of Immunology*, **126**, 385–387.

BESEDOVSKY, H. O., DEL REY, A., SORKIN, E. & DINARELLO, C. A. (1986) Immunoregulatory feedback between interleukin-1 and glucocorticoid hormones. *Science*, **233**, 652–654.

BOLTON, C. & FLOWER, R. J. (1989) The effects of the anti-glucocorticoid RU38486 on steroid-mediated suppression of experimental allergic encephalomyelitis (EAE) in the Lewis rat. *Life Sciences*, **45**, 97–104.

BYRON, K. A., VARIGOS, G. & WOOTTON, A. (1992) Hydrocortisone inhibition of human interleukin-4. *Immunology*, **77**, 624–626.

COHEN, J. J. & DUKE, R. C. (1984) Glucocorticoid activation of a calcium-dependent endonuclease in thymocyte nuclei leads to cell death. *Journal of Immunology*, **132**, 38–42.

COHEN, J. J., DUKE, R. C., FADOK, V. A. & SELLINS, K. S. (1992) Apoptosis and programmed cell death in immunity. *Annual Review of Immunology*, **10**, 267–293.

COMPTON, M. M. & CIDLOWSKI, J. A. (1986) Rapid *in vivo* effects of glucocorticoids on the integrity of rat lymphocyte genomic deoxyribonucleic acid. *Endocrinology*, **118**, 38–45.

CORRIGAN, C. J. & KAY, A. B. (1992) T cells and eosinophils in the pathogenesis of asthma. *Immunology Today*, **13**, 501–507.

CULPEPPER, J. A. & LEE, F. (1985) Regulation of IL-3 expression by glucocorticoids in cloned murine T lymphocytes. *Journal of Immunology*, **135**, 3191–3197.

CUPPS, T. R. & FAUCI, A. S. (1982) Corticosteroid-mediated immunoregulation in man. *Immunological Reviews*, **65**, 133–155.

DAYNES, R. A. & ARANEO, B. A. (1989) Contrasting effects of glucocorticoids on the capacity of T cells to produce the growth factors interleukin 2 and interleukin 4. *European Journal of Immunology*, **19**, 2319–2325.

DEL PRETE, G., MAGGI, E., PARRONCHI, P., CHRETIEN, I., TIRI, D., MACCHIA, M., RICCI, J., BANCHEREAU, J., DE VRIES, J. & ROMAGNANI, S. (1988) IL-4 is an essential factor for the IgE synthesis induced *in vitro* by human T cell clones and their supernatants. *Journal of Immunology*, **140**, 4193–4198.

DESCOTES, J. (1990) *Drug-Induced Immune Diseases*, Amsterdam: Elsevier.

FAUCI, A. S. (1978/9) Mechanisms of the immunosuppressive and anti-inflammatory effects of glucocorticoids. *Journal of Immunopharmacology*, **1**, 1–25.

FINKELMAN, F. D., KATONA, I. M., MOSMANN, T. R. & COFFMAN, R. L. (1988a) IFN-γ regulates the isotypes of Ig secreted during *in vivo* humoral immune responses. *Journal of Immunology*, **140**, 1022–1027.

FINKELMAN, F. D., KATONA, I. M., URBAN, J. F., JR, HOLMES, J., OHARA, J., TUNG, A. S., SAMPLE, J. G. & PAUL, W. E. (1988b) IL-4 is required to generate and sustain *in vivo* IgE responses. *Journal of Immunology*, **141**, 2335–2341.

FISCHER, A. & KONIG, W. (1991) Influence of cytokines and cellular interactions on the glucocorticoid-induced Ig (E,G,A,M) synthesis of peripheral blood mononuclear cells. *Immunology*, **74**, 228–233.

FLAVAHAN, N. A., SLIFMAN, N. R., GLEICH, G. J. & VANHOUTTE, P. M. (1988) Human eosinophil major basic protein causes hyperreactivity of respiratory smooth muscle. *American Review of Respiratory Disease*, **138**, 685–688.

GARVY, B. A., TELFORD, W. G., KING, L. E. & FRAKER, P. J. (1993) Glucocorticoids and irradiation-induced apoptosis in normal murine bone marrow B-lineage lymphocytes as determined by flow cytometry. *Immunology*, **79**, 270–277.

GESSANI, S., MCCANDLESS, S. & BAGLIONI, C. (1988) The glucocorticoid dexamethasone inhibits synthesis of interferon by decreasing the level of its mRNA. *Journal of Biological Chemistry*, **263**, 7454–7457.

GILLIS, S., CRABTREE, G. R. & SMITH, K. A. (1979) Glucocorticoid-induced inhibition of T cell growth factor production. II. The effect on the *in vitro* generation of cytolytic T cells. *Journal of Immunology*, **123**, 1632–1638.

GLEICH, G. J. (1990) The eosinophil and bronchial asthma: current understanding. *Journal of Allergy and Clinical Immunology*, **85**, 422–436.

GONZALO, J. A., GONZALEZ-GARCIA, A., MARTINEZ, A. C. & KROEMER, G. (1993) Glucocorticoid-mediated control of the activation and clonal deletion of peripheral T cells *in vivo*. *Journal of Experimental Medicine*, **177**, 1239–1246.

GRABSTEIN, K., DOWER, S., GILLIS, S., URDAL, D. & LARSEN, A. (1986) Expression of interleukin 2, interferon gamma and IL-2 receptor by human peripheral blood lymphocytes. *Journal of Immunology*, **136**, 4503–4508.

GRUBER, J., SGONC, R., HY, Y. H., BEUG, H. & WICK, G. (1994) Thymocyte apoptosis induced by elevated endogenous corticosterone levels. *European Journal of Immunology*, **24**, 1115–1121.

GUNDEL, R. H., LETTS, L. G. & GLEICH, G. J. (1991) Human eosinophil major basic protein induces airway constriction and airway hyperresponsiveness in primates. *Journal of Clinical Investigation*, **87**, 1470–1473.

HALL, N. R. & GOLDSTEIN, A. L. (1984) Endocrine regulation of host immunity. The role of steroids and thymosin. In FENICHEL, R. L. & CHIRIGOS, M. A. (Eds), *Immune Modulation Agents and their Mechanisms*, pp. 533–563, New York: Marcel Dekker.

HENDERSON, L. L., LARSON, J. B. & GLEICH, G. J. (1973) Effect of corticosteroids on seasonal increases in IgE antibody. *Journal of Allergy and Clinical Immunology*, **52**, 352–357.

JOHANSSON, S. G. O. & JUHLIN, L. (1970) Immunoglobulin E in 'healed' atopic dermatitis and after treatment with corticosteroids and azathioprine. *British Journal of Dermatology*, **82**, 10–13.

KELSO, A. & MUNCK, A. (1984) Glucocorticoid inhibition of lymphokine secretion by alloreactive T lymphocyte clones. *Journal of Immunology*, **133**, 784–791.

KERN, J. A., LAMB, R. J., REED, J. C., DANIELE, R. P. & NOWELL, P. C. (1988) Dexamethasone inhibition of interleukin-1 beta-production by human monocytes: post-transcriptional mechanisms. *Journal of Clinical Investigation*, **81**, 237–244.

KIMBER, I. & DEARMAN, R. J. (1994) Immune responses to contact and respiratory allergens. In DEAN, J. H., LUSTER, M. I., MUNSON, A. E. & KIMBER, I. (Eds), *Immunotoxicology and Immunopharmacology*, 2nd Edn, pp. 663–679, New York: Raven Press.

KIZAKI, H., TADAKUMA, T., ODAKA, C., MURAMATSU, J. & ISHIMURA, Y. (1989) Activation of a suicide process of thymocytes through DNA fragmentation by calcium ionophores and phorbol esters. *Journal of Immunology*, **143**, 1790–1794.

KLEBL, F. H., WEBER, G., KALDEN, J. R. & NUSSLEIN, H. G. (1994) *In vitro* and *in vivo* effect of glucocorticoids on IgE and IgG subclass secretion. *Clinical and Experimental Allergy*, **24**, 1022–1029.

KNUDSEN, P. J., DINARELLO, C. A. & STROM, T. B. (1987) Glucocorticoids inhibit transcriptional and post-transcriptional expression of interleukin-1 in U937 cells. *Journal of Immunology*, **139**, 4129–4134.

KROEMER, G. (1995) The pharmacology of T cell apoptosis. *Advances in Immunology*, **58**, 211–296.

KUHN, R., RAJEWSKY, K. & MULLER, W. (1991) Generation and analysis of interleukin-4 deficient mice. *Science*, **254**, 707–710.

KUSNECOV, A. W. & RABIN, B. S. (1994) Stressor-induced alterations of immune function: mechanisms and issues. *International Archives of Allergy and Immunology*, **105**, 107–121.

LAITINEN, L. A., LAITINEN, A. & HAAHTELA, T. (1992) A comparative study of the effects of an inhaled corticosteroid, budesonide and a β_2-agonist, terbutaline, on airway inflammation in newly diagnosed asthma: a randomized, double-blind, parallel-group controlled trial. *Journal of Allergy and Clinical Immunology*, **90**, 34–42.

LEW, W., OPPENHEIM, J. J. & MATSUSHIMA, K. (1988) Analysis of the suppression of IL-α and IL-1β production in human peripheral blood mononuclear adherent cells by a glucocorticoid hormone. *Journal of Immunology*, **140**, 1895–1902.

MACPHEE, I. A. M., ANTONI, F. A. & MASON, D. W. (1989) Spontaneous recovery of rats from experimental allergic encephalomyelitis is dependent on regulation of the immune system by endogenous adrenal corticosteroids. *Journal of Experimental Medicine*, **169**, 431–445.

MASON, D. W., MACPHEE, I. A. M. & ANTONI, F. A. (1990) The role of the neuroendocrine system in determining genetic susceptibility to experimental allergic encephalomyelitis in the rat. *Immunology*, **70**, 1–5.

MCCONKEY, D. J. & ORRENIUS, S. (1989) 2,3,7,8-Tetrachlorodibenzo-*p*-dioxin (TCDD) kills glucocorticoid-sensitive thymocytes *in vivo. Biochemistry and Biophysics Research Communications*, **160**, 1003–1008.

MCCONKEY, D. J., HARTZELL, P., DUDDY, S. K., HAKANSSON, H. & ORRENIUS, S. (1988) 2,3,7,8-Tetrachlorodibenzo-*p*-dioxin kills immature thymocytes by Ca^{2+}-mediated endonuclease activation. *Science*, **242**, 256–259.

MCCONKEY, D. J., HARTZELL, P., NICOTERA, P. & ORRENIUS, S. (1989) Calcium-activated DNA fragmentation kills immature thymocytes. *FASEB Journal*, **3**, 1843–1849.

MCCONKEY, D. J., JONDAL, M. B. & ORRENIUS, S. G. (1994) Chemical-induced apoptosis in the immune system. In DEAN, J. H., LUSTER, M. I., MUNSON, A. E. & KIMBER, I. (Eds), *Immunotoxicology and Immunopharmacology*, 2nd Edn, pp. 473–485, New York: Raven Press.

MIGLIORATI, G., PAGLIACCI, C., MORACA, R., CROCICCHIO, F., NICOLETTI, I. & RICCARDI, C. (1992) Interleukins modulate glucocorticoid-induced thymocyte apoptosis. *International Journal of Clinical Laboratory Research*, **21**, 300–307.

MIGLIORATI, G., NICOLETTI, I., D'ADAMIO, F., SPRECA, A., PAGLIACCI, C. & RICCARDI, C. (1994) Dexamethasone induces apoptosis in mouse natural killer cells and cytotoxic T lymphocytes. *Immunology*, **81**, 21–26.

MIYASHITA, T. & REED, J. C. (1992) *bcl*-2 Gene transfer increases resistance of S49.1 and WEHI7.2 lymphoid cells to cell death and DNA fragmentation induced by glucocorticoids and multiple chemotherapeutic drugs. *Cancer Research*, **52**, 5407–5411.

MOSMANN, T. R. & COFFMAN, R. L. (1989) Heterogeneity of cytokine secretion patterns and functions of helper T cells. *Advances in Immunology*, **46**, 111–147.

MOSMANN, T. R., SCHUMACHER, J. H., STREET, N. F., BUDD, R., O'GARRA, A., FONG, T. A. T., BOND, M. W., MOORE, K. W. M., SHER, A. & FIORENTINO, D. F. (1991) Diversity of cytokine synthesis and function of mouse $CD4^+$ T cells. *Immunological Reviews*, **123**, 209–229.

MUNCK, A., GUYRE, P. M. & HOLBROOK, N. J. (1984) Physiological functions of glucocorticoids in stress and their relation to pharmacological actions. *Endocrine Reviews*, **5**, 25–47.

NAZARETH, L. V., HARBOUR, D. V. & THOMPSON, E. B. (1991) Mapping of the human glucocorticoid receptor for leukaemic cell death. *Journal of Biological Chemistry*, **266**, 12976–12980.

NIETO, M. A. & LOPEZ-RIVAS, A. (1989) IL-2 protects T lymphocytes from glucocorticoid-induced DNA fragmentation and cell death. *Journal of Immunology*, **143**, 4166–4170.

NIETO, M. A., GONZALES, A., LOPEZ-RIVAS, A., DIAZ-ESPADA, F. & GAMBON, F. (1990) IL-2 protects against anti-CD3-induced cell death in human medullary thymocytes. *Journal of Immunology*, **145**, 1364–1368.

NIETO, M. A., GONZALEZ, A., GAMBON, F., DIAZ-ESPADA, F. & LOPEZ-RIVAS, A. (1992) Apoptosis in human thymocytes after treatment with glucocorticoids. *Clinical and Experimental Immunology*, **88**, 341–344.

NUSSLEIN, H. G., TRAG, T., WINTER, M., DIETZ, A. & KALDEN, J. R. (1992) The role of T cells and the effect of hydrocortisone on interleukin-4-induced IgE synthesis by non-T cells. *Clinical and Experimental Immunology*, **90**, 286–292.

NUSSLEIN, H. G., WEBER, G. & KALDEN, J. R. (1994) Synthetic glucocorticoids potentiate IgE synthesis. *Allergy*, **49**, 365–370.

OWENS, G. P., HAHN, W. E. & COHEN, J. J. (1991) Identification of mRNAs associated with programmed cell death in murine thymocytes. *Molecular and Cell Biology*, **11**, 4177–4188.

Peers, S. H., Duncan, G. S., Flower, R. J. & Bolton, C. (1995) Endogenous corticosteroids modulate lymphoproliferation and susceptibility to experimental allergic encephalomyelitis in the Brown Norway rat. *International Archives of Allergy and Immunology*, **106**, 20–24.

Pene, J., Rousset, F., Briere, F., Chretien, I., Paliard, X., Banchereau, J., Spits, H. & De Vries, J. E. (1988) IgE production by normal human B cells induced by alloreactive T cell clones is mediated by IL-4 and suppressed by IFN-γ. *Journal of Immunology*, **141**, 1218–1224.

Posey, W. C., Nelson, H. S., Branch, B. & Pearlmann, D. S. (1978) The effects of acute corticosteroid therapy for asthma on serum immunoglobulin levels. *Journal of Allergy and Clinical Immunology*, **62**, 340–348.

Sapolsky, R., Rivier, C., Yamamoto, G., Plotsky, P. & Vale, W. (1987) Interleukin-1 stimulates the secretion of hypothalamic corticotropin-releasing factor. *Science*, **238**, 522–524.

Scheinman, R. I., Cogswell, P. C., Lofquist, A. K. & Baldwin, A. S. Jr (1995) Role of transcriptional activation of IκBα in mediation of immunosuppression by glucocorticoids, *Science*, **270**, 283–286.

Schleimer, R. P. & Bochner, B. S. (1994) The effects of glucocorticoids on human eosinophils. *Journal of Allergy and Clinical Immunology*, **94**, 1202–1213.

Schwartz, L. M. & Osborne, B. A. (1993) Programmed cell death, apoptosis and killer genes. *Immunology Today*, **14**, 582–590.

Schwartzman, R. A. & Cidlowski, J. A. (1994) Glucocorticoid-induced apoptosis of lymphoid cells. *International Archives of Allergy and Immunology*, **105**, 347–354.

Settipane, G. A., Pudupakkham, R. K. & McGowan, J. H. (1978) Corticosteroid effect on immunoglobulins. *Journal of Allergy and Clinical Immunology*, **62**, 162–166.

Spreafico, F., Allegrucci, M., Merendino, A. & Luini, A. (1985) Chemical immunodepressive drugs: their action on the cells of the immune system and immune mediators. In Dean, J. H., Luster, M. I., Munson, A. E. & Amos, H. (Eds), *Immunotoxicology and Immunopharmacology*, pp. 179–192, New York: Raven Press.

Stam, W. B., Van Oosterhout, A. J. M. & Nijkamp, F. P. (1993) Pharmacologic modulation of T_{H1}- and T_{H2}-associated lymphokine production. *Life Sciences*, **53**, 1921–1934.

Sternberg, E. M., Hill, J. M., Chrousos, G. P., Kamilaris, T., Listwak, S. J., Gould, P. W. & Wilder, R. I. (1989) Inflammatory mediator-induced hypothalamic-pituitary adrenal axis activation is defective in streptococcal cell wall arthritis-susceptible Lewis rats. *Proceedings of the National Academy of Sciences, USA*, **86**, 2374–2378.

Vacca, A., Martinotti, S., Screpanti, I., Maroder, M., Felli, M. P., Farina, A. R., Gismondi, A., Santoni, A., Frati, I. & Gullino, A. (1990) Transcriptional regulation of interleukin 2 gene by glucocorticoid hormones. Role of steroid receptor and antigen-responsive 5 flanking sequences. *Journal of Biological Chemistry*, **265**, 8075–8080.

Vacca, A., Felli, M. P., Farina, A. R., Martinotti, S., Maroder, M., Screpanti, I., Meco, D., Petrangeli, E., Frati, L. & Gulino, A. (1992) Glucocorticoid receptor-mediated suppression of the cooperativity between nuclear factor of activated T cells and AP-1 enhancer elements. *Journal of Experimental Medicine*, **175**, 637–646.

Whitmore, S. E. (1995) Delayed systemic allergic reactions to corticosteroids. *Contact Dermatitis*, **32**, 193–198.

Wiegers, G. J., Croiset, G., Reul, J. M. H. M., Holsboer, F. & de Kloet, E. R. (1993) Differential effects of corticosteroids on rat peripheral blood T lymphocyte mitogenesis *in vivo* and *in vitro*. *American Journal of Physiology*, **265**, E825–E830.

Wright, A. P., Zilliacus, J., McEwan, I. J., Dahlman-Wright, K., Almlof,

T., Carlstedt-Duke, J. & Gustafsson, J. A. (1993) Structure and function of the glucocorticoid receptor. *Journal of Steroid Biochemistry and Molecular Biology*, **47**, 11–19.

Wu, C. Y., Fargeas, C., Nakajima, T. & Delespesse, G. (1991) Glucocorticoids suppress the production of interleukin 4 by human lymphocytes. *European Journal of Immunology*, **21**, 2465–2467.

Wyllie, A. H. (1980) Glucocorticoid-induced thymocyte apoptosis is associated with endogenous endonuclease activation. *Nature*, **284**, 555–556.

Zieg, G., Lack, G., Harbeck, R. J., Gelfand, E. W. & Leung, D. Y. M. (1994) *In vivo* effects of glucocorticoids on IgE production. *Journal of Allergy and Clinical Immunology*, **94**, 222–230.

Zubiaga, A. M., Munoz, E. & Huber, B. T. (1992) IL-4 and IL-2 selectively rescue Th cell subsets from glucocorticoid-induced apoptosis. *Journal of Immunology*, **149**, 107–112.

SECTION FIVE

Clinical Interfaces: Adrenal and Corticosteroid Involvement in Human Toxicology and Extrapolation to Man

11

Corticosteroid Adverse Effects and Drug Interactions in Man

P. F. D'ARCY

The Queen's University of Belfast

11.1 Introduction

It is salutary to consider that, although the corticosteroids have been in major clinical use for over 40 years, it is still necessary on occasion to remind practitioners of their inherent dangers. They are very much two-edged weapons.

In the early days of corticosteroid research, corticosteroids were classified as either glucocorticoids or mineralocorticoids. The distinction was not clear-cut since many of the then available corticoids had both properties. Research since those days has tended to produce more specific compounds, although all glucocorticoids still have some mineralocorticoid action and vice versa. The two classes differ in their toxicity spectrum and therefore a brief description of their differences is appropriate in this context. Tables 11.1 and 11.2 show examples of both types of corticoids.

11.2 Glucocorticoids

Grahame-Smith & Aronson (1984) have produced an excellent summary of the properties of glucocorticoids. They have effects on protein, carbohydrate and fat metabolism; they induce the mobilization of proteins and amino acids from skeletal muscle, skin and bone. Enzymes involved in gluconeogenesis are induced by glucocorticoids and the mobilized amino acids are converted in the liver to glucose and then glycogen. Large doses of glucocorticoids cause hyperglycaemia and are diabetogenic. Body fat is redistributed by glucocorticoids leading to obesity, 'buffalo hump' and 'moon face'.

Glucocorticoids have anti-inflammatory and immunosuppressant properties. They prevent capillary vasodilatation and increased vascular permeability which normally lead to tissue oedema and swelling. They have inhibitory effects on the cellular components of the acute inflammatory response, inhibiting the migration of leucocytes and phagocytic activity. They inhibit certain aspects of chronic inflammation including capillary and fibroblast proliferation and the deposition of collagen. Their immunosuppressant effects are mediated via inhibition of lymphocyte functions; the response of both B-cells and T-cells to antigens are suppressed and this results in impairment of humoral and cellular immunity.

Table 11.1 Corticosteroids with major glucocorticoid effects

Beclomethasone dipropionate (50)	Fluprednisolone (2)[b]
Betamethasone (25)	Hydrocortisone (cortisol) (1)
Budesonide (1)[a]	Medrysone
Ciclomethasone	Meprednisone
Cortisone acetate (0.8)	Methylprednisolone (5)
Cortivazol (10)[b]	Paramethasone acetate (10)
Deflazacort (2)[b]	Prednisolone (4)
Dexamethasone (25)	Prednisone (4)
Flumethasone pivalate	Prednylidene
Flunisolide (12.8)[a]	Tixocortol pivalate
Fluoromethalone acetate	Triamcinolone (5)

Note: Unless otherwise stated, figures in parentheses indicate anti-inflammatory potency relative to hydrocortisone. Agents without potency comparators are mainly used by topical application.
[a] Systemic potency as measured in thymus involution test in rodents.
[b] Systemic potency as assessed by dosage comparisons.

Table 11.2 Corticosteroids with major mineralocorticoid effects

Aldosterone
Deoxycortone acetate
Fludrocortisone acetate

The potency of some glucocorticoids relative to hydrocortisone are shown in Table 11.1.

11.3 Mineralocorticoids

Mineralocorticoids (Table 11.2), as their names implies, have major actions on mineral and electrolyte metabolism. They act on the distal convoluted tubule of the kidney, enhancing sodium and water reabsorption and potassium and hydrogen ion excretion (Grahame-Smith & Aronson 1984).

In this chapter, discussions will be orientated towards two major headings: first, the adverse effects of corticosteroids, and second, their drug interactions with other therapy. Both these types of adverse events are largely predetermined and predictable from their pharmacological and metabolic actions. Reviews on adverse effects with corticosteroids have been published by Dahl (1985) (topical), Stead & Cook (1989) (inhaled), Seal & Compton (1986), O'Donnell (1989) and Barnes (1995) (inhaled).

11.4 Adverse Effects of Corticosteroids

Adverse effects are directly related to the dosage and duration of treatment. Two categories of toxic effects occur in the therapeutic use of corticosteroids. First, there are those effects resulting from continued doses in excess of normal physiological requirements (hypercorticism), and second, there are those effects resulting from withdrawal of the extended use of higher doses (hypocorticism).

11.5 Hypercortism

11.5.1 General Effects

With prolonged treatment, with all dosage forms, all the features of Cushing's syndrome may occur: moon face, buffalo hump, supraclavicular fat pads, obesity, striae, acne, hirsutism, bruising, and osteoporosis with an increased risk of spontaneous fractures. There is also an increased susceptibility to infection, hyperglycaemia and glycosuria and enhancement of latent diabetes mellitus, and fluid and electrolyte disturbances which may worsen hypertension and evoke cardiac failure in susceptible individuals. An effect on tissue repair is manifest in delayed wound healing and increased liability to infection. Acute psychotic reactions, most commonly occurring in women and middle-aged patients, are associated with the use of glucocorticoids.

Corticosteroids may interfere with the normal anti-inflammatory responses and this may result in suppression of the clinical signs and symptoms of peptic ulcer disease and ulcer perforation, although studies on the association between corticosteroids have produced conflicting results (Anonymous 1987, Piper *et al.* 1991, Guslandi & Tittobello 1992). There may be myopathy characterized by weakness of the proximal musculature of arms and legs and associated shoulder and pelvic muscles, and respiratory muscles (Decramer & Stas 1992), and hypercoagulability of blood with thromboembolic episodes. A brief review of some of these adverse events is given below.

11.5.2 Bones and Joints

Corticosteroid-induced avascular necrosis of bone, together with its diagnosis and treatment, has been reviewed by Nixon (1984), Lukert & Raisz (1990) and Capell (1992). It is one of the most disabling complications of therapy and is found in patients with a variety of disease states. Even short courses of high-dose corticosteroids may be associated with its development. Rizzato & Montemurro (1993) have shown that exogenous corticosteroid-induced bone loss is fully reversible in patients under 45 years of age after steroid treatment for sarcoidosis was withdrawn. This report of reversibility needs confirmation in elderly people where the capacity for recovery of bone mass could be reduced.

Laan *et al.* (1993) have shown that glucocorticoid-induced bone loss may vary in different parts of the skeleton, with the anterior cortical rim of the vertebral body being more susceptible to the steroid's effects than other regions. The 'calcium-sparing' use of deflazacort appears to offer a degree of protection against both decreased calcium absorption and bone loss (Aicardi *et al.* 1993, Gennari 1993). Laan *et al.* (1992) have described vertebral osteoporosis in rheumatoid arthritis patients treated with low-dose corticosteroids. Treatment of corticosteroid-induced osteoporosis has included the use of oestrogens and androgens, sodium fluoride, calcitonins and bisphosphonates.

11.5.3 Eyes

11.5.3.1 Glaucoma

Repeated use of glucocorticoids or their continued application to the eye may be followed by an increase in intraocular pressure. The extent to which this happens depends on the age of the patient and his/her genetic make-up. Fortunately, the raised pressure is usually totally reversible upon withdrawal of the corticosteroid. Recent reports have implicated periocular steroid and topical steroid drops in the genesis of raised intraocular pressure and glaucomatous damage (Nielsen & Sorensen 1978, Vie 1990, Aggarwal *et al.* 1993, Butcher *et al.* 1994) in addition to the formation of cataracts (Costagliola *et al.* 1989).

The onset of raised intraocular pressure in man is usually a matter of weeks after local, and months after systemic, therapy (David & Berkowitz 1969) This occurs in ~30% of patients receiving local therapy and in a lower percentage when the steroid is given systemically. Severe increases of intraocular pressure resembling those of acute glaucoma have been reported, and cupping of the optic discs and visual-field defects produced by the raised pressure are similar to those of open-angle glaucoma. The changes are usually reversible providing treatment is withdrawn.

The mechanism of steroid-induced ocular hypertension may involve increased aqueous production, but an increase in resistance to the outflow tract seems to be the most important contributing factor. The trabecular network, which separates the anterior chamber from the canal of Schlemmn, contains collagen strands and a single layer of endothelial cells. Corticosteroids may cause swelling of the collagen strands by increasing viscosity and water-binding capacity of the mucopolysaccharides. This would block the outflow tract and increase resistance. Other effects of topical corticosteroids that could raise outflow resistance include increased vasoconstriction and pupil dilatation, both being a potentiation of the normal sympathetic tone of the eye. The use of topical steroids applied to the face may evoke ocular hypertension or glaucome. Aggarwal *et al.* (1993) have reported five such cases which demonstrate the potentially blinding complications of topical facial steroids.

11.5.3.2 Posterior subcapsular cataracts

Black and co-workers in 1960 first described posterior subcapsular cataracts (PSC) in rheumatoid arthritis patients on long-term corticosteroid therapy. Since then there have been numerous articles in the literature reporting varying indices of the association of PSC and long-term corticosteroid therapy in adults. In contrast to glaucoma, reports related to lens changes following local therapy have been few, whereas the majority of reports have been with systemic dosage (Tripathi *et al.* 1992).

There is some evidence to suggest that PSC complications of systemic corticosteroid treatment are more common in children (Braver *et al.* 1967), with lower dosage and shorter periods of therapy than is common with adults (Fürst *et al.* 1966). Kaye *et al.* (1993) have examined ocular complications of long-term, low-dose prednisone therapy in children and found no evidence of predisposition to higher intraocular pressure than control children, although they were more likely to develop PSCs.

With regard to the aetiology of the disease, steroid-induced cataract is almost always bilateral; the lesion usually occupies the polar region of the posterior cortex, just within the posterior lens capsule. It extends forward into the cornea irregularly but its borders are sharply defined (Oglesby *et al.* 1961). Vision is not usually impaired early in the development of a steroid-induced cataract, and slit-lamp examination is necessary for early detection. It has been suggested that the most important factor in steroid-induced PSC cataract formation in adults may be variability in individual susceptibility to the side-effects of corticosteroids and that possibly constitutional (genetic) factors may be important dominants (Škala & Prchal 1980).

11.5.3.3 Exophthalmos

It has been known for some time that corticosteroids will induce ophthalmos in experimental animals (Williams 1953, 1955, Aterman & Greenberg 1954). The first reports of steroid-induced exophthalmos in humans came from Slansky *et al.* (1976). They reported four cases associated with high doses of prednisone of from 3 to 12 year's duration. In one of these cases there was a reduction in exophthalmos when the prednisone dosage was reduced from 25 to 5 mg per day. In these patients the exophthalmos accompanied other known complications of prolonged corticosteroid therapy including hyperglycaemia, posterior subcapsular cataract, and elevated intraocular tension.

11.5.4 Growth

In children, growth retardation occurs and epiphyseal closure may be delayed (Grahame-Smith & Aronson 1984, Fletcher 1986, Polito & Di Toro 1992). This has also occurred with doses of inhaled steroids higher than 400 μg daily (Priftis *et al.* 1991, Wolthers & Pedersen 1992). Hughes (1987) has reviewed the effects of corticosteroids on the growth of children. Gibson *et al.* (1993) have assessed growth retardation after dexamethasone by knemometry.

11.5.5 Infections and Infestations

It is by now so well known that corticosteroids can suppress immune function and increase susceptibility to, and the severity of, any bacterial, fungal, viral or parasitic infection, that there is little new that can be added to the warnings that have been expressed so frequently in the past. Some examples of increased susceptibility to infection are provided below.

11.5.5.1 Tuberculosis

Tuberculosis is a major bacterial infection of concern in this respect, and it has long been known that manifestations of tuberculous infection or aggravation of existing tuberculosis, as well as reactivation of completely quiet disease, can occur during treatment with corticosteroids (Espersen 1963, Horne 1990, McGowan *et al.* 1992).

11.5.5.2 *Bacterial meningitis*

Corticosteroids have been used for their anti-inflammatory effects in the hope of controlling oedema in bacterial meningitis. However, an editorial in the *Lancet* (1982) reviewed the use of corticosteroids in meningitis and concluded that there was no unequivocal evidence of benefit from their use in pyrogenic meningitis. Indeed, it was suggested that harm might possibly be done by lessening CSF penetration by penicillins since they only cross the blood–brain barrier well when the meninges are inflamed (Hieber & Nelson 1977). The general opinion is that more study is needed in adults and neonates, as well as in children, to assess fully the benefits of corticosteroids in bacterial meningitis and to weigh these benefits against their risk (Bahal & Nahata 1991, McGowan *et al.* 1992).

11.5.5.3 *Viral infections*

Since the pioneering work of Findlay & Howard (1952) on the influence of corticosteroids and corticotrophin on the multiplication of many viruses, it has been known that viral infections may run a very severe course in patients on corticosteroid treatments. Infections with herpes virus seem especially dangerous (Duckworth 1973, Charasse *et al.* 1992). In his review on corticotrophin and corticosteroids, Erill (1991) has emphasized that corticosteroid treatment in patients receiving a renal transplant may increase susceptibility to infection and he cited a report of fatal hepatic necrosis caused by disseminated type 5 adenovirus infection in a recipient of a kidney from a 2-year-old victim of drowning (Norris *et al.* 1989).

Chickenpox (varicella) is of current concern since it is normally a minor illness, however, it may be fatal in patients who are immunosuppressed, whatever the cause. There are ~30 fatalities annually from chickenpox in the UK, one-third of which are associated with immunosuppressions. Manifestations of fulminant illness include pneumonia, hepatitis and disseminated intravascular coagulation. A rash is not necessarily a prominent feature (Committee on Safety of Medicines/Medicines Control Agency 1994).

Systemic corticosteroids substantially increase the risk of severe chickenpox (Dowell & Bresee 1993), and the risk of severe herpes zoster is also likely to be increased by corticosteroids. All patients taking systemic corticosteroids for purposes other than replacement should be regarded at risk, unless they have had chickenpox. Currently, there is no good evidence that topical, inhaled or rectal preparations are associated with an increased risk of severe chickenpox (Committee on Safety of Medicines/Medicines Control Agency 1994).

11.5.5.4 *Fungal infections*

Fungal infections may run a more severe course in corticosteroid-treated patients and such patients frequently have a greater susceptibility to them. Erill (1991) cited a report of two cases of fatal disseminated aspergillosis developing in patients who received a short course of high-dose corticosteroid therapy prior to liver transplantation (Brems *et al.* 1988). Apart from systemic fungal infections, it has long been known that topical infections due to yeasts are also a hazard during long-term corticosteroid treatment; for example, Dennis & Itken (1964) reported monilial infection after the use of an aerosol containing dexamethasone in asthmatic patients. Five of 25 treated patients developed infection of the oropharanx, and two an infection of the larynx with *Candida albicans.*

Fungal infections feature predominantly among the opportunistic organisms which can cause disease during the development of infections with HIV virus. The fungi encountered are the same as those found in other situations characterized by depressed cell-mediated immunity.

Corticosteroids have been used to reduce the inflammatory response in *Pneumonocystis carinii* pneumonia (PCP) (a common complication of AIDS) and reduce the likelihood of death (Bozzette *et al.* 1990, Kovacs & Masur 1990, National Institutes of Health 1990, McGowan *et al.* 1992) although other groups have expressed reservations (Chmel 1990, Sattler 1991). Bernstein *et al.* (1994) have described a case of a patient admitted with severe PCP whose treatment included high-dose steroids. Overwhelming cryptococcal disease developed, contributing to his death. Mahaffey *et al.* (1993) have described two cases of PCP in AIDS treated with adjuvant corticosteroid therapy which led to overwhelming coccidiomycosis.

PCP may be associated with various non-AIDS immunodeficiency states. For example, Bernstein *et al.* (1993) have reported two patients with ulcerative colitis who developed PCP during high-dose corticosteroid therapy. Varma *et al.* (1993) have reported a case of invasive pulmonary aspergillosis and nocardiosis in an immunocompromised host following high-dose, prolonged corticosteroid therapy for glomerulonephritis.

11.5.5.5 *Malaria*

Glucocorticoids, such as dexamethasone, can reduce the vasogenic oedema associated with some types of cerebral disease, and because of this they have been used extensively since the 1960s in the treatment of cerebral malaria (*Plasmodium falciparum*) on the assumption that cerebral oedema is a consistent feature of that disease. Only a few investigators argued against this uncritical acceptance of the use of corticosteroids and a review in the *British Medical Journal* by Hall in 1976 concluded that a controlled trial of corticosteroids in falciparum malaria was long overdue because the value had not been established. In such a controlled study, Warrell *et al.* (1982) compared the steroid with placebo and showed that dexamethasone significantly increased the duration of coma and the incidence of complications including pneumonia and gastrointestinal bleeding. It was therefore concluded that dexamethasone was deleterious in cerebral malaria and that it should no longer be used.

11.5.6 **Psychotic Reactions**

Euphoria is extremely common with corticosteroid treatment and though minor mood elevation could well reflect relief from physical distress, there is still an appreciable incidence of drug-induced euphoria. Patients may also develop severe mental complications and the risk of psychosis has been judged to be $\sim 5\%$ in patients on daily doses of 40 mg or more of prednisone (Boston Collaborative Drug Surveillance Program, 1972, Hall *et al.* 1979). Psychoses usually resolve with dosage reduction or controlled withdrawal of the steroid although antipsychotic medication may be indicated if the symptoms are severe or prolonged (Klein 1992, Travlos & Hirsch 1993). Mental disturbances have been reported after a brief dental administration of dexamethasone (MacKay & Eisendrath 1992).

11.5.7 Skin Conditions

Corticosteroids have been associated with skin manifestations since their early use. For example, rosacea-like lesions and skin atropy have occurred in patients using topical hydrocortisone which had hitherto been regarded as a relatively safe form of treatment (Guin 1981). Skin thinning and purpura occurred in patients receiving inhaled steroids (Capewell *et al.* 1990, Shuttleworth *et al.* 1990). Oral and topical steroids are known to produce acne (Plewig & Kligman 1973, Kaidbey & Kligman, 1974), and inappropriate use of topical steroids on the face can result in a condition which simulates acne and perioral dermatitis (Plewig & Kligman 1973). Hypersensitivity to topical steroids is becoming increasingly recognized; patients with stasis dermatitis and leg ulceration appear to be especially sensitive (Wilkinson & English 1992).

In general, the more potent a topical steroid the more severe the unwanted effects (Drug and Therapeutics Bulletin 1995). It is not only systemic and topical steroids that may cause skin problems since high-dose inhaled corticosteroids are known to have systemic effects on the skin; Hughes *et al.* (1992) have described a 75-year-old patient who developed acne whilst using an inhaled glucocorticoid.

11.6 Hypocortism

Suppression of the pituitary-adrenal axis occurs inevitably to a greater or lesser extent with all corticosteroid usage. Withdrawal of the steroid after treatment, unless this is done gradually to allow recovery of normal adrenocortical function, will then result in hypoadrenalism, which may result in Addisonian crisis.

In stressful episodes, the adrenal cortex is unable to supply the endogenous corticosteroids that are required of it. Surgical and anaesthetic trauma are perhaps the best examples of controlled stressful episodes and historically the first report came in 1952 when a 34-year-old man, who had been treated for 18 months with cortisone for rheumatoid arthritis, had a cup arthroplasty on his hip. He developed irreversible shock and died (Fraser *et al.* 1952). There have been many such examples since (see review by Fletcher 1968).

Corticosteroids can cross the placental barrier and cause foetal adrenal failure; early cases have been reviewed by Bongiovanni & McPadden (1960) and Oppenheimer (1964). Children may be particularly sensitive to hypoadrenalism. Zwaan *et al.* (1992) reported a case of acute adrenal insufficiency in a 7-year-old asthmatic girl after discontinuation of inhaled corticosteroid therapy. The use of spacers may improve safety with inhaled corticosteroids. They not only improve efficacy of asthma treatment, but they may also reduce adverse effects including suppression of the pituitary-adrenal axis (Prahl & Jensen 1987, Brown *et al.* 1990,

Table 11.3 Interactions involving corticosteroids

Anaesthetics	Possible hypotension unless corticosteroid cover is given. Interaction includes course of steroid treatment in the past 2 months (Feldman 1963).
Antibacterials	Rifampicin is a potent P450 enzyme inducer; it accelerates the metabolism of corticosteroids and reduces their clinical efficacy (Venkatesan 1992). Acute adrenal crisis has been precipitated by rifampicin in patients with adrenal insufficiency (Elansary & Earis 1983), and induction of microsomal enzyme systems may be enough to compromise even patients with mildly impaired cortisol production). Critical hypotension has also developed in non-Addisonian patients within 7–10 days of starting rifampicin therapy (Boss 1983).
Anticoagulants	Corticosteroids may potentiate the effects of warfarin and can induce gastric ulceration with dangerous haemorrhage in anticoagulated patients (Koch-Weser & Sellers 1971). Ulceration may go unnoticed due to the euphoric effects of corticosteroids.
Antidiabetics	Antagonism of hypoglycaemic effects. Corticosteroids have intrinsic hyperglycaemic activity and may induce diabetes mellitus or may upset the control of the established diabetic patient (Taylor 1986, O'Byrne & Feely 1990).
Anti-emetics	Dexamethasone makes a significant contribution to the efficacy of ondansetron in the control of acute cisplatinum induced emesis (Smyth *et al.* 1991).
Anti-epileptics	Carbamazepine, phenobarbitone, phenytoin and primidone are hepatic microsomal P450 enzyme inducers and will accelerate the metabolism of corticosteroids and reduce their effects (Werk *et al.* 1969, Boylan *et al.* 1976, McLelland & Jack 1978).
Antihypertensives	The mineralocorticoid effects of corticosteroids, Na^+ and water retention may antagonize the effects of concomitant hypotensive therapy (Reynolds 1993b).
Aspirin	Corticosteroids decrease the blood salicylate concentration by increasing the glomerular filtration rate. Decreasing corticosteroid dosage in patients on aspirin may result in increased serum salicylate levels with the possibility of salicylism (Klinenberg & Miller 1965). Corticosteroids and aspirin are both ulcerogenic (Emmanuel & Montgomery 1971).
Bile acid-binding resins	Colestipol causes a significant impairment of oral hydrocortisone absorption (Nekl & Aron 1993).

Table 11.3 (Cont.)

Cardiacglycosides	Hypokalaemia due to corticosteroids potentiates the effect of cardiac glycosides (Reynolds 1993a).
Chlorpromazine	Chlorpromazine reduces gut motility and may enhance the absorption of corticosteroids (Forrest *et al.* 1970).
Cyclosporin	A mutual inhibition of metabolism occurs between cyclosporin and corticosteroids increasing the plasma concentration of both agents (Yee & McGuire 1990). Increased cyclosporin plasma levels increase the danger of nephrotoxicity (Calne 1980, Boogaerts *et al.* 1982, Durrant *et al.* 1982, Langhoff & Madsen 1983; Klintmalm & Säwe 1984, Öst *et al.* 1985).
Diuretics	Mineralocorticoids promote Na^+ and water retention and antagonize diuretic effects; acetazolamide, frusemide, other loop diuretics and thiazides increase the risk of hypokalaemia. Potassium-losing diuretics given together with corticosteroids may produce serious hypokalaemia (Reynolds 1993b).
Oral contraceptives	Oestrogens enhance the anti-inflammatory action of corticosteroids and also retard the metabolism of cortisol possibly by its increased binding to globulin (Spangler *et al.* 1969). Contraceptive failure has been reported in women using intra-uterine devices and using corticosteroid therapy (Anonymous 1983). There have been several reports of an enhanced effect of corticosteroids in women receiving oestrogens or oral contraceptives; the dose of corticosteroid may need to be reduced (Anonymous 1983, Back & Orme 1990).
Rifampicin	Induction of cytochrome P450 reduces the half-life of cortisol and steroid requirements are increased four-fold in Addison's disease (Edwards *et al.* 1974). Antitubercular rifampicin therapy in patients with renal transplants caused increased metabolism of corticosteroids and onset of signs of graft rejection (Buffington *et al.* 1976). Treatment of nephrotic syndrome by prednisolone in a child also receiving riampicin plus isoniazid failed and improved when the two antitubercular agents were stopped (Hendrickse *et al.* 1979, McAllister *et al.* 1983). The plasma clearance of prednisolone increased by 45% when rifampicin was given (McAllister *et al.* 1983), and prednisolone dosage had to be increased by 93% when asthmatic patients also took rifampicin (Powell-Jackson *et al.* 1983).

Table 11.3 (*cont.*)

Smallpox vaccination	Methotrexate plus corticosteroids can be fatal; smallpox vaccination is contra-indicated with this combination (Haim & Alroy 1967).
Sympathomimetics	Corticosteroids increase the risk of hypokalaemia if high doses are given with high doses of bambuterol, fenoterol, pirbuterol, reproterol, rimiterol, ritodrine, salbutamol, salmeterol, terbutaline and tulobuterol (British National Formulary 1994).
Ulcer-healing drugs	Comcomitant administration of corticosteroids, especially mineralocorticoids, with carbenoxolone increases the risk of hypokalaemia. Pseudoaldosteronism is a possible risk of therapy with carbenoxolone (Pinder *et al.* 1976).

Keeley 1992). Helfer & Rose (1989) have reviewed corticosteroids and adrenal suppression.

11.7 Drug Interactions Involving Corticosteroids

The spectrum of drug–drug interactions involving corticosteroids is limited and the mechanisms involved are largely well understood. For example, glucocorticoids can induce cytochromes P450 CYP3A3, 3A4, 3A5 and 3A7, and may compete for metabolism with a number of drug substances including nifedipine, cyclosporin, erythromycin, lidocaine, midazolam, quinidine and warfarin which are metabolized by the same enzyme systems (Gonzalez 1993). It is not surprising therefore that interactions between some of these drugs and individual corticosteroids have been reported in the clinic. Such interactions normally occur after oral or parenteral dosage since generally interactions do not occur when the steroids are used topically or by inhalation.

The pharmacokinetics of corticosteroids can be modified by concomitant use of P450-inducing (eg with phenytoin or rifampicin) or -inhibiting drugs (eg cimetidine) and these interactions are largely predictable. The glucocorticoid actions of corticosteroids will influence the actions of insulin and oral hypoglycaemics, and mineralocorticoid effects can, by Na^+ and water retention, reduce the effects of hypotensives and greatly influence the adverse effects of concomitant diuretics or other potassium-losing drugs. These effects may underlie significant clinical interactions. Many of these interactions were discovered during the early use of steroids and are now well recognized. Table 11.3 summarizes some of these established interactions. Generally, such interactions do not occur with inhaled corticosteroids or topical preparations unless dosage is high and prolonged. Interactions are largely confined to oral or systemically administered steroids.

References

AGGARWAL, R. K., POTAMITIS, T., CHONG, N. H. V., GUARRO, M., SHAH, P. & KETERPAL, S. (1993) Extensive visual loss with topical facial steroids. *Eye*, **7**, 664–666.

AICARDI, G., BENSO, L., VIGNOLO, M., TERRAGNA, A., VERRINA, E., CORDONE, G., COPPO, R., SERNIA, O., SARDELLA, M. L. & DI BATTISTA, E. (1993) Dose-dependent effects of deflazacort and prednisone on growth and skeletal maturation. *British Journal of Rheumatology*, **32**(Suppl. 2), 39–43.

ANONYMOUS (1983) Corticosteroids and oral contraceptives. *Drug Interaction News*, **3**, 48.

ANONYMOUS (1987) Do corticosteroids cause peptic ulcers? *Drug and Therapeutics Bulletin*. **25**, 41–43.

ATERMAN, K. & GREENBERG, S. M. (1954) Experimental exophthalmos produced by cortisone. *Archives of Ophthalmology*, **51**, 822–831.

BACK, D. J. & ORME, ML'E. (1990) Pharmacokinetic drug interactions with oral contraceptives. *Clinical Pharmacokinetics*, **18**, 472–484.

BAHAL, N. & NAHATA, M. C. (1991) The role of corticosteroids in infants and children with bacterial meningitis. *DICP Annals of Pharmacotherapy*, **25**, 542–545.

BARNES, P. J. (1995) Inhaled glucocorticoids for asthma. *New England Journal of Medicine*, **332**, 868–875.

BERNSTEIN, B., FLOMENBERG, P. & LETZER, D. (1994) Disseminated cryptococcal disease complicating steroid therapy for *Pneumocystis carinii* pneumonia in a patient with AIDS. *Southern Medical Journal*, **87**, 537–538.

BERNSTEIN, C. N., KOLODNY, M., BLOCK, E. & SHANAHAN, F. (1993) *Pneumocystis carinii* pneumonia in patients with ulcerative colitis treated with corticosteroids. *American Journal of Gastroenterology*, **88**, 574–577.

BLACK, R. L., OGLESBY, R. B., von SALLMANN, L. & BUNIM, J. J. (1960) Posterior subcapsular cataracts induced by corticosteroids in patients with rheumatoid arthritis. *Journal of the American Medical Association*, **174**, 166–171.

BONGIOVANNI, A. M. & MCPADDEN, A. J. (1960) Steroids during pregnancy and possible fetal consequences. *Fertility and Sterility*, **11**, 181–186.

BOOGAERTS, M. A., ZACHEE, P. & VERWILGHEN, R. L. (1982) Cyclosporin, methylprednisolone and convulsions. *Lancet*, **ii**, 1216–1217.

BOSS, G. (1983) Rifampicin and adrenal crisis. *British Medical Journal*, **287**, 52.

BOSTON COLLABORATIVE DRUG SURVEILLANCE PROGRAM (1972) Acute adverse reactions to prednisone in relation to dosage. *Clinical Pharmacology and Therapeutics*, **13**, 694–698.

BOYLAN, J. J., OWEN, D. S., JR & CHIN, J. B. (1976) Phenytoin interference with dexamethasone. *Journal of the American Medical Association*, **235**, 803–804.

BOZZETTE, S. A., SATTLER, F. R., CHIU, J., WU, A. W., GLUCKSTEIN, D., KEMPER, C., BARTOK, A., NIOSI, J., ABRAMSON, I., COFFMAN, J. *et al.* (21 authors) (1990) A controlled trial of early adjunctive treatment with corticosteroids for *Pneumocystis carinii* pneumonia in the acquired immunodeficiency syndrome. *New England Journal of Medicine*, **323**, 1451–1457.

BRAVER, D. A., RICHARDS, R. D. & GOOD, T. A. (1967) Posterior subcapsular cataracts in steroid-treated children. *Archives of Ophthalmology*, **77**, 161–162.

BREMS, J. J., HIATT, J. R., KLEIN, A. S., HART, J., EL-KHOURY, G., WINSTON, D., MILLIS, J. M. & BUSUTTIL, R. W. (1988) Disseminated aspergillosis complicating orthotopic liver transplantation for fulminant hepatic failure refractory to corticosteroid therapy. *Transplantation*, **46**, 479–481.

BRITISH NATIONAL FORMULARY (1994) Sympathomimetics, $Beta_2$, *British National Formulary*, 28th Edn, p. 527, London: Pharmaceutical Press.

BROWN, P., BLUNDELL, G., GREENING, A. & CROMPTON, G. (1990) Do large volume spacer devices reduce the systemic effects of high-dose inhaled corticosteroids? *Thorax*, **45**, 736–739.

BUFFINGTON, G. A., DOMINGUEZ, J. H., PIERING, W. F., HERBERT, L. E., KAUFFMAN, H. M., Jr & LEMANN, J., JR (1976) Interaction of rifampicin and glucocorticoids: adverse effects on renal allograph function. *Journal of the American Medical Association*, **236**, 1958–1960.

BUTCHER, J. M., AUSTIN, M., MCGALLIARD, J. & BOURKE, R. D. (1994) Bilateral cataracts and glaucoma induced by long term use of steroid eye drops. *British Medical Journal*, **309**, 43.

CALNE, R. Y. (1980) Cyclosporin. *Nephron*, **26**, 57–63.

CAPELL, H. (1992) Selected side-effects. 5. Steroid therapy and osteonecrosis. *Prescribers' Journal*, **32**, 32–34.

CAPEWELL, S., REYNOLDS, S., SHUTTLEWORTH, D., EDWARDS, C. & FINLAY, A. (1990) Purpura and dermal thinning associated with high-dose inhaled corticosteroid. *British Medical Journal*, **300**, 1548–1551.

CHARASSE, C., LIEGAUX, J. M., LE TULZO, Y., CAMUS, C., THOMAS, R. CARTIER, F., RAMEE, M. P. & ANDRE, P. (1992) Fatal co-infection with pulmonary tuberculosis and herpes simplex virus I pneumonia under steroid treatment. *European Respiratory Journal*, **5**, 108–111.

CHMEL, H. (1990) *Pneumocystis carinii* pneumonia: the steroid dilemma. *Archives of Internal Medicine*, **150**, 1793–1794.

COMMITTEE ON SAFETY OF MEDICINES/MEDICINES CONTROL AGENCY (1994) Severe chickenpox associated with systemic corticosteroids. *Current Problems in Pharmacovigilance*, **20**, 1–2.

COSTAGLIOLA, C., CATA-GIOVANNELLI, B., PICCIRILLO, A. & DELFINO, M. (1989) Cataracts associated with long-term topical steroids. *British Journal of Dermatology*, **12**, 472–473.

DAHL, M. G. C. (1985) Hazards of topical steroid therapy. *Adverse Drug Reaction Bulletin*, **No. 115 (December)**, 428–431.

DAVID, D. S. & BERKOWITZ, J. S. (1969) Ocular effects of topical and systemic corticosteroids. *Lancet*, **ii**, 149–151.

DECRAMER, M. & STAS, K. J. (1992) Corticosteroid-induced myopathy involving respiratory muscles in patients with chronic obstructive pulmonary disease or asthma. *American Review of Respiratory Disease*, **146**, 800–802.

DENNIS, M. & ITKEN, I. H. (1964) Effectiveness and complications of aerosol dexamethasone phosphate in severe asthma. *Journal of Allergy*, **35**, 70–76.

DOWELL, S. F. & BRESEE, J. S. (1993) Severe varicella associated with steroid use. *Pediatrics*, **92**, 223–228.

DRUG AND THERAPEUTICS BULLETIN (1995) Once-a-day topical corticosteroids. *Drug and Therapeutics Bulletin*, **33**, 21–22.

DUCKWORTH, R. (1973) Acute peridontal conditions – their pathogenesis and management. *British Dental Journal*, **135**, 168–169.

DURRANT, S., CHIPPING, P. M., PALMER, S. GORDON-SMITH, E.C. (1982) Cyclosporin A, methylprednisolone and convulsions. *Lancet*, **ii**, 829–830.

EDITORIAL (1982) Steroids in bacterial meningitis – helpful or harmful? *Lancet*, **i**, 1164.

EDWARDS, O. M., COURTENAY-EVANS, R. J., GALLEY, J. M. HUNTER, J. & TAIT, A. D. (1974) Changes in cortisol metabolism following rifampicin therapy. *Lancet*, **ii**, 549–551.

ELANSARY, E. H. & EARIS, J. E. (1983) Rifampicin and adrenal crisis. *British Medical Journal*, **286**, 1861–1862.

EMMANUEL, J. H. & MONTGOMERY, R. D. (1971) Gastric ulcer and anti-arthritic drugs. *Postgraduate Medical Journal*, **47**, 227–232.

ERILL, S. (1991) Corticotrophins and corticosteroids. In DUKES, M. N. G. & ARONSON, J. K. (Eds), *Side Effects of Drugs Annual*, pp. 418–425, Amsterdam: Elsevier.

ESPERSEN, E. (1963) Corticosteroids and pulmonary tuberculosis. Activation of four cases. *Acta Tuberculosea et Pneumologica Belgica*, **43**, 1–8.

FELDMAN, S. A. (1963) Effect of changes in electrolytes, hydration and pH upon the reactions to muscle relaxants. *British Journal of Anaesthesia*, **35**, 546–551.

FINDLAY, G. M. & HOWARD, E. M. (1952) The effects of cortisone and adrenocorticotrophic hormone on poliomyelitis and on other virus infections. *Journal of Pharmacy and Pharmacology*, **4**, 37–42.

FLETCHER, A. P. (1986) Drug-induced endocrine dysfunction. In D'ARCY, P. F. & GRIFFIN, J. P. (Eds), *Iatrogenic Diseases*, 3rd Edn, pp. 358–381, Oxford: Oxford University Press.

FORREST, F. M., FORREST, I. S. & SERRA, M. T. (1970) Modification of chlorpromazine metabolism by some other drugs frequently administered to psychotic patients. *Journal of Biological Psychiatry*, **2**, 53–58.

FRASER, C. G., PREUSS, F. S. & BIGFORD, W. D. (1952) Adrenal atrophy and irreversible shock associated with cortisone therapy. *Journal of the American Medical Association*, **149**, 1542–1543.

FÜRST, C., SMILEY, W. K. & ANSELL, B. M. (1966) Steroid cataract. *Annals of Rheumatic Diseases*, **25**, 364–367.

GENNARI C. (1993) Differential effect of glucocorticoids on calcium absorption and bone mass. *British Journal of Rheumatology*, **32**, (Suppl. 2), 11–14.

GIBSON, A. T., PEARSE, R. G. & WALES, J. K. (1993) Growth retardation after dexamethasone administration: assessment of knemometry. *Archives of Diseases in Childhood*, **69**, (5 Spec. No.), 505–509.

GONZALEZ, F. J. (1993) Cytochrome P450 in human. In SCHENKMAN J. B. & GREIM, H. (Eds), *Cytochrome P450*, pp. 239–257, Heidelberg: Springer.

GRAHAME-SMITH, D. G. & ARONSON, J. K. (1984) Corticosteroids, In *Oxford Textbook of Clinical Pharmacology and Drug Therapy*, pp. 664–669, Oxford: Oxford University Press.

GUIN, J. D. (1981) Complications of topical hydrocortisone. *Journal of the American Academy of Dermatology*, **4**, 417–422.

GUSLANDI, M. & TITTOBELLO, A. (1992) Steroid ulcers, a myth revisited. *British Medical Journal*, **304**, 655–656.

HAIM, S. & ALROY, G. (1967) Methotrexate in psoriasis. *Lancet*, **i**, 1165.

HALL, A. P. (1976) The treatment of malaria. *British Medical Journal*, **1**, 323–328.

HALL, R. C., POPKIN, M. K., STICKNEY, S. K. & GARDNER, E. R. (1979) Presentation of the steroid psychoses. *Journal of Nervous and Mental Disease*, **167**, 229–236.

HELFER, E. L. & ROSE, L. I. (1989) Corticosteroids and adrenal suppression: characterizing and avoiding the problem. *Drugs*, **38**, 838–845.

HENDRICKSE, W., MCKIERNAN, J., PICKUP, M. & LOWE, J. (1979) Rifampicin-induced non-responsiveness to corticosteroid treatment in nephrotic syndrome. *British Medical Journal*, **1**, 306.

HIEBER, J. P. & NELSON, J. D. (1977) A pharmacologic evaluation of penicillin in children with purulent meningitis. *New England Journal of Medicine*, **297**, 410.

HORNE, N. W. (Ed.) (1990) *Modern Drug Treatment of Tuberculosis*, 7th Edn, London: Chest, Heart and Stroke Association.

HUGHES, I. A. (1987) Steroids and growth. *British Medical Journal*, **295**, 683–684.

HUGHES, J. R., HIGGINS, E. M. & DU VIVER, A. W. P. (1992) Acne associated with inhaled glucocorticosteroids. *British Medical Journal*, **305**, 1000.

KAIDBEY, K. H. & KLIGMAN, A. M. (1974) The pathogenesis of topical steroid acne. *Journal of Investigative Dermatology*, **62** 31–36.

KAYE, L. D., KALENAK, J. W., PRICE, R. L. & CUNNINGHAM, R. (1993) Ocular implications of long-term prednisone therapy in children. *Journal of Pediatric Ophthalmology and Strabismus*, **30**, 142–144.

KEELEY, D. (1992) Large volume plastic spacers in asthma: should be used more. *British Medical Journal*, **305**, 598–599.

KLEIN, J. F. (1992) Adverse psychiatric effects of systemic glucocorticoid therapy (Review). *American Family Physician*, **46**, 1469–1474.

KLINENBERG, J. R. & MILLER, F. (1965) Effect of corticosteroids on blood salicylate concentrations. *Journal of the American Medical Association*, **194**, 601–604.

KLINTMALM, G. & SÄWE, J. (1984) High dose methylprednisolone increases plasma cyclosporin levels in renal transplant recipients. *Lancet*, **i**, 731.

KOCH-WESER, J. & SELLERS, E. M. (1971) Drug interactions with coumarin anticoagulants. *New England Journal of Medicine*, **285**, 487–498, 547–558.

KOVACS, J. A. & MASUR, H. (1990) Are corticosteroids beneficial as adjunctive therapy for pneumocystis pneumonia in AIDS? *Annals of Internal Medicine*, **113**, 1–3.

LAAN, R. F., VAN RIEL, P. L., VAN ERNING, L. J., LEMMENS, J. A., RUIJS, S. H. & VAN DE PUTTE, L. B. (1992) Vertebral osteoporosis in rheumatoid arthritis patients: effect of low dose prednisone therapy. *British Journal of Rheumatology*, **31**, 91–96.

LAAN, R. F., BUIJS, W. C., VAN ERNING, L. J., LEMMENS, J. A., CORSTENS F. H., RUIJS, S. H., VAN DE PUTTE, L. B. & VAN RIEL, P. L. (1993) Differential effects of glucocorticoids on cortical appendicular and cortical vertebral bone mineral content. *Calcified Tissue International*, **52**, 5–9.

LANGHOFF, E. & MADSEN, S. (1983) Rapid metabolism of cyclosporin and prednisone in a kidney transplant patient receiving tuberculostatic treatment. *Lancet*, **ii**, 1031.

LUKERT, B. P. & RAISZ, L. G. (1990) Glucocorticoid-induced osteoporosis: pathogenesis and management. *Annals of Internal Medicine*, **112**, 352–364.

MACKAY, S. & EISENDRATH, S. (1992) Adverse reaction to dental corticosteroids. *General Dentistry*, **40**, 136–138.

MAHAFFEY, K. W., HIPPENMEYER, C. L., MANDEL, R. & AMPEL, N.M. (1993) Unrecognized coccidiomycosis complicating *Pneumocystis carinii* pneumonia in patients infected with the human immunodeficiency virus and treated with corticosteroids. A report of two cases. *Archives of Internal Medicine*, **153**, 1496–1498.

MCALLISTER, W. A. C., THOMPSON, P. J., AL-HABET, S. M. & ROGERS, H. J. (1983) Rifampicin reduces effectiveness and bioavailability of prednisolone. *British Medical Journal*, **286**, 923–925.

MCGOWAN, J. E., CHESNEY, P. J., CROSSLEY, K. B. & LAFORCE, F. M. (1992) Report by the Working Group on Steroid Use. Antimicrobial Agents Committee, Infectious Diseases Society of America: guidelines for the use of systemic glucocorticoids in the management of selected infections. *Journal of Infectious Diseases*, **165** 1–13.

MCLELLAND, J. & JACK, W. (1978) Phenytoin/dexamethasone interaction: a clinical problem. *Lancet*, **i**, 1096–1097.

NATIONAL INSTITUTES OF HEALTH - UNIVERSITY OF CALIFORNIA (1990) Expert Panel for Corticosteroids as Adjunctive Therapy for Pneumocystis Pneumonia. Consensus statement on the use of corticosteroids as adjunctive therapy for pneumocystis pneumonia in the acquired immunodeficiency syndrome, *New England Journal of Medicine*, **323**, 1500–1504.

NEKL, K. E. & ARON, D. C. (1993) Hydrocortisone-colestipol interaction. *Annals of Pharmacotherapy*, **27**, 980–981.

NIELSEN, N. & SORENSEN, P. N. (1978) Glaucoma induced by application of corticosteroids to the periorbital region. *Archives of Dermatology*, **114**, 953–954.

NIXON, J. E. (1984) Early diagnosis and treatment of steroid induced avascular necrosis of bone. *British Medical Journal*, **288**, 741–744.

NORRIS, S. H., BUTLER, T. C., GLASS, N. & TRAN, R. (1989) Fatal hepatic necrosis caused by disseminated type 5 adenovirus in a renal transplant recipient. *American*

Journal of Nephrology, **9**, 101–105.

O'BYRNE, S. & FEELY, J. (1990) Effects of drugs on glucose tolerance in non-insulin dependent-diabetics (Part 1). *Drugs*, **40**, 6–18.

O'DONNELL, J. (1989) Adverse effects of corticosteroids. *Journal of Pharmacy Practice*, **11**, 256–266.

OGLESBY, R. B., BLACK, R. L., VON SALLMANN, L. & BUNIM, J.J. (1961) Cataracts in rheumatoid arthritis patients treated with corticosteroids. *Archives of Ophthalmology*, **66**, 519–523.

OPPENHEIMER, E. H. (1964) Lesions in the adrenals of an infant following maternal corticosteroid therapy. *Bulletin of the Johns Hopkins Hospital*, **114**, 145–151.

ÖST, L., KLINTMALM, G. & RIGDÉN, O. (1985) Mutual interaction between prednisolone and cyclosporine in renal transplant patients. *Transplantation Proceedings*, **17**, 1252–1255.

PINDER, R. M., BROGDEN, R. N., SAWYER, P. R., SPEIGHT, T. M., SPENCER, R. & AVERY, G. S. (1976) Carbenoxolone: a review of its pharmacological properties and therapeutic efficacy in peptic ulcer disease. *Drugs*, **11**, 245–307.

PIPER, J. M., RAY, W. A., DAUGHERTY, J. R. & GRIFFIN, M. R. (1991) Corticosteroid use and peptic ulcer disease: role of nonsteroidal anti-inflammatory drugs. *Annals of Internal Medicine*, **114**, 735–740.

PLEWIG, G. & KLIGMAN, A. M. (1973) Induction of acne by topical steroids. *Archiv Für Dermatologische Forschung*, **247** 29–52.

POLITO, C. & DI TORO, R. (1992) Delayed pubertal growth spurt in glomerulopathic boys receiving alternate-day prednisone. *Child Nephrology and Urology*, **12**, 202–207.

POWELL-JACKSON, P. R., GRAY, B. T., HEATON, R. W., COSTELLO, J. F., WILLIAMS, R. & ENGLISH, J. (1983) Adverse effect of rifampicin on steroid-dependent asthma. *American Review of Respiratory Disease*, **128**, 307–310.

PRAHL, P. & JENSEN, T. (1987) Decreased adrenocortical suppression utilising the nebuhaler for inhalation of steroid aerosols. *Clinical Allergy*, **17**, 393–398.

PRIFTIS, K., EVERARD, M. L. & MILNER, A. D. (1991) Unexpected side-effects of inhaled steroids: a case report. *European Journal of Pediatrics*, **150**, 448–449.

REYNOLDS, J. E. F. (Ed.) (1993a) Digoxin: precautions. In *Martindale. The Extra Pharmacopoeia*, 30th Edn, pp. 666–667, London: Pharmaceutical Press.

REYNOLDS, J. E. F. (Ed.) (1993b) Precautions for corticosteroids. In *Martindale, The Extra Pharmacopoeia*, 30th Ed, p. 714, London: Pharmaceutical Press.

RIZZATO, G. & MONTEMURRO, L. (1993) Reversibility of exogenous corticosteroid-induced bone loss. *European Respiratory Journal*, **6**, 116–119.

SATTLER, F. R. (1991) Who should receive corticosteroids as adjunctive treatment for *Pneumocystis carinii* pneumonia? *Chest*, **99**, 1058–1061.

SEAL, J. P. & COMPTON, M. R. (1986) Side-effects of corticosteroid agents. *Medical Journal of Australia*, **144**, 139–142.

SHUTTLEWORTH, D., EDWARDS, C., CAPEWELL, S., REYNOLDS, S. & FINDLAY, A. Y. (1990) Inhaled corticosteroids and skin thinning. *British Journal of Dermatology*, **122**, 268.

SKALA, H. W. & PRCHAL, J. T. (1980) Effect of corticosteroids on cataract formation. *Archives of Ophthalmology*, **98**, 1773–1777.

SLANSKY, H. H., KOLBERT, G. & GARTNER, S. (1976) Exophthalmos induced by steroids. *Archives of Ophthalmology*, **77**, 579–581.

SMYTH, J. F., COLEMAN, R. E., NICOLSON, M., GALLMEIER, W. M., LEONARD, R. C. F., CORNBLEET, M. A., ALLAN, S. G., UPADHYAYA, B. K. & BRUNTSCH, U. (1991) Does dexamethasone enhance control of acute cisplatin induced emesis by ondansetron? *British Medical Journal*, **303** 1423–1426.

SPANGLER, A. S., ANTONIADES, H. N., SOTMAN, S. L. & INDERBITIZIN, T. M. (1969) Enhancement of the anti-inflammatory actions of hydrocortisone by estrogen. *Journal of*

Clinical Endocrinology and Metabolism, **29**, 650–655.

STEAD, R. J. & COOK, N. J. (1989) Adverse effects of inhaled corticosteroids. *British Medical Journal*, **298**, 403–404.

TAYLOR, R. (1986) Drugs and glucose tolerance. *Adverse Drug Reactions Bulletin*, **No. 127 (December)**, 452–455.

TRAVLOS, A. & HIRSCH, G. (1993) Steroid psychosis: a cause of confusion on the acute spinal cord injury unit. *Archives of Physical Medicine and Rehabilitation*, **74**, 312–315.

TRIPATHI, R. C., KIPP, M. A., TRIPATHI, B. J., KIRSCHNER, B. S., BORISUTH, N. S., SHEVELL, S. K. & ERNEST, J. T. (1992) Ocular toxicity of prednisone in pediatric patients with inflammatory bowel disease. *Lens Eye Toxicity Research*, **9**, 469–482.

VENKATESAN, K. (1992) Pharmacokinetic drug interactions with rifampicin. *Clinical Pharmacokinetics*, **22**, 47–65.

VARMA, P. P., CHUGH, S., GUPTA, K. L., SAKHUJA, V. & CHUGH, K.S. (1993) Invasive pulmonary aspergillosis and nocardiosis in an immunocompromised host. *Journal of the Associations of Physicians of India*, **41**, 237–238.

VIE, R. (1990) Glaucoma and amaurosis associated with longterm application of topical corticosteroids to the eyelids. *Acta Dermato-Venereologica (Stockholm)*, **60**, 541–542.

WARRELL, D. A., LOOAREESUWAN, S., WARRELL, M. J., KASEMSARN, P., INTARAPRAESERT, R., BUNNAG, D. & HARINASUTA, T. (1982) Dexamethasone proves deleterious in cerebral malaria. A double-blind trial in 100 comatose patients. *New England Journal of Medicine*, **306**, 313–319.

WERK, E. E., CHOI, Y., SHOLITON, L., OLINGER, C. & HAGUE, N. (1969) Interference in the effect of dexamethasone by diphenylhydantoin. *New England Journal of Medicine*, **281**, 32–34.

WILKINSON, S. M. & ENGLISH, J. S. (1992) Hydrocortisone sensitivity: clinical features of fifty-nine cases. *Journal of the American Academy of Dermatology*, **25**, 683–687.

WILLIAMS, A. W. (1953) Exophthalmos in cortisone-treated experimental animals. *British Journal of Experimental Pathology*, **34**, 621–624.

WILLIAMS, A. W. (1995) Pathogenesis of cortisone-induced exophthalmos in guinea-pigs. *British Journal of Experimental Pathology*, **36**, 245–247.

WOLTHERS, O. D. & PEDERSEN, S. (1992) Controlled study of linear growth in asthmatic children during treatment with inhaled glucocorticosteroids. *Pediatrics*, **89**, 839–842.

YEE, G. C. & MCGUIRE, T.R. (1990) Pharmacokinetic drug interactions with cyclosporin (Part II). *Clinical Pharmacokinetics*, **19**, 400–415.

ZWAAN, C. M., ODINK, R. J. H., DELEMARRE-VAN DE WAAL, H. A., DANKERT-ROELSE, J. E. & BOKMA, J. A. (1992) Acute adrenal insufficiency after discontinuation of inhaled corticosteroid therapy. *Lancet*, **340**, 1289–1290.

12

Adverse Drug Reactions and the Adrenal Glands

A Pharmacoepidemiological Approach

RONALD D. MANN

Drug Safety Research Unit, Southampton

12.1 A Pharmacoepidemiological Approach

Pharmacoepidemiology is the study of the use of, and the effects of, drugs in large numbers of people [1]. It involves the use of hypothesis-generating techniques which signal possible adverse drug reactions (ADRs) and hypothesis-testing techniques which, in formal studies, confirm or refute such signals. The data presented in this chapter originate from the UK but the same messages appear in the data of the Food and Drugs Administration (FDA) in the USA and the World Health Organization's (WHO) Collaborating Centre at Uppsala, Sweden.

In the UK there are two principal methods of generating hypotheses regarding possible ADRs: these are the yellow card scheme run by the Committee on Safety of Medicines (CSM) and Prescription-Event Monitoring (PEM) run by the Drug Safety Research Unit (DSRU) at Southampton. There are also two principal methods of hypothesis-testing: these are the studies conducted, using record-linkage, by the Medicines Evaluation and Monitoring Organization (MEMO) in Tayside, Scotland, and studies using the computerized general practitioner research database now known as GPRD but originally known as the VAMP database. These hypothesis-testing systems will not be further described as the data to be discussed originate from the database of the CSM.

The yellow card scheme [2] began in 1964 in response to the thalidomide disaster which produced an estimated 10 000 deformed babies in the countries in which this drug was widely used in early pregnancy [3]. Yellow cards (in various forms) are widely available to all prescribers in the UK; advice on the reporting of suspected ADRs is equally widely available. The great advantage of the scheme is that it allows the reporting of suspected reactions occurring with any drug on the UK market and throughout the whole of the lifetime of each licensed medicine. The great disadvantage is that there is gross under-reporting so that the number of reports is only a very poor indicator of the true incidence of the reaction. Nevertheless, the scheme is of proven worth and will, for the foreseeable future, represent the only affordable means of detecting rare ADRs.

PEM relies on the fact that all National Health Service (NHS) prescriptions in England are sent, by the dispensing pharmacists, to a central Prescription Pricing

Authority (PPA). The DSRU receives, under confidential cover, electronic copies of all of the prescriptions for the newly-marketed drugs which are being monitored. Questionnaires (green forms) are then, after an appropriate interval, posted to the doctors who originated the prescriptions for each individual patient. These green forms seek a number of items of information, including a dated account of any events which may have occurred in the individual subjects. These data, when computerized, are linked with the prescribing data in order to look for events or clusters of events which might represent ADRs. The great advantage of the method is that it allows rapid monitoring of the early 'real world' experience of the drug and it does this in large cohorts of patients. The principal disadvantage is that not all of the doctors complete and return the green forms so that response rates vary from approximately 50 to 75%.

The monitoring of the safety of medicines requires all of these methods which serve different purposes because of their contrasting strengths and weaknesses. However, the pharmacoepidemiological data of greatest interest in respect of ADRs and the adrenals are those of the yellow card scheme.

In keeping with this pharmacoepidemiological approach, this chapter will be concerned only with the clinical literature and will omit consideration of the very extensive literature on animal and laboratory studies. It will also largely omit reports on the treatment of neoplasia affecting the adrenal glands for in most of these studies the effects are intended and do not represent ADRs.

12.2 Anatomy and Physiology of the Adrenal Glands

The adult adrenal glands each weigh 4–5 g. They are situated at the upper poles of the kidneys. The structure and principal hormonal products are summarized in Table 12.1.

Before birth the adrenals are much larger and the cortex is dominated by an innermost fetal zone which degenerates soon after birth. The zona glomerulosa (the outermost active layer of the cortex) synthesizes and secretes mineralocorticoids, principally aldosterone. The zona fasciculata produces cortisol (hydrocortisone). The zona reticularis and the fetal zone synthesize the weak adrenal androgens, principally DHEA (dehydroepiandrosterone) and its sulphate. The differentiation of function between the three zones of the cortex depends on the presence or absence of a few specific enzymes. Oestrogens are known to impair cortisol synthesis

Table 12.1 The hormonal products of the adrenal glands

Gross structure	Constituent parts	Hormonal products
Cortex	Capsule	–
	Zona glomerulosa	Aldosterone
	Zona fasciculata	Cortisol
	Zona reticularis	DHEA
Medulla		Adrenaline

Note: DHEA, Dehydroepiandrosterone.

(although there has been only one yellow card between 1990 and 1995 reporting adrenal insufficiency in association with the use of oestrogens; in this case the drug cited was ethinyloestradiol and the outcome was fatal).

The mineralocorticoids of the adrenal are physiologically important. The enzyme renin is secreted by the juxtaglomerular apparatus of the kidney. Renin acts upon a circulating substrate of hepatic origin to release the decapeptide, angiotensin I. Angiotensin I is then converted by angiotensin-converting enzymes (ACE) in the tissues, including the lung, to produce the active octapeptide, angiotensin II. The synthesis and secretion of aldosterone are controlled by a number of factors, principally the concentration of angiotensin II and the extracellular sodium and potassium concentrations. Aldosterone itself acts on intracellular mineralocorticoid receptors in the distal renal tubule to promote sodium uptake and potassium excretion.

12.3 Pathways to Adverse Reactions Affecting the Adrenals

These pathways provide potential for important drug interactions and adverse drug events. The angiotensin-converting enzyme inhibitors (ACE inhibitors), such as captopril, enalapril, lisinopril and ramipril, inhibit the conversion of angiotension I to angiotensin II. They are used in the treatment of hypertension and heart failure and, although effective and generally well tolerated, they can cause very rapid hypotension in some patients. For this reason ACE inhibitor therapy in patients with severe heart failure should be initiated in hospital; initiation in hospital is also recommended for patients with mild to moderate heart failure in circumstances which include pre-existing electrolyte disturbances, hypotension, high-dose diuretic therapy, renal impairment and age 70 years or greater. Heart failure patients can also suffer dangerous hyperkalaemia (especially if potassium-sparing diuretics have been used). Profound first-dose hypotension and substantial deterioration of renal function have also been fairly frequently described in these patients but are not always directly drug-attributable. Thus it is essential to be familiar with the prescribing information when using drugs of this important and very useful class.

The glucocorticoids, principally cortisol (hydrocortisone) in man, markedly affect glucose homeostasis; they also have catabolic effects on muscle and they enhance hepatic gluconeogenesis. They inhibit the immune system, the processes of repair and the synthesis of collagen in bones and soft tissues. Their endocrine effects are very diverse for they affect the action of many other hormones. If they are absent the response of stress of any kind is grossly diminished in a way that is of considerable clinical importance.

In the adult the weak adrenal androgens seem to be of only very limited importance.

The adrenal medulla receives a sympathetic nerve supply which causes the secretion of adrenaline in response to stress. The blood supply of the medulla comes largely from the adrenal cortex and is rich in cortisol. The last stage of the synthesis of adrenaline is cortisol-dependent so that pituitary corticotrophin (ACTH) and adrenal cortisol insufficiency results in adrenaline deficiency. Glucocorticoids and adrenaline interact at a number of target tissues and in asthma synthetic analogues of these hormones are used together.

The trophic hormones driving adrenal activity can also have profound clinical effects. Chronic ACTH excess produces hypertrophy of the adrenal cortex and alterations in the blood vessels. Multiple nodules can develop in the adrenal glands and become almost autonomous; they can also undergo malignant change. Chronic ACTH insufficiency results in adrenal atrophy, the clinical picture being mainly of glucocorticoid, rather than mineralocorticoid, deficiency.

12.4 Diseases of the Adrenal Cortex and Relevant Adverse Drug Reactions

12.4.1 *Adrenocortical Insufficiency*

Causative conditions include ACTH deficiency, congenital enzyme disorders and acute or chronic destruction of the adrenal. Acute destruction (eg in meningococcal septicaemia) produces the Waterhouse-Friderichsen syndrome which may be associated with massive circulatory collapse and death. Chronic destruction of the adrenals is most frequently due to tuberculosis or autoimmune disease and produces the features of Addison's disease.

Thomas Addison (1793–1860), an English physician at Guy's Hospital, London, UK, was a contemporary of Bright and Hodgkin. His monograph 'On the constitutional and local effects of disease of the suprarenal capsule' was published in 1855 and represents the beginning of endocrinology. The key clinical features of Addison's disease (as it was named by Trousseau in Paris) are pigmentation in the mouth, skin creases and pressure areas; hypotensive crises accompanied by salt and water loss, hyperkalaemia, hypercalcaemia and hypoglycaemia; and associated vitiligo, myxoedema or pernicious anaemia.

Drugs can cause adrenocortical insufficiency: glucocorticoid therapy is a major cause of secondary insufficiency and is a substantial clinical problem; glucocorticoid metabolism can be enhanced by drugs such as rifampicin and carbamazepine.

Yellow card reports to the CSM from 1964 to 1995 inclusive provide the totals for adrenal insufficiency given in Table 12.2.

The numbers given in Table 12.2 are small and, of course, one or two reports are seldom helpful, but where the evidence is a little stronger most of the reports are associated with the systemic or topical use of glucocorticoids. The glucocorticoids mentioned in Table 12.2 are prednisolone, clobetasol, cortisone, beclomethasone, betamethasone and prednisone.

The glucocorticoids, as a class, include fludrocortisone (the mineralocorticoid activity of which is so high that its anti-inflammatory activity is of no clinical relevance); hydrocortisone and cortisone (which have marked mineralocorticoid effects so that they are not used systemically for long-term anti-inflammatory purposes though hydrocortisone is used for adrenal replacement therapy and intravenously in emergencies such as anaphylaxic shock and topically in the management of inflammatory skin conditions); prednisolone and prednisone (both of which have predominantly glucocorticoid activity although prednisone is only active after conversion to prednisolone in the liver); betamethasone and dexamethasone (which have very high glucocorticoid and very low mineralocorticoid activities so that they are used when water retention would be a disadvantage) and beclomethasone (which, like some of the esters of betamethasone, exerts a marked topical effect on the skin and lungs so that it is used in dermatology and in the management

Table 12.2 Reports to the CSM on adrenal insufficiency

	Single products		Multiple products	
Drug	Total	Fatal	Total	Fatal
Prednisolone	10	4	0	0
Clobetasol	8	0	0	0
Cortisone	6	5	0	0
Beclomethasone	5	0	1	0
Etomidate	5	0	0	0
Rifampicin	3	0	0	0
Betamethasone	2	1	1	0
Busulphan	2	1	0	0
Ketoconazole	2	0	0	0
Prednisone	2	0	0	0
Salbutamol	1	0	1	0
Others[a]	28	2	5	2
Total	74	13	8	2

Note: [a] Single reports only with each drug. Source: CSM (personal communication).

of asthma by inhalation). There is a very extensive literature on adrenal suppression (and related investigations) associated with the use of these steroids given orally, systematically or intramuscularly [4–32] in adults or children. Similar studies involving intra-articular, periocular, extradural and intrathecal use have also been reported [33–38]. Comparable observations have been made following the inhalational [39–64] and topical or cutaneous [65–75] use of these agents. Clobetasol occupies the second position in Table 12.2. This is a highly potent steroid used in restricted amounts for the short-term treatment only of severe, resistant, inflammatory skin disorders unresponsive to less potent corticosteroids. Studies of this drug and its potentiality for causing adrenal suppression have been reported not only to the CSM but also elsewhere [76–77].

Etomidate, an anaesthetic induction agent, also appears in Table 12.2. It has been known since 1983 that this agent can have an undesirable suppressant effect on adrenocortical function [78–97]. Other reports have appeared suggesting that various anaesthetic agents can, in some circumstances, have similar effects [98–101].

The effect of rifampicin, (the subject of three reports in the CSM data provided in Table 12.2) on steroid metabolism has already been mentioned. This agent is now regarded as a key component of almost all antituberculous regimens. The drug induces hepatic enzymes which accelerate the metabolism of several drugs, including corticosteroids. The degree of adverse effect can reach the proportions of an Addisonian crisis [102–106].

Ketoconazole is noted above as being the subject of two reports to the CSM. This is an imidazole antifungal better absorbed than most drugs of its therapeutic class but associated with a well-established risk of fatal hepatotoxicity. The literature is, however, fairly persuasive [107–118] that the drug can cause adrenal suppression and there are reports of its use in the treatment of Cushing's syndrome.

Busulphan (an alkylating agent most frequently used in chronic myeloid leukaemia) and salbutamol (an anti-asthma drug) complete Table 12.2. In both instances confounding by indication or the effects of co-administered agents may account for the infrequent reports to the CSM (although the early report on structure–activity relationships for ulcerogenic and adrenocorticolytic effects of alkyl nitriles, amines and thiols of Szabo and Reynolds [119] is of interest in this regard).

Other drugs which do not appear in Table 12.2 but which can cause adrenal suppression are aminoglutethimide (which has largely replaced adrenalectomy in postmenopausal women with advanced breast cancer [120–133] and trilostane, which has similar uses. Both are powerful adrenal antagonists capable of producing a 'medical adrenalectomy' so that replacement therapy may be needed. These drugs have also been used in treating advanced cancer of the prostate and in selected cases of Cushing's syndrome. Medroxyprogesterone acetate and megestrol acetate are progestogens largely used as second- or third-line therapy in breast cancer; glucocorticoid effects at high dose may lead to a cushingoid effect but adrenal suppression can also occur [134–136]. Tetracosactrin, an analogue of corticotrophin (ACTH), is used to test adrenocortical function; ACTH is no longer commercially available in the UK but the literature contains reports of studies of adrenal responsiveness to ACTH and its analogues [137–140]. A drug not much met with in developed countries is Suramin, a trypanocide used in the treatment of the early stages of African trypanosomiasis and as an antihelmintic in the treatment of onchocerciasis; it has been tried in the treatment of AIDS. It is given by slow intravenous injection and can cause severe immediate reactions. There have been a number of reports of adrenal insufficiency associated with its use [141–145]. Cyproterone acetate is an anti-androgen used in the treatment of severe hypersexuality and sexual deviation in the male. It is also used in prostatic cancer and in the treatment of severe acne and hirsutism in women. Precautions in its use include monitoring adrenocortical function regularly. Reports are concerned with secondary adrenal insufficiency due to the drug and with the site of action of its anti-adrenal steroidogenic effect; a number of authors have also explored its use in different indications [146–157].

Adrenal suppression can, of course, reach the proportions of an Addisonian crisis with many of these agents. Acute adrenal insufficiency after the discontinuation of inhaled steroids has been reported, as has primary adrenocortical failure masked by the administration of exogenous steroids [158]. Acute adrenal failure can follow haemorrhage into the adrenal and such bleeding has been reported following the use of ACTH [159–163] or anticoagulants [164–173].

12.4.2 *Adrenocortical Excess*

Excess activity of the adrenal cortex (Cushing's syndrome) ie either ACTH-dependent, ACTH-independent or of a mixed aetiology which can include the alcohol-induced pseudo-Cushing's syndrome. ACTH-dependent causes include those of pituitary origin (eg the pituitary basophil adenoma of Cushing's disease), those due to ACTH secretion from ectopic sites (eg oat cell carcinoma, pancreatic carcinoma or phaeochromocytoma) and the iatrogenic effects of ACTH therapy. Doctors, by using drugs which produce adrenocortical excess, probably cause a large proportion of cases of Cushing's syndrome. ACTH-independent causes include adrenal adenoma or carcinoma and excessive glucocorticoid therapy. The administration of such therapy is the other way that doctors cause Cushing's syndrome.

Harvey Cushing (1869–1939), a US neurosurgeon, is credited with having reduced the mortality rate in brain surgery from 90 to 8%. He published his monograph on 'The pituitary body and its disorders' in 1912. One of his accomplishments was his two-volume study of 'The life of Sir William Osler' (1926). This study, one of the very finest of medical biographies, won him the Pulitzer Prize.

The main clinical features of Cushing's syndrome are central obesity, 'moon face', 'buffalo hump', unenlarged wrists and ankles, proximal myopathy, osteoporosis, abdominal striae and deficient wound healing, depression and psychosis, hypertension and associated diabetes mellitus. Many of these features appear alone or together when glucocorticoids (eg hydrocortisone) are administered in excessive quantities.

Yellow card reports to the CSM from 1964 to 1995 inclusive provide the totals for suspected cases of Cushing's syndrome and hyperadrenalism given in Table 12.3.

Again, one or two cases are seldom reliably informative. The significant role played by the glucocorticoids is to be expected. Beclomethasone, clobetasol, betamethasone and prednisolone together account for 23 of the 38 yellow card reports given in Table 12.3 for drugs used as single (non-combination) products. The literature contains many studies of hyperadrenalism resulting from the systemic [174–185] or topical [186–202] use of glucocorticoids; studies involving ACTH and its effects in this context are also numerous [203–207].

Medroxyprogesterone (Depo-Provera) is a long-acting progestogen given by injection and, whilst most often used as a contraceptive, it also finds uses in some forms of malignant disease and in the management of menstrual disorders. The literature contains three reports of Cushing's syndrome induced by medroxyprogesterone and a study of the possible place of the drug in male contraception [208–211].

Ethinyloestradiol completes the drugs individually listed in Table 12.3; the two relevant reports are concerned with combination products. In addition to this information, studies are available in the literature reporting increased adrenocortical

Table 12.3 Reports to the CSM on hyperadrenalism

	Single products		Multiple products	
Drug	Total	Fatal	Total	Fatal
Beclomethasone	8	0	0	0
Clobetasol	7	0	0	0
Betamethasone	6	0	0	0
Medroxyprogesterone	4	0	0	0
Ethinyloestradiol	0	0	2	0
Prednisolone	2	0	0	0
Others[a]	11	0	2	0
Total	38	0	4	0

Note: [a] Single reports only with each drug. Source: CSM (personal communication).

responsiveness to exogenous ACTH in oral contraceptive users; the effect of oral contraceptive treatment of the serum concentration of dehydroisoandrosterone sulphate, and the effect of these contraceptives on adrenal cortical function [212–215]. However, it seems unlikely that oral contraceptives cause hyperadrenalism or Cushing-like states.

Mixed pictures have been reported and Young *et al.* from Liverpool, UK have entered a case of unrecognized Cushing's syndrome and adrenal suppression due to topical clobetasol propionate into the literature [216]. Alcohol can also produce a well-known pseudo-Cushing's syndrome [217–222]. A very large number of additional drugs have also been occasionally reported as causing hyperadrenalism but it is difficult to be sure that the association is causative in many of these situations.

12.5 Disorders of Aldosterone Secretion

The basic structures of the mineralocorticoid system (renin, angiotensin I and II and aldosterone) have already been briefly mentioned. In addition, the heart is now known to be involved in salt and water homeostasis through the peptide atrial naturietic (ANP) which is produced by the endocrine cells of the right atrium; ANP acts on the renal tubule to cause a sodium diuresis.

12.5.1 *Hypoaldosteronism*

This may be part of overall adrenal cortical deficiency or it may occur in patients with a normal output of glucocorticoids. Aldosterone deficiency results in sodium depletion, hypotension and hyperkalaemia. The condition may be primary (due to an enzyme deficiency late in the aldosterone pathway), or secondary (due to isolated renin deficiency). The latter occurs in diabetes and in autonomic neuropathy, since beta-adrenergic stimulation normally provokes the secretion of renin.

12.5.2 *Hyperaldosteronism*

This may be primary (Conn's syndrome) or secondary. Primary aldosteronism is a rare cause of mild hypertension and hypokalaemia. In Chinese patients, aldosteronism can cause hypokalaemic paralysis which can be associated with adrenal adenoma or hyperplasia.

Secondary hyperaldosteronism is common in patients with heart failure, severe hypertension, cirrhosis or nephrosis. Increased renin production leads to sodium retention and potassium secretion. Rarely, the condition is due to a renin-secreting tumour of the kidney.

The conditions are treated with an aldosterone antagonist, such as spironolactone, or with drugs that act on the distal tubule sodium pump, such as amiloride.

The literature includes a discussion regarding pseudo-aldosteronism associated with the topical use of 9-alpha-fluoroprednisolone [223–227]. In addition, glycyrrhizin, liquorice-containing laxatives and carbenoxolone have been reported to induce prolonged pseudo-aldosteronism [228–230].

12.6 Diseases of the Adrenal Medulla

12.6.1 Adrenal Medullary Insufficiency

This condition is of importance in hypopituitarism and generalized autonomic neuropathy (the Shy-Drager syndrome) in which they may be impairment of the response to hypoglycaemia.

12.6.2 Adrenal Medullary Excess

Phaeochromocytomas are tumours of the adrenal medulla. They are rare causes of hypertension. Most tumours of this type secrete noradrenaline and cause episodes of headache, sweating, palpitations, flushing and anxiety. Reports that imipramine, pethedine, phenothiazine or metoclopramide can provoke a phaeochromocytomal hypertensive or shock crisis have appeared in the literature. It has also been reported that haemorrhagic necrosis of a phaeochromocytoma can be associated with phentolamine administration. There have also been reports of pulmonary oedema following propranolol therapy in cases of phaeochromocytoma [231–239].

12.7 Disadvantages of the Corticosteroids

The corticosteroids readily produce type A adverse reactions, overdose or prolonged use exaggerating the physiological effects of the drug administered.

Glucocorticoid effects include diabetes, osteoporosis, avascular necrosis of the femoral head, mental disturbances (including serious psychoses, sometimes with severe depression involving a risk of suicide, and sometimes with euphoria), muscle wasting (which can produce proximal myopathy), peptic ulceration (which may proceed to haemorrhage or perforation) and Cushing's syndrome. In such patients withdrawal must be gradual in order to avoid the onset of adrenal insufficiency. In children there may be suppression of growth, and in pregnancy there may be adverse effects on the development of the adrenals in the child.

It is of importance that administration of these steroids may suppress the clinical signs of infection so that septicaemia may be masked and tuberculosis allowed to advance rapidly and extensively prior to diagnosis. Chickenpox and measles pose particular problems in non-immune patients given corticosteroids and, unless they have had chickenpox, all patients receiving oral or parenteral corticosteroids for purposes other than replacement should be considered to be at risk of severe chickenpox. Passive immunization with varicella-zoster immunoglobulin is needed in non-immune exposed patients receiving systemic corticosteroids or given these agents within the preceeding 3 months. There is no good evidence that risks of this kind are associated with the use of topical, inhaled or rectal corticosteroids but it is *essential that clinicians attending non-immune corticosteroid-treated patients exposed to chickenpox or measles should consult the CSM warnings* given in the British National Formulary (BNF).

Both systemic and topical adverse effects can occur when corticosteroids are applied to the skin. The topical effects can include irreversible striae atrophicae, increased hair growth, perioral dermatitis in young women, local acne and cutaneous depigmentation. Topical corticosteroids, although they have important uses in the eye, can cause severe problems: in the presence of infection due to herpes simplex virus they aggravate the condition and may cause loss of sight or even of the eye; in predisposed patients use may also lead to steroid glaucoma.

Adrenal suppression can result from the long-term use of corticosteroids which, by depressing the secretion of corticotrophin, may lead to adrenal atrophy. In such patients the withdrawal of steroids must be very gradual in order to avoid precipitating acute adrenal insufficiency, hypotension and death. Adrenal atrophy may persist for months or years and any illness, stress or surgical intervention can precipitate acute adrenal insufficiency. Thus all patients taking corticosteroids should carry 'steroid cards'. These warn doctors that the patient is, or has been, receiving steroids. Reintroduction of steroid therapy may be needed to cover intercurrent illnesses, emergencies or operations. The information on the steroid card is particularly essential for anaesthetists.

Mineralocorticoid effects include sodium and water retention, potassium loss and hypertension. They are most marked with fludrocortisone; intermediate with cortisone, hydrocortisone, corticotrophin and tetracosactrin; less marked with betamethasone and dexamethasone; and are minimal with methylprednisolone, prednisolone, prednisone and triamcinolone.

12.8 Conclusion

The adrenal glands can be affected by many drugs which can either suppress or enhance the functions of these important endocrine structures. This chapter has been concerned with a pharmacoepidemiological approach to these problems but it should be read in association with systematic reviews of the subject [240].

References

1. Strom, B. L. (1994) What is pharmacoepidemiology? In Strom B. L. (Ed.), *Pharmacoepidemiology*, 2nd Edn, pp. 3–13, Chichester: John Wiley.
2. Mann, R. D. & Rawlins, M. D. (1993) Spontaneous adverse drug reaction reporting systems. In Mann, R. D., Rawlins, M. D. & Auty, R. M. (Eds), *A Textbook of Pharmaceutical Medicine: Current Practice*, pp. 323–330, Carnforth: Parthenon.
3. Mann, R. D. (1984) (Ed.) Aspects of modern drug use. In *Modern Drug Use: An Enquiry Based on Historical Principles*, Lancaster: MTP Press.
4. Nowakowska-Jankiewicz, B. & Zieleniewski, J. (1986) The effect of hydrocortisone and dexamethasone on the mitotic activity of regenerating adrenal cortex. *Acta Medica Polona*, **27**(3–4), 97–99.
5. Radermecker, M. (1978) Fonction surrenalienne chez l'asthmatique traite pendant deux ans par aerosols de beclomethasone. *Acta Tuberculosea et Pneumologica Belgica*, **69**(2), 107–112.
6. Anonymous (1980) Corticosteroids and hypothalamic-pituitary-adrenocortical function. *British Medical Journal*, **280**, 813–814.
7. Lufkin, E. G., Wahner, H. W. & Bergstralh, E. J. (1988) Reversibility of steroid-induced osteoporosis. *American Journal of Medicine*, **85**(6), 887–888.
8. Romano, P. E., Traisman, H. S. & Green, O. C. (1977) Fluorinated corticosteroid toxicity in infants. *American Journal of Ophthalmology*, **84**(2), 247–50.
9. Carson, T. E., Daane, T. A. & Weinstein, R. L. (1975) Long-term intramuscular administration of triamcinolone acetonide. Effect on the hypothalmic-pituitary-adrenal axis. *Archives of Dermatology*, **111**(12), 1585–1587.

10. MIKHAIL, G. R., SWEET, L. C. & MELLINGER, R. C. (1977) Long-term intramuscular administration of triamcinolone acetonide (Letter). *Archives of Dermatology*, **113**(1), 111.
11. FABRIS, C., MARTANO, C., PRADUI, G. & SANTORO, M. A. (1979) Division palatine et corticotherapie au cours de la grossesse. *Archives Francaises de Pediatrie*, **36**(10), 1046–1048.
12. GUEST, G., RAPPAPORT, R., PHILIPPE, F. & THIBAUD, E. (1983) Survenue de vergetures graves chez des adolescents atteints d'hyperplasie surrenale congenitale par defaut de 21-hydroxylation et traites par la dexamethasone. *Archives Francaises de Pediatrie*, **40**(6), 453–456.
13. JOB, J. C., MUNCK, A., CHAUSSAIN, J. L. & CANLORBE, P. (1985) Traitement de l'hyuperplasie surrenale virilisante chez les adolescents. Emploi et inconvenients de la dexamethasone. *Archives Francaises de Pediatrie*, **42**(9), 765–769.
14. DOWNIE, W. W., DIXON, J. S., LOWE, J. R., RHIND, V. M. *et al.* (1978) Adrenocortical suppression by synthetic cortiosteroid drugs: a comparative study of prednisolone and betamethasone. *British Journal of Clinical Pharmacology*, **6**(5), 397–399.
15. NEWTON, R. W., BROWNING, M. C., IQBALM, J., PIERCY, N. *et al.* (1978) Adrenocortical suppression in workers manufacturing synthetic glucocorticoids. *British Medical Journal*, **1**(6105), 73–74.
16. PIEZZI, R. S. & MIRANDA, J. C. (1981) Effect of dexamethasone on the neonatal adrenal medulla. *Cell and Tissue Research*, **220**(1), 213–217.
17. HALBERG, F., CORNELISSEN, G., TARQUINI, B., BENVENUTI, M. *et al.* (1984) Timing of medical diagnosis and treatment: clino-circadian quantification of suppression by dexamethasone of the adrenal cortical cycle in healthy men. *Chronobiologia*, **11**(1), 43–50.
18. SHARP, A. M., HANDELSMAN, D. J., RISTUCCIA, R. M. & TURTLE, J. R. (1982) Dexamethasone suppression of adrenocortical function. *Clinical Chemistry*, **28**(6), 1333–1334.
19. PRAHL, P. (1991) Adrenocortical suppression following treatment with beclomethasone and budesonide. *Clinical and Experimental Allergy*, **21**(1), 145–146.
20. ANSELL, B. M. (1991) Overview of the side effects of corticosteroid therapy. *Clinical and Experimental Rheumatology*, **9**(Suppl. 6), 19–20.
21. MUNRO, D. D. (1976) Topical corticosteroid therapy and its effect on the hypothalamic-pituitary-adrenal axis. *Dermatological* **152**(Suppl. 1) 173–180.
22. BROGDEN, R. N., PINDER, R. M., SAWYER, P. R., SPEIGHT, T. M. *et al.* (1976) Hydrocortisone 17-butyrate: a new topical corticosteroid preliminary report. *Drugs*, **12**(4), 249–257.
23. CLISSOLD, S. P. & HEEL, R. C. (1984) Budesonide. A preliminary review of its pharmacodynamic properties and therapeutic efficacy in asthma and rhinitis. *Drugs*, **28**(6), 485–518.
24. HELFER, E. L. & ROSE, L. I. (1989) Corticosteroids and adrenal suppression. Characterizing and avoiding the problem. *Drugs*, **38**(5), 838–845.
25. PULLAR, T. & STURROCK, R. D. (1983) Adrenal response in rheumatoid arthritis treated with long-term steroids. *European Journal of Rheumatology and Inflammation*, **6**(2), 187–191.
26. GRABNER, W. (1977) Zur Problematik der Kortikoid-induzierten NNR-Insuffizienz bei chirurgischen Eingriffen. *Fortschritte der Medizin*, **95**(30), 1866–1868.
27. GUILLEMANT, J. & GUILLEMANT, S. (1982) Effect of dexamethasone on steroidogenesis and on adrenocortical cyclic AMP and cyclic GMP responses to ACTH. *Hormone and Metabolic Research*, **14**(10), 547–550.
28. SAITO, E., ICHIKAWA, Y. & HOMMA, M. (1979) Direct inhibitory effect of dexamethasone on steroidogenesis of human adrenal *in vivo*. *Journal of Clinical Endocrinology and Metabolism*, **48**(5), 861–863.

29. HAMILTON, B. P., ZADIK, Z., EDWIN, C. M., HAMILTON, J. H. *et al.* (1979) Effect of adrenal suppression with dexamethasone in essential hypertension. *Journal of Clinical Endocrinology and Metabolism*, **48**(5), 848–853.
30. ASO, M. & SHIMAO, S. (1983) Systemic effects of 0.064% betamethasone dipropionate ointment and cream by simple application – with emphasis upon the suppression of adrenocortical functions. *Journal of Dermatology*, **10**(1), 55–62.
31. WEBB, D. R. (1977) Beclomethasone in steroid-dependent asthma. Effective therapy and recovery of hypothalamo-pituitary-adrenal function. *JAMA*, **238**(14), 1508–1511.
32. SPIEGEL, R. J., VIGERSKY, R. A., OLIFF, A. I., ECHELBERGER, C. K. *et al.* (1979) Adrenal suppression after short-term corticosteroid therapy. *Lancet*, **1**(8117), 630–633.
33. ESSELINCKX, W., KOLANOWSKI, J. & NAGANT DE DEUXCHAISNES, C. (1982) Adrenocortical function and responsiveness to tetracosactrin infusions after intra-articular treatment with triamcinolone acetonide and hydrocortisone acetate. *Clinical Rheumatology*, **1**(3), 176–184.
34. WEISS, A. H. (1989) Adrenal suppression after corticosteroid injection of periocular hemangiomas. *Americal Journal of Ophthalmology*, **107**(5), 518–522.
35. JACOBS, S., PULLAN, P. T., POTTER, J. M. & SHENFIELD, G. M. (1983) Adrenal suppression following extradural steroids. *Anaesthesia*, **38**(10), 953–956.
36. EDMONDS, L. C., VANCE, M. L. & HUGHES, J. M. (1992) Morbidity from paraspinal depo corticosteroid injections for analgesia: Cushing's syndrome and adrenal suppression. *Anesthesia and Analgesia*, **72**(6), 820–822.
37. MARSHALL, K. A. & CHAPLAN, S. R. (1991) Adrenal suppression and paraspinal corticosteroids. *Anesthesia and Analgesia*, **74**(5), 773.
38. CHERNOW, B., VIGERSKY, R., O'BRIAN, J. T. & GEORGES, L. P. (1982) Secondary adrenal insufficiency after intrathecal steroid administration. *Journal of Neurosurgery*, **56**(4), 567–570.
39. BISGAARD, H., PEDERSEN, S., DAMKJAER-NIELSEN, M. & OSTERBALLE, O. (1991) Adrenal function in asthmatic children treated with inhaled budesonide. *Acta Paediatrica Scandinavica*, **80**(2), 213–217.
40. KONIG, P. (1993) The risks and benefits of inhaled anti-inflammatory therapy in children. *Agents and Actions Supplements*, **49**, 181–188.
41. DROSZCZ, W., LECH, B., MALUNOWICZ, E. & PIOTROWSKA, B. (1980) Factors influencing adrenocortical suppression during long-term triamcinolone acetonide therapy in asthma. *Annals of Allergy*, **44**(3), 174–176.
42. KWONG, F. K., SUE, M. A. & KLAUSTERMEYER, W. B. (1987) Corticosteroid complications in respiratory disease. *Annals of Allergy*, **58**(5), 326–330.
43. STEAD, R. J. & COOKE, N. J. (1989) Adverse effects of inhaled corticosteroids. *British Medical Journal*, **298**(6671), 403–404.
44. WONG, J. & BLACK, P. (1992) Acute adrenal insufficiency associated with high dose inhaled steroids. *British Medical Journal*, **304**(6839), 1415.
45. SPITZER, S. A., KAUFMAN, H., KOPLOVITZ, A., TOPILSKY, M. *et al.* (1976) Beclomethasone dipropionate and chronic asthma. The effect of long-term aerosol administration on the hypothalamic-pituitary-adrenal axis after substitution for oral therapy with corticosteroids. *Chest*, **70**(1), 38–42.
46. YERNAULT, J. C., LECLERCQ, R., SCHANDEVYL, W., VIRASORO, E. *et al.* (1977) The endocrinometabolic effects of beclomethasone dipropionate in asthmatic patients. *Chest*, **71**(6), 698–702.
47. ALTMAN, L. C., FINDLAY, S. R., LOPEZ, M., LUKACSKO, P. *et al.* (1992) Adrenal function in adult asthmatics during long-term daily treatment with 800, 1200 and 1600 micrograms triamcinolone acetonide. Multicenter study. *Chest*, **101**(5), 1250–1256.
48. PRAHL, P. & JENSEN, T. (1987) Decreased adreno-cortical suppression utilizing the Nebuhaler for inhalation of steroid aerosols. *Clinical Allergy*, **17**(5), 393–398.

49. Vlasses, P. H., Ferguson, R. K., Koplin, J. R., Clementi, R. A. *et al.* (1981) Adrenocortical function after chronic inhalation of fluocortinbutyl and beclomethasone dipropionate. *Clinical Pharmacology and Therapeutics*, **29**(5), 643–649.
50. Boe, J. & Thulesius, O. (1975) Treatment of steroid-dependent asthma with beclomethaxone dipropionate administered by aerosol. *Current Therapeutic Research, Clinical and Experimental*, **17**(5), 460–466.
51. Molema, J., Lammers, J. W., van Herwaarden, C. L. & Folgering, H. T. (1988) Effects of inhaled beclomethasone dipropionate on beta 2-receptor function in the airways and adrenal responsiveness in bronchial asthma. *European Journal of Clinical Pharmacology*, **34**(6), 577–583.
52. Jennings, B. H., Andersson, K. E. & Johansson, S. A. (1991) Assessment of systemic effects of inhaled glucocorticoids: comparison of the effects of inhaled budesonide and oral prednisolone on adrenal function and markers of bone turnover. *European Journal of Clinical Pharmacology*, **40**(1), 77–82.
53. Priftis, K., Everard, M. L. & Milner, A. D. (1991) Unexpected side-effects of inhaled steroids: a case report. *European Journal of Pediatrics*, **150**(6), 448–449.
54. Daman-Willems, C. E., Dinwiddie, R., Grant, D. B., Rivers, R. P. *et al.* (1994) Temporary inhibition of growth and adrenal suppression associated with the use of steroid nose drops. *European Journal of Pediatrics*, **153**(9), 632–634.
55. Gordon, A., McDonald, C. F., Thomson, S. A., Frame, M. H. *et al.* (1987) Dose of inhaled budesonide required to produce clinical suppression of plasma cortisol. *European Journal of Respiratory Disease*, **71**(1), 10–14.
56. Feiss, G., Morris, R., Rom, D., Mansfield, L. *et al.* (1992) A comparative study of the effects of intranasal triamcinolone acetonide aerosol (ITAA) and prednisone on adrenocortical function. *Journal of Allergy and Clinical Immunology*, **89**(6), 1151–1156.
57. Vaz, R., Senior, B., Morris, M. & Binkiewicz, A. (1982) Adrenal effects of beclomethasone inhalation therapy in asthmatic children. *Journal of Pediatrics*, **100**(4), 660–662.
58. Konig, P. & Goldstein, D. E. (1982) Adrenal function after beclomethasone inhalation therapy (Letter). *Journal of Pediatrics*, **101**(4), 646–647.
59. British Thoracic and Tuberculosis Association (1975) Inhaled corticosteroids compared with oral prednisone in patients starting long-term corticosteroid therapy for asthma. A controlled trial by the British Thoracic and Tuberculosis Association. *Lancet*, **2**(7933), 469–473.
60. Brown, H. M. (1986) Nocturnal adrenal suppression in children inhaling beclomethasone dipropionate (Letter). *Lancet*, **1**(8492), 1269.
61. Wales, J. K., Barnes, N. D. & Swift, P. G. (1991) Growth retardation in children on steroids for asthma (Letter). *Lancet*, **338**(8781), 1535.
62. Wahn, U., Rauh, W., Lipinski, C., Klett, M. *et al.* (1978) Nebennierenrindenfunktion nach Langzeitbehandlung mit Beclometasondipropionat (BDP) beim kindlichen Asthma Bronchiale. *Monatsschr-Kinderheilkd*, **126**(2), 87–89.
63. Duchene, A. & Chadenas, D. (1990) Insuffisance cortico-surrenale apres corticotherapie prolongee en aerosols. *Presse Medicale*, **19**(8), 381.
64. Ninan, T. K., Reid, I. W., Carter, P. E., Smail, P. J. *et al.* (1993) Effects of high doses of inhaled corticosteroids on adrenal function in children with severe persistent asthma. *Thorax*, **48**(6), 599–602.
65. Nilsson, J. E. & Gip, L. J. (1979) Systemic effects of local treatment with high doses of potent corticosteroids in psoriatics. *Acta Dermato-Venereologica*, **59**(3), 245–248.
66. Gruenberg, J. C. & Mikhail, G. R. (1976) Percutaneous adrenal suppression with topically applied corticosteroids. *Archives of Surgery*, **111**(10), 1165.
67. Wendt, H. & Reckers, R. (1976) Systemische Wirkung von Diflucortolonvalerianat nach dermaler Applikation. *Arzneimittel-Forschung*, **26**(7b), 1509–1513.

68. WENDT, H. (1977) Wirkung von Flucortin-butylester auf die NNR-Funktion des Menschen. *Arzneimittel-Forschung*, **27**(11a), 2236–2238.
69. DOWNIE, D. A., CAINS, G. D. & PEEK, R. D. (1978) The effect of betamethasone diproprionate cream 0.05% (Dirposone) on adrenocortical function. *Australasian Journal of Dermatology*, **19**(3), 114–117.
70. MUNRO, D. D. (1976) The effect of percutaneously absorbed steroids on hypothalamic-pituitary-adrenal function after intensive use in in-patients. *British Journal of Dermatology*, **94**(Suppl. 12), 67–76.
71. ALLENBY, C. F., MAINM, R. A., MARSDEN, R. A. & SPARKES, C. G. (1975) Effect on adrenal function of topically applied clobetasol propionate (Dermovate). *British Medical Journal*, **4**(5997), 619–621.
72. CARRUTHERS, J. A., AUGUST, P. J. & STAUGHTON, R. C. (1975) Observations on the systemic effect of topical clobetasol propionate (Dermovate). *British Medical Journal*, **4**(5990), 203–204.
73. KECSKES, A., HEGER-MAHN, D., KUHLMANN, R. K. & LANGE, L. (1993) Comparison of the local and systemic side effects of methylprednisolone aceponate and mometasone furoate applied as ointments with equal antiinflammatory activity. *Journal of the American Academy of Dermatology*, **29**(4), 576–580.
74. ASO, M. (1983) The effects of potent topical corticosteroids on adrenocortical function. *Journal of Dermatology*, **10**(2), 145–149.
75. ASO, M. & SHIMAO, S. (1983) Systemic effects of 0.064% betamethasone dipropionate ointment and cream by simple application – with emphasis upon the suppression of adrenocortical functions. *Journal of Dermatology*, **10**(1), 55–62.
76. ORTEGAM, E., BURDICK, K. H. & SEGRE, E. J. (1975) Adrenal suppression by clobetasol priopionate (Letter). *Lancet*, **1**(7917), 1200.
77. ANONYMOUS (1986) Clobetasol – a potent new topical corticosteroid. *Medical Letter on Drugs and Therapeutics*, **28**(715), 57–59.
78. GOVERDE, H. J., BOIDIN, M. P., ZANDSTRA, D. F. & AGOSTON, S. (1985) Inhibition of corticosteroidogenesis by etomidate. *Acta Anaesthesiologica Belgica*, **36**(4), 375–380.
79. BOIDIN, M. P. (1986) Can etomidate cause an Addisonian crisis? *Acta Anaesthesiologica Belgica*, **37**(3), 165–170.
80. MEHTA, M. P., DILLMAN, J. B., SHERMAN, B. M., GHONEIM, M. M. *et al.* (1985) Etomidate anesthesia inhibits the cortisol response to surgical stress. *Acta Anaesthesiologica Scandinavica*, **29**(5), 486–489.
81. WANSCHER, M., TONNESEN, E., HUTTEL, M. & LARSEN, K. (1985) Etomidate infusion and adrenocortical function. A study in elective surgery. *Acta Anaesthesiologica Scandinavica*, **29**(5), 483–485.
82. DECOSTER, R., HELMERS, J. H. & NOORDUIN, H. (1985) Effect of etomidate on cortisol biosynthesis: site of action after induction of anaesthesia. *Acta Endocrinologica (Copenhagen)*, **110**(4), 526–531.
83. STUTTMANN, R., ALLOLIO, B., BECKER, A., DOEHN, M. *et al.* (1988) Etomidat versus Etomidat und Hydrokortison zur Narkoseeinleitung bei abdominalchirurgischen Eingriffen. *Anaesthesist*, **37**(9), 576–582.
84. VANACKER, B., WIEBALCK, A., VAN-AKEN, H., SERMEUS, L. *et al.* (1993) Induktionsqualitat und Nebennierenrindenfunktion. Ein klinischer Vergleich von Etomiedat – Lipuror und Hypnomidate. *Anaesthesist*, **42**(2), 81–89.
85. SEITZ, W., FRITZ, K., LUBBE, N., GRAMBOW, D. *et al.* (1985) Suppression der Nebennierenrinde durch Infusion von Etomidat wahrend Allgemeinanasthesie. *Anast-Intensivther Notfallmed*, **20**(3), 125–130.
86. WAGNER, R. L. & WHITE, P. F. (1984) Etomidate inhibits adrenocortical function in surgical patients. *Anesthesiology*, **61**(6), 647–651.
87. NATHAN, N., VANDROUX, J. C. & FEISS, P. (1991) Role de la vitamine C sur les

effets corticosurrenaliens de l'etomidate. *Annales Francaises D Anesthesie et de Reanimation*, **10**(2), 329–332.
88. PREZIOSI, P. & VACCA, M. (1982) Etomidate and corticotrophic axis. *Archives Internationales de Pharmacodynamie et de Therapie*, **256**(2), 308–310.
89. LAMBERT, A., MITCHELL, R. & ROBERTSON, W. R. (1985) Effects of propofol, thiopentone and etomidate on adrenal steroidogenesis *in vitro*. *British Journal of Anaesthesia*, **57**(5), 505–508.
90. FELLOWS, I. W., BASTOW, M. D., BYREN, A. J. & ALLISON, S. P. (1983) Adrenocortical suppression in multiply injured patients: a complication of etomidate treatment. *British Medical Journal (Clinical Research Edition)*, **287**(6408), 1835–1837.
91. FRY, D. E. & GRIFFITHS, H. (1984) The inhibition of etomidate of the 11 beta-hydroxylation of cortisol. *Clinical Endocrinology (Oxford)*, **20**(5), 625–629.
92. CONNER, C. S. (1984) Etomidate and adrenal suppression. *Drug Intelligence and Clinical Pharmacology*, **18**(5), 393–394.
93. HINDS, C. J. (1984) Etomidate and adrenocortical function (Letter). *Intensive Care Medicine*, **10**(5), 268–269.
94. DORR, H. G., KUHNLE, U., HOLTHAUSEN, H., BIDLINGMAIER, F. *et al.* (1984) Etomidate: a selective adrenocortical 11 beta-hydroxylase inhibitor. *Klin Wochenschr*, **62**(21), 1011–3.
95. ALLOLIO, B., STUTTMANN, R., LEONHARD, U., FISCHER, H., *et al.* (1984) Adrenocortical suppression by a single induction dose of etomidate, *Klinische Wochenschrift*, **62**(21), 1014–1017.
96. SPIEGEL, R. J., VIGERSKY, R. A., OLIFF, A. I., ECHELBERGER, C. K. *et al.* (1979) Adrenal suppression after short-term corticosteroid therapy. *Lancet*, **1**(8117), 630–633.
97. PREZIOSI, P. & VACCA, M. (1988) Adrenocortical suppression and other endocrine effects of etomidate. *Life Sciences*, **42**(5), 477–489.
98. MEHTA, S. & BURTON, P. (1975) Adrenocortical function related to Althesin anaesthesia and surgery in man. *Anaesthesia*, **30**(2), 170–173.
99. ABDULLA, W., KRENNRICH, G., CORDES, U. & DRAUSE, U. (1981) Verhalten der sympathoadrenalen und adrenocorticalen Aktivitat bei Enflurannarkose und Operation. *Anaesthesist*, **30**(10), 493–496.
100. APPEL, E., DUDZIAK, R., PALM, D. & WNUK, A. (1979) Sympathoneuronal and sympathoadrenal activation during ketamine anesthesia. *European Journal of Clinical Pharmacology*, **16**(2), 91–95.
101. GOTHERT, M., DORN, W. & LOEWENSTEIN, I. (1976) Inhibition of catecholamine release from the adrenal medulla by halothane. Site and mechanism of action. *Naunyn-Schmiedebergs Archives of Pharmacology*, **294**(3), 239–249.
102. ELANSARY, E. H. & EARIS, J. E. (1983) Rifampicin and adrenal crisis. *British Medical Journal (Clinical Research Edition)*, **286**(6381), 1861–1862.
103. KYRIAZOPOULOU, V., PARPAROUSI, O. & VAGENAKIS, A. G. (1984) Rifampicin-induced adrenal crisis in Addisonian patients receiving corticosteroid replacement therapy. *Journal of Clinical Endocrinology and Metabolism*, **59**(6), 1204–1206.
104. HEY, A. A., CONAGLEN, J. V., ESPINER, E. A. & THORNLEY, P. E. (1983) Rifampicin-induced adrenal crisis (Letter). *New Zealand Medical Journal*, **96**(744), 988–989.
105. EDIGER, S. K. & ISLEY, W. L. (1988) Rifampicin-induced adrenal insufficiency in the acquired immunodeficience syndrome: difficulties in diagnosis and treatment. *Postgraduate Medical Journal*, **64**(751), 405–406.
106. WILKINS, E. C., HNIZDO, E. & COPE, A. (1989) Addisonian crisis induced by treatment with rifampicin. *Tubercle*, **70**(1), 69–73.
107. BEST, T. R., JENKINS, J. K., MURPHY, F. Y., NICKS, S. A. *et al.* (1987) Persistent adrenal insufficiency secondary to low-dose ketoconazole therapy. *American Journal of*

Medicine, **82**(2 Spec. No.), 676–680.

108. PONT, A., GRAYBILL, J. R., CRAVEN, P. C., GALGIANI, J. N. *et al.* (1984) High-dose ketoconazole therapy and adrenal and testicular function in humans. *Archives of Internal Medicine*, **144**(11), 2150–2153.
109. KHOSLA, S., WOLFSON, J. S., DEMERJIAN, Z. & GODINE, J. E. (1989) Adrenal crisis in the setting of high-dose ketoconazole therapy. *Archives of Internal Medicine* **149**(4), 802–804.
110. LAMBERT, A., MITCHELL, R. & ROBERTSON, W. R. (1986) The effect of ketoconazole on adrenal and testicular steroidogenesis *in vitro*. *Biochemical Pharmacology*, **35**(22), 3999–4004.
111. BRADBROOK, I. D., GILLIES, H. C., MORRISON, P. J., ROBINSON, J. *et al.* (1985) Effects of single and multiple doses of ketoconazole on adrenal function in normal subjects. *British Journal of Clinical Pharmacology*, **20**(2), 163–165.
112. MCCANCE, D. R., HADDEN, D. R., KENNEDY, L., SHERIDAN, B. *et al.* (1987) Clinical experience with ketoconazole as a therapy for patients with Cushing's syndrome. *Clinical Endocrinology*, **27**(5), 593–599.
113. KHANDERIA, U. (1991) Use of ketoconazole in the treatment of Cushing's syndrome. *Clinical Pharmacology*, **10**(1), 12–13.
114. WINQUIST, E. W., LASKEY, J., CRUMP, M., KHAMSI, F. *et al.* (1995) Ketoconazole in the management of paraneoplastic Cushing's syndrome secondary to ectopic adrenocorticotropin production. *Journal of Clinical Oncology*, **13**(1), 157–164.
115. BRITTON, H., SHEHAB, Z., LIGHTNER, E., NEW, M. *et al.* (1988) Adrenal response in children receiving high doses of ketoconazole for systemic coccidioidomycosis. *Journal of Pediatrics*, **112**(3), 488–492.
116. TUCKER, W. S., JR, SNELL, B. B., ISLAND, D. P. & GREGG, C. R. (1985) Reversible adrenal insufficiency induced by ketoconazole. *JAMA*, **253**(16), 2413–2414.
117. WEBER, M. M., LUPPA, P. & ENGELHARDT, D. (1989) Inhibition of human adrenal androgen secretion by ketoconazole. *Klinische Wochenschrift*, **67**(14), 707–712.
118. WHITE, M. C. & KENDALL-TAYLOR, P. (1985) Adrenal hypofunction in patients taking ketoconazole (Letter). *Lancet*, **1**(8419), 44–45.
119. SZABO, S. & REYNOLDS, E. C. (1975) Structure–activity relationships for ulcerogenic and adrenocorticolytic effects of alkyl nitriles, amines, and thiols. *Environmental Health Perspectives*, **11**, 135–140.
120. WELLS, S. A., JR, SANTEN, R. J., LIPTON, A., HAAGENSEN, S. E., JR *et al.* (1978) Medical adrenalectomy with aminoglutethimide; clinical studies in postmenopausal patients with metastatic breast carcinoma. *Annals of Surgery*, **187**(5), 475–484.
121. ROBINSON, M. R. (1980) Aminoglutethimide: medical adrenalectomy in the management of carcinoma of the prostate. A review after 6 years. *British Journal of Urology*, **52**(4), 328–329.
122. HARRIS, A. L. (1985) Could aminoglutethimide replace adrenalectomy? *Breast Cancer Research and Treatment*, **6**(3), 201–211.
123. SANTEN, R. J. & WELLS, S. A., JR (1980) The use of aminoglutethimide in the treatment of patients with metastatic carcinoma of the breast. *Cancer*, **46**(Suppl. 4), 1066–1074.
124. MURRAY, R. M. & PITT, P. (1982) Aminoglutethimide in tamoxifen-resistant patients: The Melbourne Experience. *Cancer Research*, **42**(Suppl. 8), 3437s–3441s.
125. HAVLIN, K. A. & TRUMP, D. L. (1988) Aminoglutethimide: theoretical considerations and clinical results in advanced prostate cancer. *Cancer Treatment and Research*, **39**, 83–96.
126. KOELMEYER, T. D., STEPHENS, E. J. & WOOD, H. F. (1978) Experience with 6-aminoglutethimide in the treatment of metastatic breast cancer. *Clinical Oncology*, **4**(4), 323–327.
127. RICCIARDI, M. P., PELLEGRINI, A., GRIECO, M., MARRONI, P. *et al.* (1992)

Effects of stress on adrenal steroidogenesis blockade by aminoglutethimide. *In Vivo*, **6**(1), 103–106.

128. SANTEN, R. J., WELLS, S. A., RUNIC, S., GUPTA, C. *et al.* (1977) Adrenal suppression with aminoglutethimide. I. Differential effects of aminoglutethimide on glucocorticoid metabolism as a rationale for use of hydrocortisone. *Journal of Clinical Endocrinology and Metabolism*, **45**(3), 469–479.
129. SAMOJLIK, E., SANTEN, R. J. & WELLS, S. A. (1977) Adrenal suppression with aminoglutethimide. II. Differential effects of aminoglutethimide on plasma androstenedione and estrogen levels. *Journal of Clinical Endocrinology and Metabolism*, **45**(3), 480–487.
130. MISBIN, R. I., CANARY, J. & WILLARD, D. (1976) Aminoglutethimide in the treatment of Cushing's syndrome. *Journal of Clinical Pharmacology*, **16**(11–12), 645–651.
131. SANFORD, E. J. DRAGO, J. R., ROHNER, T. J., JR, SANTEN, R. *et al.* (1976) Aminoglutethimide medical adrenalectomy for advanced prostatic carcinoma, *Journal of Urology*, **115**(2), 170–174.
132. MURRAY, R. M., PITT, P. & JERUMS, G. (1981) Medical adrenalectomy with aminoglutethimide in the management of advanced breast cancer. *Medical Journal of Australia*, **1**(4), 179–181.
133. ANONYMOUS (1981) Aminoglutethimide. *Medical Letter on Drugs and Therapeutics*, **23**(16), 71–72.
134. VAN-VEELEN, H., WILLEMSE, P. H., SLEIJFER, D. T., VAN DER PLOEG, E. *et al.* (1985) Mechanism of adrenal suppression by high-dose medroxyprogesterone acetate in breast cancer patients. *Cancer Chemotherapy and Pharmacology*, **15**(2), 167–170.
135. VAN-VEELEN, H., WILLEMSE, P. H., SLEIJFER, D. T., SLUITER, W. J. *et al.* (1985) Endocrine effects of medroxyprogesterone acetate: relation between plasma levels and suppression of adrenal steroids in patients with breast cancer. *Cancer Treatment Reviews*, **69**(9), 977–983.
136. VAN-VEELEN, H., HOUWERZIJL, J., RODING, T. J., TJABBES, T. *et al.* (1982) Oral high-dose medroxyprogesterone acetate causes adrenal suppression in patients with breast cancer. *European Journal of Cancer and Clinical Oncology*, **18**(10), 1035–1036.
137. KOLANOWSKI, J., ESSELINCKX, W., NAGANT-DE-DEUXCHAISNES, C. & CRABBE, J. (1977) Adrenocortical response upon repeated stimulation with corticotrophin in patients lacking endogenous corticotrophin secretion. *Acta Endocrinologica*, **85**(3), 595–660.
138. PAYET, N., LEHOUX, J. G. & ISLER, H. (1980) Effect of ACTH on the proliferative and secretory activities of the adrenal glomerulosa. *Acta Endocrinologica*, **93**(3), 365–374.
139. KRUYT, N. & ROLLAND, R. (1982) Cortisol, 17 alpha-OH, progesterone and androgen responses to a standardized ACTH-stimulation in different stages of the normal mestrual cycle. *Acta Endocrinologica*, **96**(2), 243–251.
140. WEST, S. H. & BASSETT, J. R. (1992) A circadian rhythm in adrenal responsiveness to ACTH is not confirmed by *in vitro* studies. *Acta Endocrinologica*, **126**(4), 363–368.
141. FEUILLAN, P., RAFFELD, M., STEIN, C. A., LIPFORD, N. *et al.* (1987) Effects of suramin on the function and structure of the adrenal cortex in the cynomolgus monkey. *Journal of Clinical Endocrinology and Metabolism*, **65**(1), 153–158.
142. DORFINGER, K., VIERHAPPER, H., WILFING, A., CZERNIN, S. *et al.* (1991) The effects of suramin on human adrenocortical cells *in vitro*: suramin inhibits cortisol secretion and adrenocortical cell growth. *Metabolism*, **40**(10), 1020–1024.
143. PY, A., DAIROU, F. & DE-GENNES, J. L. (1987) Toxicite surrenale de la suramine. *Presse Medicale*, **16**(39), 1979.
144. DORFINGER, K., NIEDERLE, B., VIERHAPPER, H., ASTRID, W. *et al.* (1991) Suramin and the human adrenocortex: results of experimental and clinical studies. *Surgery*, **110**(6), 1100–1105.

145. LaRocca, R. V., Cooper, M. R., Uhrich, M., Danesi, R. *et al.* (1991) Use of suramin in treatment of prostatic carcinoma refractory to conventional hormonal manipulation. *Urologic Clinics of North America*, **18**(1), 123–129.
146. Schurmeyer, T., Graff, J., Senge, T. & Nieschlag, E. (1986) Effect of oestrogen or cyproterone acetate treatment on adrenocortical function in prostate carcinoma patients. *Acta Endocrinologica*, **113**(3), 360–367.
147. Migally, N. (1979) Effect of cyproterone acetate on the structure of the adrenal cortex. *Archives of Andrology*, **2**(2), 109–115.
148. Savage, D. C. & Swift, P. G. (1981) Effect of cyproterone acetate on adrenocortical function in children with precocious puberty. *Archives of Disease in Childhood*, **56**(3), 218–222.
149. Bhargava, A. S., Kapp, J. F., Poggel, H. A., Heinick, J. *et al.* (1981) Effect of cyproterone acetate and its metabolites on the adrenal function in man, rhesus monkey and rat. *Arzneimettelforschung*, **31**(6), 1005–1009.
150. Lambert, A., Mitchell, R. M. & Robertson, W. R. (1985) On the site of action of the anti-adrenal steroidogenic effect of cyproterone acetate. *Biochemical Pharmacology*, **34**(12), 2091–2095.
151. de la Torre, B., Noren, S., Hedman, M. & Diczfalusy, E. (1979) Effect of cyproterone acetate (CPA) on gonadal and adrenal function in men. *Contraception*, **20**(4), 377–396.
152. Von Muhlendahl, K. E., Weber, B., Muller-Hess, R., Korth-Schutz, S. *et al.* (1977) Nebennierenrindeninsuffizienz bei Cyproteronacetatbehandlung. *Deutsche Medizinische Wochenschrift*, **102**(29), 1074.
153. Stewart, P. M., Grieve, J., Nairn, I. M., Padfield, P. L. *et al.* (1987) Lithium inhibits the action of fludrocortisone on the kidney. *Clinical Endocrinology*, **27**(1), 63–68.
154. Helge, H. & Korth-Schutz, S. (1980) Klinische Anwendung von Cyproteronacetat bei Kindern. *Gynakologe*, **13**(1), 44–53.
155. Girard, J. & Baumann, J. B. (1976) Secondary adrenal insufficiency due to cyproterone acetate (Proceedings). *Journal of Endocrinology*, **69**(3), 13P–14P.
156. Hammerstein, J., Meckies, J., Leo-Rossberg, Moltz, L. *et al.* (1975) Use of cyproterone acetate (CPA) in the treatment of acne, hirsutism and virilism. *Journal of Steroid Biochemistry*, **6**(6), 827–836.
157. von-Muhlendahl, K. E., Korth-Schutz, S., Muller-Hess, R., Helge, H. *et al.* (1977) Cyproterone acetate and adrenocortical function (Letter). *Lancet*, **1**(8022), 1160–1161.
158. Zwaan, C. M., Odink, R. J., Delemarre-van-de-Waal, H. A., Dankert-Roelse, J. E. *et al.* (1992) Acute adrenal insufficiency after discontinuation of inhaled corticosteroid therapy (Letter). *Lancet*, **340**(8830), 1289–1290.
159. Jain, C. U., Gudi, K. & Giovannielloi, J. (1990) Adrenocorticotropic hormone-induced bilateral adrenal hemorrhage (Letter). *American Journal of Roentgenology*, **154**(2), 424–425.
160. Dunlap, S. K., Meiselman, M. S., Breuer, R. I., Panella, J. S. *et al.* (1989) Bilateral adrenal hemorrhage as a complication of intravenous ACTH infusion in two patients with inflammatory bowel disease. *American Journal of Gastroentrology*, **84**(10), 1310–1312.
161. Redman, J. F. & Faas, F. H. (1976) Acute unilateral adrenal hemorrhage following ACTH administration in a patient with Cushing's syndrome. *American Journal of Medicine*, **61**(4), 533–536.
162. Simpson, W. T. (1979) Adrenal hemorrhage during the treatment of ulcerative colitis with adrenocorticotrophic hormone. *Canadian Journal of Surgery*, **22**(1), 82–83.
163. Marcus, H. I., Connon, J. J. & Stern, H. S. (1986) Bilateral adrenal hemorrhage during ACTH treatment of ulcerative colitis. Report of a case and review of the liter-

ature. *Diseases of the Colon and Rectum*, **29**(2), 130–132.

164. OGNIBENE, A. J. & MCBRIDE, H. (1987) Adrenal hemorrhage: a complication of anticoagulant therapy – a case history. *Angiology*, **38**(6), 479–483.
165. GRANRY, J. C., HOUET, J. F. & DELHUMEAU, A. (1989) Hematomes des surrenales et heparine. *Annales Francaises D Anesthesie et de Reanimation*, **8**(6), 650–655.
166. LESCHI, J. P., GOEAU-BRISSONIERE, O., COGGIA, M. & CHICHE, L. (1994) Heparin-related thrombocytopenia and adrenal hemorrhagic necrosis following aortic surgery. *Annals of Vascular Surgery*, **8**(5), 506–508.
167. HARDWICKE, M. B. & KISLY, A. (1992) Prophylactic subcutaneous heparin therapy as a cause of bilateral adrenal hemorrhage. *Archives of Internal Medicine*, **152**(4), 845–847.
168. MARCUS, A. O., DUICK, D. S., LITTLE, D. & WALDMANN, E. B. (1982) Bilateral adrenal hemorrhage during anticoagulant therapy. *Arizona Medicine*, **39**(9), 575–577.
169. ARTHUR, C. K., GRANT, S. J. MURRAY, W. K., ISBISTER, J. P. *et al.* (1985) Heparin-associated acute adrenal insufficiency (Letter). *Australian and New Zealand Journal of Medicine*, **15**(4), 454–455.
170. SOUIED, F., POURRIAT, J. L., LE-ROUX, G., HOANG, P. *et al.* (1992) Adrenal hemorrhagic necrosis related to heparin-associated thrombocytopenia (see comments), *Critical Care Medicine*, **20**(8), 1192.
171. ERNEST, D. & FISHER, M. M. (1991) Heparin-induced thrombocytopaenia complicated by bilateral adrenal haemorrhage. *Intensive Care Medicine*, **17**(4), 238–240.
172. AHMAD, S. (1979) Acute Addison's disease and heparin (Letter). *JAMA*, **241**(25), 2703.
173. LEEHEY, D., GANTT, C. & LIM, V. (1981) Heparin-induced hypoaldosteronism. Report of a case. *JAMA*, **246**(19), 2189–2190.
174. MEUNIER, P. J., DEMPSTER, D. W., EDOUARD, C., CHAPUY, M. C. *et al.* (1984) Bone histomorophometry in corticosteroid-induced osteoporosis and Cushing's syndrome. *Advances in Experimental Medicine and Biology*, **171**, 191–200.
175. GRUBB, S. R., CANTLEY, L. K., JONES, D. L. & CARTER, W. H. (1981) Iatrogenic Cushing's syndrome after intrapericardial corticosteroid therapy. *Annals of Internal Medicine*, **95**(6), 706–707.
176. O'SULLIVAN, M. M., RUMFELD, W. R., JONES, M. K. & WILLIAMS, B. D. (1985) Cushing's syndrome with suppression of the hypothalamic-pituitary-adrenal axis after intra-articular steroid injections. *Annals of Rheumatic Diseases*, **44**(8), 561–563.
177. HUGHES, J. M., HICHENS, M., BOOZE, G. W. & THORNER, M. O. (1986) Cushing's syndrome from the therapeutic use of intramuscular dexamethasone acetate. *Archives of Internal Medicine*, **146**(9), 1848–1849.
178. FOREST, M. G., LECOQ, A., SALLE, B. & BERTRAND, J. (1981) Does neonatal phenobarbital treatment affect testicular and adrenal functions and steroid binding in plasma in infancy? *Journal of Clinical Endocrinology and Metabolism*, **52**(1), 103–110.
179. LANGSTON, J. R. & KOLODNY, S. C. (1976) Cushing's syndrome associated with the intradermal injection of triamcinolone diacetate. *Journal of Oral Surgery*, **34**(9), 846–849.
180. MEIKLE, A. W., CLARKE, D. H. & TYLER, F. H. (1976) Cushing syndrome from low doses of dexamethasone. A result of slow plasma clearance. *JAMA*, **235**(15), 1592–1593.
181. HIMATHONGKAM, T. (1978) Florid Cushing's syndrome and hirsutism induced by desoximethasone. *JAMA*, **239**(5), 430–431.
182. BERGREM, H., JERVELL, J. & FLATMARK, A. (1985) Prednisolone pharmacokinetics in cushingoid and non-cushingoid kidney transplant patients. *Kidney International*, **27**(2), 459–464.
183. BRADLEY, B. S., KUMAR, S. P., MEHTA, P. N. & EZHUTHACHAN, S. G. (1994) Neonatal cushingoid syndrome resulting from serial courses of antenatal betamethasone. *Obstetrics and Gynecology*, **83**(5 Part 2), 869–872.
184. TUEL, S. M., MEYTHALER, J. M. & CROSS, L. L. (1990) Cushing's syndrome from

epidural methylprednisolone. *Pain*, **40**(1), 81–84.

185. GIRRE, C., VINCENS, M., GOMPEL, A. & FOURNIER, P. E. (1987) Syndrome de Cushing iatrogene par surdosage aigu en triamcinolone. *Therapie*, **42**(3), 317–318.
186. RUIZ-MALDONADO, R., ZAPATA, G., LORDES, T. & ROBLES, C. (1982) Cushing's syndrome after topical application of corticosteroids. *American Journal of Diseases of Children*, **136**(3), 274–275.
187. GEMME, G., RUFFA, G., BONIOLI, E., LAGORIO, V. *et al.* (1984) Picture of the month. Cushing's syndrome due to topical corticosteroids. *American Journal of Diseases of Children*, **138**(10), 987–988.
188. KIMMERLE, R. & ROLLA, A. R. (1985) Iatrogenic Cushing's syndrome due to dexamethasone nasal drops. *American Journal of Medicine*, **79**(4), 535–537.
189. DANDINE, M., LAVAUD, J., BESSON-LEAUD, M. & LIMAL, J. M. (1980) Insuffisance surrenale et synrome de Cushing liatrogene par absorption cutanee de corticoides chez un nourrisson de 8 mois. *Annales de Dermatologie et de Venereologie*, **107**(3), 191–195.
190. MAY, P., STEIN, E. J., RYTER, R. J., HIRSH, F. S. *et al.* (1976) Cushing syndrome from percutaneous absorption of triamcinolone cream. *Archives of Internal Medicine*, **136**(5), 612–613.
191. BORZYSKOWSKI, M., GRANT, D. B. & WELLS, R. S. (1976) Cushing's syndrome induced by topical steroids used for the treatment of non-bullous ichthyosiform erythroderma. *Clinical and Experimental Dermatology*, **1**(4), 337–342.
192. LAWLOR, F. & RAMABALA, K. (1984) Iatrogenic Cushing's syndrome – a cautionary tale. *Clinical and Experimental Dermatology*, **9**(3), 286–289.
193. COOK, L. J., FREINKEL, R. K., ZUGERMAN, C., LEVIN, D. L. *et al.* (1982) Iatrogenic hyperadrenocorticism during topical steroid therapy: assessment of systemic effects by metabolic criteria. *Journal of the American Academy of Dermatology*, **6**(6), 1054–1060.
194. STEVENS, D. J. (1988) Cushing's syndrome due to the abuse of betamethasone nasal drops. *Journal of Laryngology and Otology*, **102**(3), 219–221.
195. HEROMAN, W. M., BYBEE, D. E., CARDIN, J., BASS, J. W. *et al.* (1980) Adrenal suppression and cushingoid changes secondary to dexamethasone nose drops. *Journal of Pediatrics*, **96**(3 Part 1), 500–501.
196. AUGSPURGER, R. R. & WETTLAUFER, J. N. (1980) Cushing's syndrome: complication of triamcinolone injection for urethral strictures in children. *Journal of Urology*, **123**(6), 932–933.
197. ORTEGA, L. D. & GRANDE, R. G. (1979) Cushing's syndrome due to abuse of dexamethasone nasal spray (Letter). *Lancet*, **2**(8133), 96.
198. CHISHOLM, D. & CHALKLEY, S. (1994) Cushing's syndrome from an inhaled glucocorticoid (Letter). *Medical Journal of Australia*, **161**(3), 232.
199. CHALKLEY, S. M. & CHISHOLM, D. J. (1994) Cushing's syndrome from an inhaled glucocorticoid. *Medical Journal of Australia*, **160**(10), 611.
200. RONAYETTE, D., HENNEQUIN, D., BOUQUIER, J. J. & BONNETBLANC, J. M. (1978) Syndrome de Cushing induit par corticotherapie locale chez un nourrisson. *Nouvelle Presse Medicale*, **7**(5), 368.
201. REINER, M., GALEAZZI, R. L. & STUDER, H. (1977) Cushing-Syndrom und Nebennierenrinden-Suppression durch intranasale. *Schweizerische Medizinische Wochenschrift* (*Journal Suisse de Medecine*), **107**(49), 1836–1837.
202. DHEIN, S. (1986) Cushing-Syndrom nach externer Glukokortikoid-Applikation bei Psoriasis. *Zeitschrift fuer Hautkrankheiten*, **61**(3), 161–166.
203. AXELROD, J. (1977) Catecholamines: effects of ACTH and adrenal corticoids. *Annals of the New York Academy of Sciences*, **297**, 274–283.
204. KOWAL, J., HORST, I., PENSKY, J. & ALFONZO, M. (1977) A comparison of the effects of ACTH, vasoactive intestinal peptide, and cholera toxin on adrenal cAMP and

steroid synthesis. *Annals of the New York Academy of Sciences*, **297**, 314–328.

205. Brownie, A. C., Gallant, S., Paul, D. P. & Bergon, L. L. (1977) Effect of ACTH on cytochrome P450 systems of the adrenal cortex. *Annals of the New York Academy of Sciences*, **297**, 349–360.
206. Krishnan, K. R., Ritchie, J. C., Manepalli, A. N., Nemeroff, C. B. *et al.* (1988) Adrenocortical sensitivity to ACTH in humans. *Biological Psychiatry*, **24**(1), 105–108.
207. Krishnan, K. R., Ritchie, J. C., Saunders, W. B., Nemeroff, C. B. *et al.* (1990) Adrenocortical sensitivity to low-dose ACTH administration in depressed patients. *Biological Psychiatry*, **27**(8), 930–933.
208. Siminoski, K., Goss, P. & Drucker, D. J. (1989) The Cushing syndrome induced by medroxyprogesterone acetate. *Annals of Internal Medicine*, **111**(9), 758–760.
209. Learoyd, D. & McElduff, A. (1990) Medroxyprogesterone induced Cushing's syndrome. *Australian and New Zealand Journal of Medicine*, **20**(6), 824–825.
210. Merrin, P. K. & Alexander, W. D. (1990) Cushing's syndrome induced by medroxyprogesterone. *British Medical Journal*, **301**(6747), 345.
211. Roger, M., Griboul, G., Delanoe, D., Fougeyrollas, B. *et al.* (1984) Contraception masculine. Fonctions gonadique et surrenalienne chez les hommes traites par la medroxyprogesterone. *Pathologie Biologie (Paris)*, **32**(8), 895–898.
212. Fujimoto, V. Y., Villanueva, A. L., Hopper, B., Moscinski, M. *et al.* (1986) Increased adrenocortical responsiveness to exogenous ACTH in oral contraceptive users. *Advances in Contraception*, **2**(4), 343–353.
213. Madden, J. D., Milewich, L., Parker, C. R. Jr, Carr, B. R. *et al.* (1978) The effect of oral contraceptive treatment on the serum concentration of dehydroisonandrosterone sulfate. *American Journal of Obstetrics and Gynecology*, **132**(4), 380–384.
214. Amin, E. S., El Sayed, M. M., El Gamel, B. A. & Nayel, S. A. (1980) Comparative study of the effect of oral contraceptives containing 50 microgram of estrogen and those containing 20 microgram of estrogen on adrenal cortical function. *American Journal of Obstetrics and Gynecology*, **137**(7), 831–833.
215. Feldman, J. M. & Roche, J. (1976) Effect of oral contraceptives on platelet monoamine oxidase, monoamine excretion, and adrenocortical function. *Clinical Pharmacology and Therapeutics*, **20**(6), 670–675.
216. Young, C. A., Williams, I. R. & MacFarlane, I. A. (1991) Unrecognized Cushing's syndrome and adrenal suppression due to topical clobetasol propionate. *British Journal of Clinical Practice*, **45**(1), 61–62.
217. Cobb, C. F. & Van Thiel, D. H. (1982) Mechanism of ethonal-induced adrenal stimulation. *Alcoholism, Clinical and Experimental Research*, **6**(2), 202–206.
218. Weinberg, J. (1989) Prenatal ethanol exposure alters adrenocortical development of offspring. *Alcoholism, Clinical and Experimental Research*, **13**(1), 73–83.
219. Smalls, A. G., Kloppenborg, P. W., Njo, K. T., Knoben, J. M. *et al.* (1976) Alcohol-induced Cushingoid syndrome. *British Medical Journal*, **2**(6047), 1298.
220. Rees, L. H., Besser, G. M., Jeffcoate, W. J., Goldie, D. J. *et al.* (1977) Alcohol-induced pseudo-Cushing's syndrome. *Lancet*, **1**(8014), 726–728.
221. Jeffcoate, W. (1993) Alcohol-induced pseudo-Cushing's syndrome. *Lancet*, **341**(8846), 676–677.
222. Kirkman, S. & Nelson, D. H. (1988) Alcohol-induced pseudo-Cushing's disease: a study of prevalence with review of the literature. *Metabolism*, **37**(4), 390–394.
223. Praga, M., Maiz, E., Mijares, R. P. & Lopez-de-Novales, E. (1983) Pseudoaldosteronism due to 9-alpha-fluoroprednisolone creams (Letter). *Clinical Nephrology*, **20**(6), 321.
224. Moran, J. M., De-Sancho, J. L., Abaigar, P., Amenabar, J. *et al.* (1984) Pseudoaldosteronism due to 9-alpha-fluoroprednisolone creams (Letter). *Clinical Nephrology*, **22**(6), 323–324.

225. MONTOLIU, J., BOTEY, A., TRILLA, A. & REVERT, L. (1984) Pseudoprimary aldosteronism from the topical application of 9-alpha-fluorprednisolone to the skin. *Clinical Nephrology*, **22**(5), 262–266.
226. BROGNOLI, M., DOREGATTI, C., RONCALI, M. & ORLANDINI, G. (1985) Pseudoprimary hyperaldosteronism from topical application of 9-alpha-fluorprednisolone on the skin (Letter). *Clinical Nephrology*, **24**(6), 315–316.
227. VITA, G., BARTOLONE, S., SANTORO, M., TOSCANO, A. *et al.* (1987) Hypokalemic myopathy in pseudogyperaldosteronism induced by fluoroprednisolone-containing nasal spray. *Clinical Nephrology*, **6**(2), 80–85.
228. TAKEDA, R., MORIMOTO, S., UCHIDA, K., NAKAI, T. *et al.* (1979) Prolonged pseudoaldosteronism induced by glycyrrhizin. *Endocrinology*, **26**(5), 541–547.
229. ARMANINI, D., SCALI, M., ZENNARO, M. C., KARBOWIAK, I. *et al.* (1989) The pathogenesis of pseudohyperaldosteronism from carbenoxolone. *Journal of Endocrinological Investigation*, **12**(5), 337–341.
230. AXELROD, J. (1977) Catecholamines: effects of ACTH and adrenal corticoids. *Annals of the New York Academy of Sciences*, **297**, 275–283.
231. MIURA, Y. & YOSHINAGA, K. (1988) Doxazosin: a newly developed, selective alpha 1-inhibitor in the management of patients with pheochromocytoma. *American Heart Journal*, **116**(6 Part 2), 1785–1789.
232. FERGUSON, K. L. (1994) Imipramine-provoked paradoxical pheochromocytoma crisis: a case of cardiogenic shock. *American Journal of Emergency Medicine*, **12**(2), 190–192.
233. CECAT, P., SONNENFELD, H., BOITTIAUX, P., SCHERPEREEL, P. *et al.* (1978) Preparation a l'intervention des pheochromocytomes par les alpha-bloquants. *Anesthesia and Analgesia*, **35**(2), 219–226.
234. ABE, M., ORITA, Y., NAKASHIMA, Y. & NAKAMURA, M. (1984) Hypertensive crisis induced by metoclopramide in patient with pheochromocytoma. *Angiology*, **35**(2), 122–128.
235. VAN WAY, C. W., FARACI, R. P., CLEVELAND, H. C., FOSTER, J. F. *et al.* (1976) Hemorrhagic necrosis of pheochromocytoma associated with phentolamine administration. *Annals of Surgery*, **184**(1), 26–30.
236. LAWRENCE, C. A. (1978) Pethidine-induced hypertension in phaeochromocytoma. *British Medical Journal*, **1**(6106), 149–150.
237. WARK, J. D. & LARKINS, R. G. (1978) Pulmonary oedema after propranolol therapy in two cases of phaeochromocytoma. *British Medical Journal*, **1**(6124), 1395–1396.
238. BOULOUX, P. M., FEATHERSTONE, R. M., CLEMENT-JONES, V., REES, L. H. *et al.* (1985) Erroneous diagnosis of phaeochromocytoma in hypertensive patient on labetalol. *Journal of the Royal Society of Medicine*, **78**(7), 588–589.
239. MONTMINY, M. & TERES, D. (1983) Shock after phenothiazine administration in a pregnant patient with a pheochromocytoma. A case report and literature review. *Journal of Reproductive Medicine*, **28**(2), 159–162.
240. ELASSER, W. & VON EICKSTEDT, K. W. (1992) Corticotrophins and corticosteroids. In DUKES, M. N. G. (Ed.), *Meyler's Side Effects of Drugs*, 12th Edn, pp. 977–994, Amsterdam: Elsevier.

13

Experimental Adrenal Toxicity

Significance of findings and relevance to man

A. D. DAYAN

St Bartholomew's Hospital Medical College, London

13.1 Introduction

The toxicologist, like any other scientist, may work analytically by dissecting mechanisms before turning to the synthesis needed to relate them to the pathogenesis of toxic effects. Or, and this type of study is no less demanding, he or she may act more like a clinician or natural historian of disease, observing man or animals affected by a substance, collating functional and structural findings and effects, and arguing back from their pattern to the likely causal processes and ultimately to the responsible toxicant.

Toxicology has yet another, and perhaps better known, facet in which empirical knowledge of effects, and often analytical understanding of pathogenic mechanisms, is combined with information about exposure to predict the likelihood and nature of the risk posed by a substance if man were exposed to it.

This complexity of experimentation, observation and prediction helps to explain why the toxicologist should approach any discussion about 'relevance' and 'significance for man' (or other species) with some reserve. In the case of the adrenal, there is need for particular circumspection since the structure of the organ is complex, its functions are manifold, and there are many hints from pharmacology and physiology of local and distant mechanisms and actions that may well be important in toxicity, but which we do not yet understand in sufficient detail to be able to make a straightforward interpretation. Despite the considerable learning presented in this book, the enquiring but not disingenuous toxicologist might still wonder, for example, whether there are still disorders to be uncovered that involve endorphins, prostaglandins and the adrenal medulla and cortex, since these ephemeral messengers are certainly secreted by the medulla and they can affect the activities of medullary and cortical cells. An analogous question might be concerned with the nature of the extreme sensitivity of adrenal cortical blood vessels and cortical cells to endotoxin, as demonstrated by the spectacular haemorrhagic necrosis found in endotoxic shock in the clinic (Waterhouse-Friedeichsen syndrome) and the laboratory. How many summated factors influence aldosterone secretion in relation to water and electrolyte metabolism? How is the differentiation effected of cortical cells towards gluco- or mineralocorticoid secretion? Once adrenal cortical cells are committed to steroidogenesis, what basic processes may switch them to progression from hyperplasia to neoplasia, with or without interruption of steroidogenesis, and so on?

The purpose of this section in brief is to point out that there are still many gaps in our understanding of the adrenal, even in one species, and relating effects seen in an experiment *in vitro* or in animals to what may happen in man must be done with caution and with as much respect for present ignorance as for current knowledge.

13.2 General Questions About the Adrenal and Toxicity

When considering the toxicology of the adrenal and its general implications in risk assessment – for this is a chief goal of much work on toxicity – there are pointed questions to be asked of any set of data or protocols intended to produce useful information, of any regulatory application, and especially of any attempt to invoke the adrenal in understanding a local or a distant toxic action.

Working backwards from consequence to cause, as is often necessary in applied toxicology, layers of increasingly focused questions concerning the evidence that the adrenal is involved in the toxic effect present themselves to the interested observer.

The clinical observer would seek answers to the following questions. What changes in the behaviour and appearance of the organism in body growth rate and proportions, and in steroid-sensitive tissues, are there that are compatible with an effect on adrenal function? Are there appropriate changes in resting or stimulated behaviour, in the external genitalia, in sexual maturity and performance, or in blood pressure and water and electrolyte balance, perhaps apparent in the urine and faeces? Has bone or subcutaneous tissue been affected by an excess of glucocorticoids? Effects on other systems can be deduced from an understanding of the physiological actions of adrenal corticoids.

The physiologist and clinical biochemist would look for changes in blood pressure, in fluid and electrolyte balance, in nitrogen and calcium balance, renal function, the state of the pituitary, and the state of the genital tissues and sexual activity. Their questions might also cover effects on the immune systems (eg an increased incidence or severity of common or unusual infections), on the nervous system (sedation, aggressiveness, learning and memory) and on development.

The pathologist will wish to know about the size, appearance and weight of the adrenals, about the straightforward light microscopic appearance of the zones and their cells, about probable secretory activity (initially perhaps as altered fat staining in the cortex or changes in chromaffin granules in the medulla, and subsequently via more refined cytochemical procedures), and about the other tissues and organs influenced by adrenal secretions, or that might affect or be affected via humoral and neural signalling systems.

The biochemist and pharmacologist will seek out their assays of endocrine and paracrine hormones, their precursors and metabolites, perhaps the receptors or enzymes, and ultimately the genes and their regulatory elements.

From any of these, or better still several supporting lines of evidence, will come acceptance that the adrenal is somehow involved in the toxic process. It is important that the types of information overlap and are mutually supportive, or at least are consistent with each other, since this not only confirms the toxic action but may well suggest its probable nature and more promising directions of enquiry at the next level of mechanistic analysis. That ultimate toxicological enquiry is likely to involve detailed investigation of the process of steroidogenesis and its control, or correspondingly of catecholamine or other neurotransmitter synthesis, release and re-uptake, and of the appropriate receptors.

The importance of evaluating the mutually supportive components of the pattern of effect applies at the laboratory level and is equally important when making a risk assessment for a compound during development or for regulatory purposes. Any discrepancy, assuming appropriate observations have been made and there is a reasonable interpretation of the data, must raise a question about the adequacy of the toxicity experiments, or about the degree of confidence in the interpretation offered. A conventional subacute or chronic toxicity test may appear to be a standardized investigation, from which only limited types of data can be obtained. In fact, if clinical, biochemical and histopathological results are combined and evaluated together, taking account of all the observed changes (or their absence) and their nature, it is possible to discern much about the detailed pathogenesis and source of a toxic action. If this information is then related to the dose applied, or more importantly to the internal dose (exposure), by taking account of toxicokinetics and metabolism, the results even of routine studies provide a powerful means for surveillance to detect an effect, and they provide a valuable minimum set of data to permit an initial analysis of the mechanism of the toxic effect. The complexity of the adrenal and its physiological regulation affords a pointed example of a further example of the need for a comprehensive assessment of every toxic response, but especially one that involves so complex an organ, and one which is influenced by many neural and endocrine mechanisms, and which, in turn, may affect many other body functions and tissues. The multiplicity of direct and indirect target organ effects and consequences, and the need, therefore, to evaluate the full pattern of toxic actions, is also apparent when the adrenal and reproductive toxicity are considered.

Toxicity affecting steroidogenesis might result in a direct action on gonadal steroid production, or it might just influence steroid metabolism in the adrenal cortex and so influence the genital tract by disturbing hypothalamo-hypophyseal activity, in addition to exposing end-organ receptors to an abnormally high concentration of steroid precursors and metabolites released from adrenal cortical cells, resulting in activation or tachyphylaxis. The end result is likely to be loss of, or at least impairment of, fertility. Should there be successful fertilization, the course of gestation and the necessary adaptive maternal responses might be impaired, and secondary virilization or feminization of the developing foetus is possible, affecting overall growth and the development of many viscera and external organs. Disorders of adrenal function during lactation or in the maturation phase after weaning may also affect growth and genital development at the functional and anatomical levels. Equally, behavioural development may be altered.

Thus, there are many ways in which adrenocortical toxicity in the maternal or paternal animals of the parental generation may affect reproduction, even culminating in persisting behavioural and anatomical changes in the F_1 generation.

13.3 The Need for Comprehensive Evaluation of Toxicity Involving the Adrenals

Sufficient has been written in the previous section to show that comprehensive functional and pathological observations are necessary if toxicity affecting the adrenal is suspected, or if there are distant physiological, biochemical or structural changes that might be due to disturbance of the quantity or nature of secretions from the adrenal cortex or medulla. This requires careful examination of clinical, laboratory and pathological findings, whether in man or animals, to reveal and define the occurrence and pattern of the toxic action.

In addition, there are many similarities between various specialized transport and synthetic pathways in the adrenal and other organs, so toxic effects on particular functions, or changes in other organs, should trigger more detailed examination of the adrenal.

As examples, consider the close relationship between the cytochrome P450 enzymes involved in steroidogenesis in cells in the adrenal cortex and the gonads, and in the liver and elsewhere, including aromatases in the brain. Inhibition or induction at one site may also affect the others to some degree, although there are sufficient differences for many responses to be quite divergent. Similarly, the metabolism of neurotransmitter catecholamines and neuropeptides in the medulla, and the diverse autonomic receptors located there, may be affected by toxic processes that are mainly apparent elsewhere in the autonomic innervation of other peripheral tissues.

Detection of toxicity affecting any component of the adrenal, or dependent on adrenal abnormalities, requires a synoptic evaluation of many different types of data. Until this has been done the real toxic hazard of a chemical will not have been adequately explored, and any risk cannot be properly evaluated.

The importance of looking both inwards at the adrenal and outwards at dependent tissues is well illustrated by several serious and even fatal illnesses of man, including acute adrenal failure secondary to various actions at the pituitary or hypothalamic level, haemodynamic disorders due to phaeochromocytomas, virilization due to abnormal corticosteroid synthesis, electrolyte disorders secondary to abnormal mineralocorticoid synthesis, and so on.

13.4 Risk Assessment and Toxicity Affecting the Adrenal

As with toxicity affecting other tissues and organs, albeit often more generalized in its consequences, risk assessment of toxicity affecting the adrenal requires recognition and characterization of the hazard, and its correlation with the dose, or better still the exposure to the substance and any toxic metabolite.

Consideration of physiological differences between species comes next, which may involve divergence in control mechanisms and receptors, in cellular and metabolic processes and their enzymes in the adrenals, and differences in the nature of the hormones produced, in addition to the responses of their target tissues. Examples include the principal production of corticosterone in the rat and of cortisol in primates, analogous differences in the detailed patterns of androgens and oestrogens in those species (and in their potency of action on genital tissues), absence of the striking clinical features of the human Cushing's syndrome in the rodent, and so on. To these must be added the conventional awareness of species differences in toxicokinetics and xenobiotic metabolism, determining exposure to the ultimate toxicant at the site of action. The final and most difficult step is the prediction of risk in man, based on knowledge of the dose–exposure–response relationship and understanding of the sources of divergence in the nature of the response.

This approach, which is conventional, should suffice for the evaluation of non-neoplastic risks, including those rare substance and processes that may affect fertility, or foetal and postnatal development via an action involving the adrenal.

Neoplasms of the adrenal cortex, as in other organs under endocrine control, usually evolve along a well-defined sequence of hyperplasia (secondary to loss of feedback inhibition), adenoma and carcinoma. The same progression, at least to adenoma formation and perhaps ultimately to carcinoma, may also be seen in man, although the development of independent autochthonous tumours in man, whether benign or malignant, is probably much less common than in the rat or mouse. Adrenocortical neoplasms do not always secrete gluco- or mineralocorticoids; and if they do, the detailed systemic physiological and pathological effects differ considerably between species. This is not as important a factor in quantitative consideration of risk as the probability that a harmful effect will occur, but it must be considered in a comprehensive account of the consequences of the risk, as in any cost–utility analysis of a candidate drug or other chemical.

Adrenal medullary neoplasms, whether the neuroblastomas of early life or the more differentiated phaeochromocytomas more common in later life, may secrete catecholamines, often with profound haemodynamic consequences. This complication appears to be more common in man than in other species, but blood pressure and other physiological consequences have not been systematically investigated in the rodent and other laboratory species.

The analyst of risks revealed in the laboratory will be aware of toxic responses peculiar to individual species – such as the ability of sugar alcohols uniquely to produce phaeochromocytomas in the rat – however, the same concern exists whenever extrapolation is attempted from animals to man.

13.5 Conclusions

Toxicity affecting the adrenal gland should be investigated and assessed in the same thorough manner as are the actions of xenobiotics on any other organ. Such a statement conceals much complexity, however, since detection of such an effect may require evaluation of many distant functional and pathological body processes and tissues, in addition to comprehensive examination of the adrenal itself, and of the other organs that control it via endocrine and neural mechanisms.

Xenobiotics may affect the adrenal directly, eg glutethimide, many antifungal imidazole derivatives and etomidate, or by disturbing its endocrine regulation via an effect, say, on ACTH secretion by the pituitary, or on fluid and acid–base control.

The need is for comprehensive attention to the features and clues to primary and more distant effects on the adrenal, or of its effects on other functions and tissues. Risk analysis additionally requires conventional accounting for species-specific factors and toxicokinetics, before a prediction can be made as to whether a given exposure will have a particular effect in another species, and the nature of such an effect.

The ultimate step is to consider the cost and utility of that risk, which once more must reflect a multiplicity of potential dysfunctions and lesions due to the complexity of the adrenals and their physiological roles.

The adrenal and its toxicological study represent an excellent model for studying integrated physiological and pathological mechanisms in health and disease, and thus for displaying the comprehensiveness required of any rigorous and comprehensive toxic risk assessment.

Index